The Responsive House

cave carving
val camonica, italy
second millenium b.c.

The Responsive House

Selected papers and discussions from
The Shirt-Sleeve Session in Responsive Housebuilding
Technologies,
held at the Department of Architecture,
Massachusetts Institute of Technology,
Cambridge, Massachusetts,
May 3-5, 1972.

Edited by Edward Allen

The MIT Press
Cambridge, Massachusetts, and London, England

Copyright © 1974 by
The Massachusetts Institute of Technology

This book was printed and bound by Semline, Inc.
in the United States of America.

Publisher's and Editor's Note.
The format of this book is intended to reflect the
informality of the Shirtsleeve Session and the frankly
experimental nature of many of the ideas presented
there. Also, by photographing the text directly from
the author's typescripts, we have avoided the time and
expense of detailed editing and composition in print.
The MIT Press   Edward Allen

Library of Congress Cataloging in Publication Data

Shirt-Sleeve Session in Responsive Housebuilding Technologies, Massachusetts Institute of Technology, 1972.
   The responsive house.

   Includes bibliographical references.
   1. House construction--Congresses. I. Allen,
Edward, 1938-   ed. II. Title.
TH4811.S516 1972   728   74-23518
ISBN 0-262-01040-2

"What of architectural beauty I now see, I know
has gradually grown from within outward, out of the
necessities and character of the indweller, who is the
only builder,--out of some unconscious truthfulness,
and nobleness, without ever a thought of the appearance;
and whatever additional beauty of this kind is
destined to be produced will be preceded by a like
unconscious beauty of life...There is some of the
same fitness in a man's building his own house that
there is in a bird's building its own nest."
--Henry David Thoreau

Quotations from Henry David Thoreau are taken from Walden, The
Writings of Henry D. Thoreau, edited by J. Lyndon Shanley, CEAA
Edition, copyright © 1971 by Princeton University Press, pages
30-49, by permission of the publisher. Photographs of participants
in The Shirt-Sleeve Session are by Roger Goldstein.

TABLE OF CONTENTS

Introduction ix
Edward Allen

Section I

**TOWARD A THEORY OF
RESPONSIVE DWELLINGS
1**

Section II

**INVOLVING PEOPLE IN
THE MAKING OF
THEIR DWELLINGS 65**

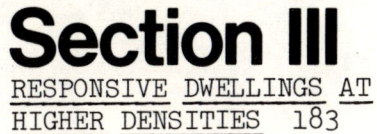

Section III
RESPONSIVE DWELLINGS AT
HIGHER DENSITIES 183

Section IV
A MAGIC HOUSING MACHINE
225

Introduction

"The Shirt-Sleeve Session is built on the idea that dwellings ought to be fitted to people, not people to dwellings; that building technologies ought to be flexible and changeable, capable of fine-grain physical conformance to patterns of human use; that people ought to be able to participate much more fully and easily, directly or through machines, in the design, construction, and later modification of their dwellings; that the few people in the world who have done work based on these or similar ideas ought to get together informally, roll up their sleeves, and share their knowledge."

"The Shirt-Sleeve Session is intended for people who have worked on the development of responsive housebuilding technologies, including technologies aimed primarily at a wide initial choice of dwelling form and those aimed at enabling easy change of dwelling form over time. The Session is meant to deal with residential structures, single or multiple, at the scale of individual rooms or dwelling units, not with schemes for neighborhoods or cities. It is for theoreticians, experimenters, and builders who have grappled with the problems inherent in the development of highly flexible, occupant-manipulable building systems." *

The Shirt-Sleeve Session grew from an increasing awareness that the last vestiges of consumer choice and occupant control are being eliminated from new housing in the United States, and that no coherent body of work is underway to develop occupant-responsive alternatives. In recent years, the cost of the individually-built house has risen so steeply that the majority of American families can no longer afford one. A massive program by the Department of Housing and Urban Development to encourage the development of industrialized housebuilding methods, begun in 1969, resulted in the submission of hundreds of well-documented proposals, from a broad spectrum of American corporations, for new residential construction systems. Of these hundreds, all but a handful were for inflexible, repetitive, boxlike systems which are capable, at best, of two or three standard configurations. Of the handful, all were for exotic dome or shell systems which are even more constraining that the boxlike systems. Not one proposed system approached the level of consumer choice and flexibility of form that is inherent in the conventional stick-built house.[1]

The surge of interest in industrialized housing which arose in response to the H.U.D. program has largely subsided. Many of the large corporations which planned to enter the housing field have quietly withdrawn, largely because of the discovery that a product as large as a house cannot be manufactured, marketed, and transported by the same methods as their more standard lines of products. Of the few corporations selected to participate in a H.U.D. demonstration project, several have declared bankruptcy and disappeared.[2]

* From "An Invitation to The Shirt-Sleeve Session on Responsive Housebuilding Technologies."

As of this writing, the two surviving trends in housebuilding technology are the "mobile home," and the "rationalized" stick-built house. In both cases, the move is toward increasing product standardization, and away from increasing consumer choice. This is not surprising, in that the two most commonly stated goals of housing industrialization are to increase production and reduce costs. The Shirt-Sleeve Session was set up to explore possible paths toward a third goal: The increase of occupant choice and control. It was not the usual academic conference. The field of discussion was ill-defined, without an established precedent of thought, research, or practice; and without the usual group of preeminent scholars or practitioners. We were a loosely-related group of about thirty persons, assembled largely through chain-letter and word-of-mouth. For three days, we shared our work and thoughts, and debated our differences. From these discussions there arose a recognizable pattern of diverse approaches. and a single significant, unanswered question about the nature of responsiveness in architecture.

The question is, simply, to _what_ would a responsive dwelling respond? What areas of human need are best answered by a change in the architecture of the immediate environment? And what sorts of responses might a building make to such needs? We found ourselves as a group incapable of defining responsiveness in housing, and, paradoxically, those who had the most difficulty in furnishing plausible examples of responsiveness were not those participants describing relatively ponderous, hard-to-change, traditional construction systems, but those who proposed technologies are capable of instant response through pneumatic or electronic means. Perhaps the closest we got to an answer was to agree that where a person could design and built for himself, using a tractable technology, the question would largely take care of itself.

The common theme which was present in all the presentations was the necessity of eliminating the middleman from the housing process. Everyone agreed that the majority of people would be better off it they could control directly the configuration of their environments, without having to resort to the hired services of architects, engineers, and builders. It was recognized that the middlemen presently serve a function of providing expertise and tools that are not commonly available, but it was felt that much of the expertise has been shrouded intentionally in a protective professional mysticism, and that the middlemen often constitute a financial and temporal barrier that deters the householder from making necessary changes in his dwelling. It was felt that housing ought to be a continual process of change and modification, rather than a static product of a high-overhead production line, and that this transformation cannot be accomplished until the dweller can work directly on his dwelling.

Anne Hollister, Edward Allen,
Neil Pinney

Daria Fisk, Dixon Bain, Hans Harms

Jim Taggart, Avery Johnson

It was recognized that the householder usually lacks information, skills, tools, and time with which to work on his dwelling. The presentation of alternative schemes for providing these commodities furnished the diversity of the Shirt-Sleeve Session, and the discussion of the schemes furnished the controversy. During the Session it appeared that we fell into two opposing camps; one advocating a return to handicraft, and the other a forward-leaping utilization of computers, plastics, and space-age technology to avoid the problems which were felt to be associated with handicraft. A retrospective attempt to diagram the relationships among the various approaches shows that this appearance was deceiving. There was no simple bipolar axis of opposing theories, but instead a rich, circular continuum of closely-related ideas, possessing of course its opposite poles, but also a finely-graded tangential linkage between the poles.

The diagrams I have drawn are imperfect. Without doubt, they leave out approaches which were not represented at the Session, and more importantly, they reflect a certain confusion of subject matter which arose in our discussions. Nicholas Negroponte noted on the last day that we had been talking about three separate topics:

1. Responsive design processes, which help the non-professional to plan or re-plan his environment.

2. Responsive construction processes, which help the non-professional to make or modify his environment more easily.

3. Responsive architecture, which actually changes itself in an attempt to accommodate the needs and desires of the occupant.

This distinction, while useful, does not eliminate the confusion entirely. Negroponte's own work has spanned all three topics, and much of the work of the other participants incorporates at least two. In making the diagrams, I have separated the first two topics into two diagrams. The third topic is found toward the lower left in both diagrams.

If the diagrams could be broken into broad categories, I would propose a tripolar model rather than a bipolar one. At the center of each is the status quo, in which the occupant is left without skills, experience, or tools with which to work on his dwelling, unless he can afford to hire professionals each time he wishes to make a change or begin anew. Toward the upper left of each diagram are the approaches which make more easily available to the occupant the resources of American industry and professionalism. Toward the right are those which help him to become his own professional, through education and simplification. Toward the lower left are the approaches which use nonfactory machines to enhance his capabilities.

APPROACHES TO INVOLVING THE
OCCUPANT IN THE DESIGN OF
HIS DWELLING

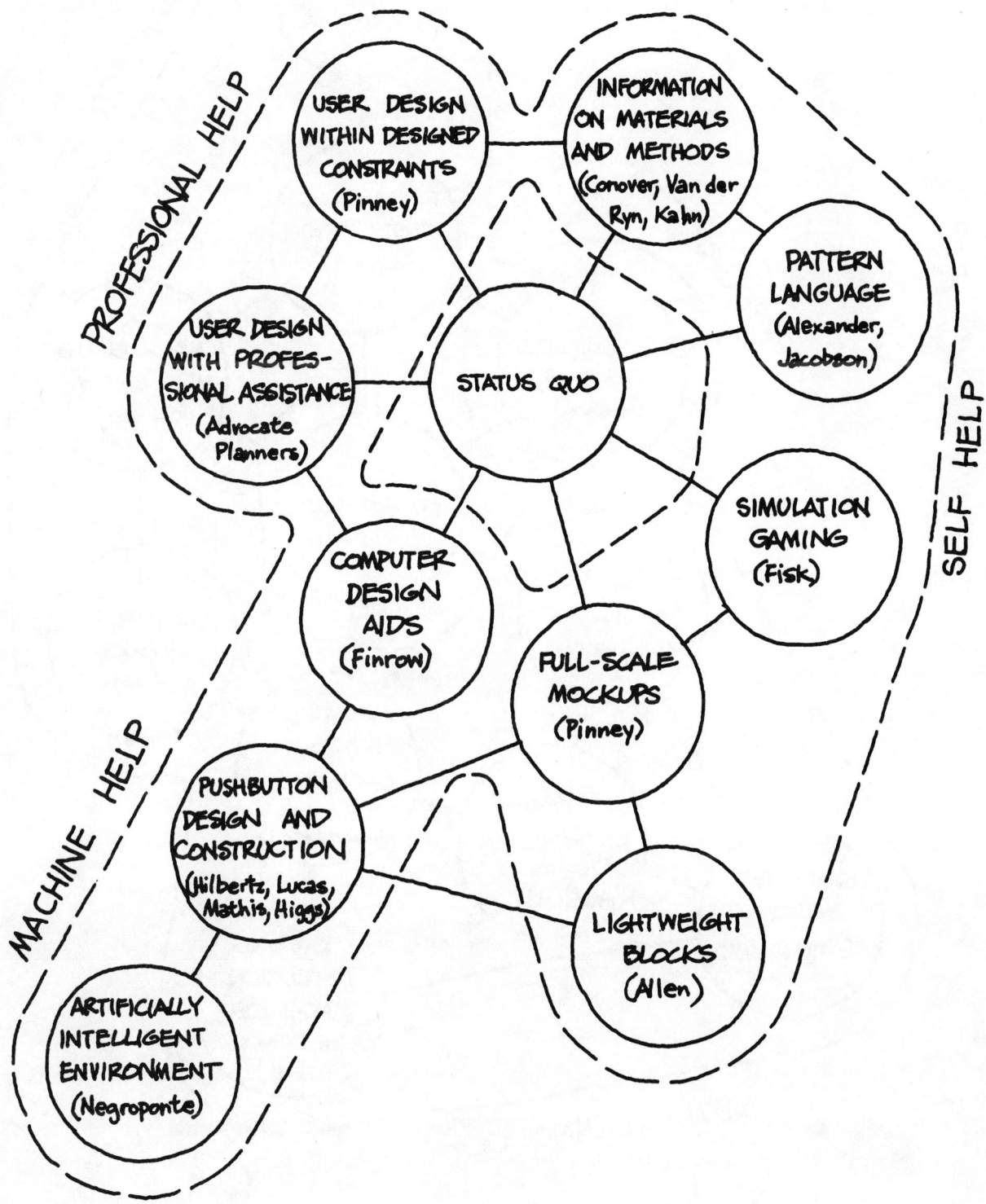

PROFESSIONAL HELP

SELF HELP

MACHINE HELP

USER DESIGN WITHIN DESIGNED CONSTRAINTS (Pinney)

INFORMATION ON MATERIALS AND METHODS (Conover, Van der Ryn, Kahn)

PATTERN LANGUAGE (Alexander, Jacobson)

USER DESIGN WITH PROFESSIONAL ASSISTANCE (Advocate Planners)

STATUS QUO

SIMULATION GAMING (Fisk)

COMPUTER DESIGN AIDS (Finrow)

FULL-SCALE MOCKUPS (Pinney)

PUSHBUTTON DESIGN AND CONSTRUCTION (Hilbertz, Lucas, Mathis, Higgs)

LIGHTWEIGHT BLOCKS (Allen)

ARTIFICIALLY INTELLIGENT ENVIRONMENT (Negroponte)

APPROACHES TO INVOLVING
THE OCCUPANT IN THE
CONSTRUCTION OF HIS
DWELLING

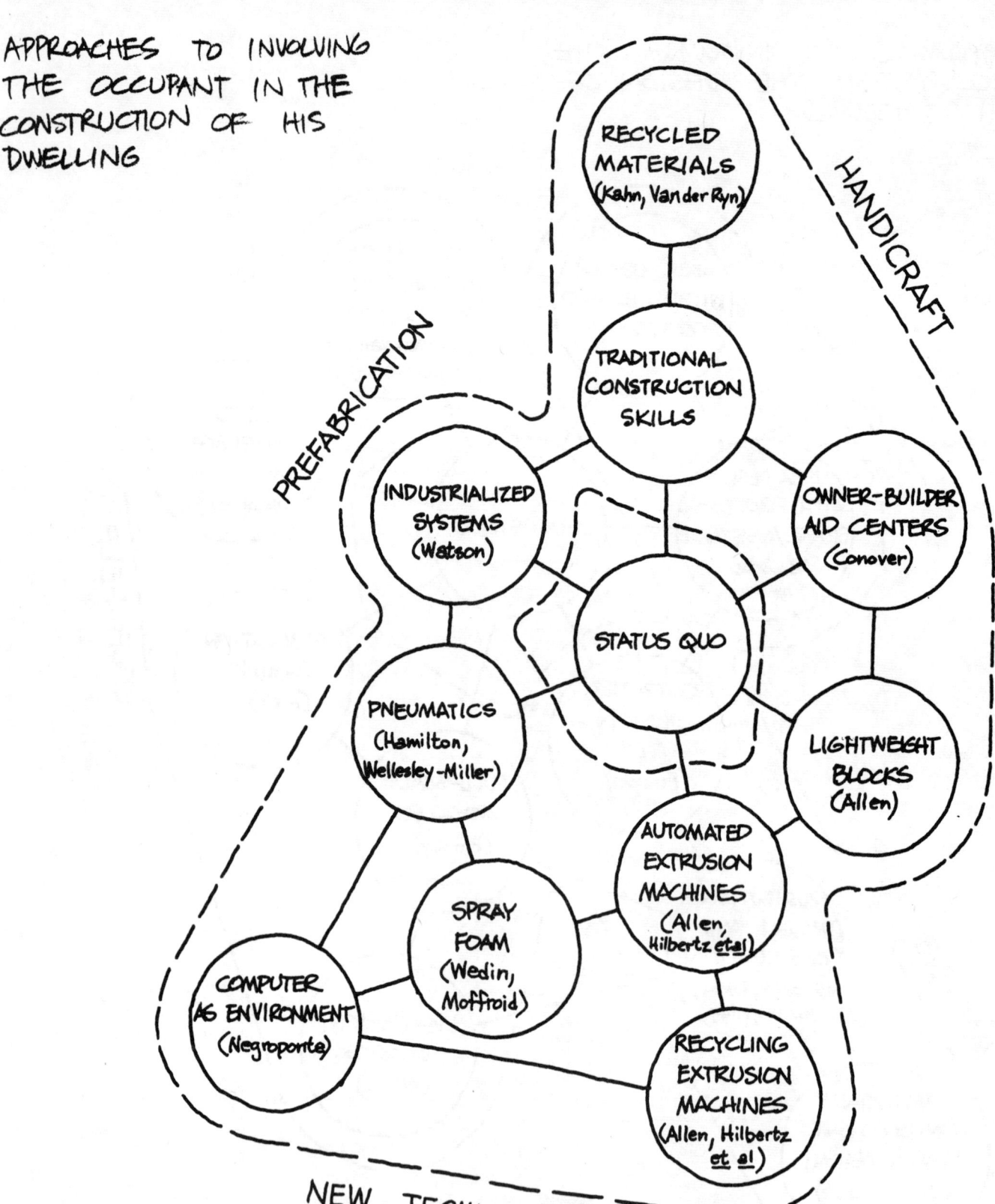

The "machine help" approach is still entirely experimental, without significant examples of actual implementation. The "industrial/professional help" approach has so far had a limited real-world impact because of the relatively large amounts of factory, capital, and professional time which it requires. The "self-help" sector alone shows much success in practice, and its success so far is confined entirely to the most conventional, laborious kinds of self-help, at the upper right of the diagrams. The extent of this kind of building activity is much larger than is commonly supposed; one new single-family house out of every five built in the United States each year is constructed by its owner.[3]

Study of the owner-builder phenomenon leads to optimism on two points. One, it indicates that a significant number of the population not only desire to take, but are able to take a strong role in the making of their own dwellings. It is tempting to speculate how much larger the proportion could become with widespread distribution of better informational materials and easier materials of construction. Two, such study reveals that owner-builders are not in it only for the money they can save. This in itself is a laudable reason to self-build, but deeper reasons emerge as well: The owner seeks the satisfaction of having provided his own house. He seeks the individualization of his house which can be achieved only with his own hands. And he seeks the confidence of knowing that he is able more fully to control the circumstances which most intimately affect his life.[4]

The goal of responsive housebuilding is to put each person in charge of his own existence, to the fullest extent that is possible. The goal of the housing industry and the government is the put each person in a physically healthful dwelling. These are complementary, mutually-reinforcing goals, not conflicting ones. Those of us who attended The Shirt-Sleeve Session feel unanimously that all industrial and governmental housing programs to date have perpetuated the self-destructive dependency of the individual on forces supposedly greater than himself. We believe that the individual can, and in perhaps the majority of cases would like to, design, build, redesign, and rebuild his dwelling as he chooses. We maintain not only that such individual control be practices within a large-scale housing effort, but that it is the only means for putting each American in a decent house. Mass-produced housing has proved desirable neither economically nor socially; one need listen only to the financial crash of many major housing producers, or the physical crash of the infamous Pruitt-Igoe public housing.

We were unable to propose whole alternatives to the present housing production system. This was, to my knowledge, the first major gathering to talk about responsive housebuilding technologies. We traded a lot of ignorance, and only a growing bit of wisdom. Many more ideas must be traded. and many more experiments must be tried, before large numbers of people will live in dwellings of their own design and construction. But we made a modest beginning. The purpose of this volume is to share our experience and our concern, in the hope that others will join in our search.

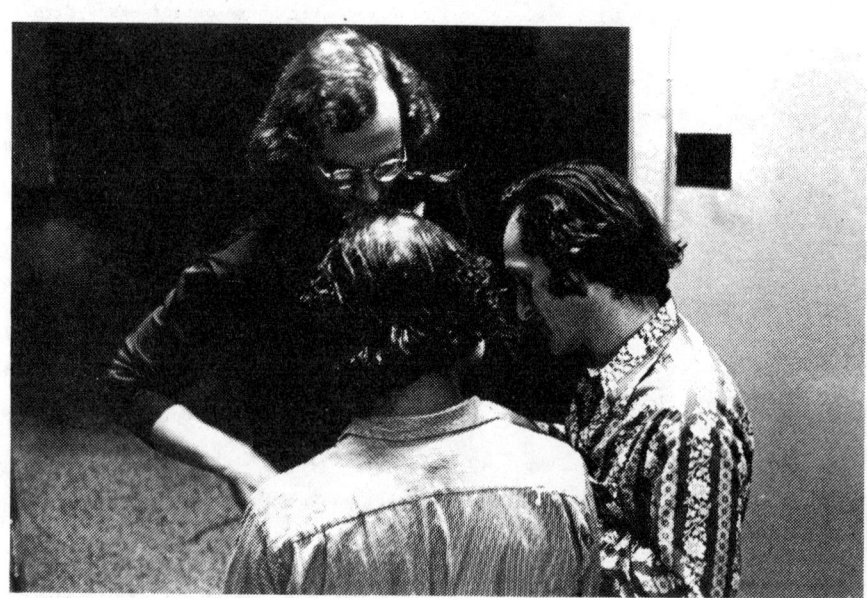

Tom Bender, Lloyd Kahn,
Sim Van der Ryn

Sean Wellesley-Miller, Blair Hamilton, Nicholas Negroponte, Anne Hollister

Sean Wellesley-Miller

David Robinson, Neil Pinney

Winslow Elliott Wedin

Max Jacobson

Daria Fisk

Dixon Bain, Max Jacobson, Huck Rorick

References

1. Housing Systems Proposals for Operation Breakthrough. U.S. Department of Housing and Urban Development, Washington, D.C., 1970.

2. James C. Hyatt, "Making Modular Units is Easy; Making Money At It Isn't So Easy," Wall Street Journal, September 7, 1972.

3. William C. Grindley, "Owner-Builders: Survivors With a Future," in Freedom to Build, John F. C. Turner and Robert Fichter, editors, New York, the MacMillan Company, 1972, p. 4.

4. William C. Grindley, The Owner Building System, A Case Study and Model of Amateur Homebuilding, M.C.P. thesis, Massachusetts Institute of Technology, 1972, pp. 9, 10, 59.

Acknowledgments

The Shirt-Sleeve Session was organized by the editor, in partnership with Huck Rorick and Sean Wellesley-Miller. Financial support was furnished by the Harvard-M.I.T. Joint Center for Urban Studies, and by the M.I.T. Department of Architecture.

Special thanks are due to the contributors to this volume, who were as responsive as the dwellings they envision, and to Ronnie Rogers, who edited the tape recordings and transcribed several major discussions.

Section I

" . . . the inhabitant, the indweller, might build truly within

and without . . . "

--Henry David Thoreau

Sean
Wellesley-Miller

Work Notes on the Need for a New Building Technology

I. Exponential Growth on a Finite Surface: The Global Context

In North America and most West European countries it is estimated that as many new buildings will be constructed by the end of the century as at present exist in them. This represents a doubling of the building stock in 30 years, or the equivalent, in the United States, of a new town for 50,000 people every seven days.

In many developing countries as many new buildings are currently needed as are at present standing in them today.

In world terms, as many or more new buildings will have to be built by the end of the century, if the minimum projected world population growth is to be adequately housed, as have been built in the whole of human history to date.

*   *   *   *   *

In developing countries this huge demand is primarily caused by population growth rates, allied to large scale urban immigration from the land to the cities. In the industrially advanced nations the causes are more complex. In the U.S. it is related to rising living standards and expectations which are partially expressed as an effective demand for better housing and services, the relative growth of the tertiary economic sector, shifts in demographic age structure, and the peaking obsolescence rates of urban building stock constructed thirty to fifty years ago during periods of rapid urban growth. Many European countries, such as Britain and Holland, also have long-standing housing shortages dating from the Second World War.

*   *   *   *   *

Whatever the cause, even in the affluent West it is doubtful if traditional building methods will be able to satisfy this demand.

Whatever new methods are developed will have large scale, long range consequences well into the next century.

This situation presents us with, at one and the same time, a challenge, a threat, and perhaps an opportunity . . .

Building programs as large as this cause social and ecological problems as well as technical and economic ones. In the U.S., the building industry is second only to agro-industry in defining our relation to, and use of, the natural environment and our impact on nature's complex but fragile ecologies. Badly planned development can have, and often has had, a deleterious effect on the natural environment, seriously damaging its carrying capacity.

Sean Wellesley-Miller is Assistant Professor of Architecture at M.I.T.

The twenty billion dollar U.S. building industry is a major consumer of many scarce resources. Domestic gas and oil reserves are estimated to last only another ten to twelve years. World supplies may well be exhausted in thirty to forty years. With six percent of the world's population, the U.S. currently consumes 34% of world energy production. 22.4% of this is devoted to lighting, heating, and cooling the building stock at a low level of efficiency. Or, put crudely, one BTU of every twelve produced globally is used to keep an American comfortable.

The values expressed in a building's design are reflected in its use. America's 67 million dwelling units are the end-point of most consumption. The average household uses some 200,000 gallons of water per annum and produces nearly three tons of garbage.

    *   *   *   *   *

Architecture is no longer limited to the innocent arrangement of spaces. Whatever new technologies we develop must be environmentally responsible as well as humanly responsive.

    *   *   *   *   *

Yesterday in Today and Today in Tomorrow: The Temporal Dimension

The average planning horizon of most architectural firms is five to ten years. The average lifetime of most buildings is in excess of thirty years. Due to their long lifetimes, buildings, more than anything else, determine what might be called the "historical future"--that part of the future that issues from and is constrained by the present. Moreover, the built environment constructed today will be, for a major part, the container and context of the "invented future"--that part of the future we can only guess at now.

    *   *   *   *   *

While the future's past inheritance can be a stabilizing influence, a source of joy and continuity in an uncertain future, it can also inflict repressive and restrictive limitations. The liberatory potential of future social and technological innovations will be conditioned by the physical environment within which they have to fit and function; the built environment we are making now, in ignorance.

    *   *   *   *   *

Even today, the difficulty of retrofitting old buildings with, say, better insulation, of upgrading their fabric or integrating new facilities, traps thousands of people in continuously deteriorating and seriously substandard environments.

In aggregate, the thermal performance of present buildings will largely predetermine future energy consumption on the basis of present efficiencies.

Or again, to what degree are the economic blackspots of the English Midlands and Northeast due to their rigid building fabric that effectively locks them out of the possibility of change?

*   *   *   *   *

The problem of time past in time future brings in its train the multiple paradoxes of gentleness and endurance, economy and response, identity and change, that confront all modern architecture.

*   *   *   *   *

Social Clothing: The Personal Dimension

In a social sense there is no such thing as a history of modern architecture, only a history of modern architects and the buildings they have made.

Mention the advent of total urbanization to the average citizen, and instead of embracing it as an opportunity for humanistic enhancement, his immediate emotional reaction is to start searching in an imaginary atlas for a desert island conveniently close to work.

An ever increasing number of people live or work in urban environments in whose design they play little or no formative role. The result is often epitomized by loss of identity, social injustice, and alienation.

Yet design participation, whether it comes via advocacy planning or computer-aided design, is as much a question of technology as it is of social and political organization or adaption of the design process. In the last resort, participation only becomes meaningful if the environment and its formative technologies are designed to support it. In the end it is a matter of control both at the community and the household level . . .

Historically we have moved from an "anthropological" architecture, to a traditional artisan architecture, to an industrial architecture, while the future seems to hold the choice between an industrialized and an eco-technical architecture; an extension or full circle . . .

In the course of these changes we have moved through material vocabularies of stone, adobe, wood, and natural fibers; to brick, timber, and tile; to steel, glass, and concrete; to foams, fabrics, alloys, plastics, and laminates . . .

. . . from a socially-evolved to a professionally-planned architecture.

*   *   *   *   *

Yet for the last century the mainstream of architecture seems to be continually one era behind the technical possibilities of its time. Thus we had regency architecture at the advent of the factory system and large-scale urbanization; the articulation of the technical vocabulary of the first half of the industrial revolution by Le Corbusier, Gropius, and other moderns in search of "an architectural symbolism for the machine age" during the rise of mass-production. Now, in the early seventies, in the era of cybernation, mainstream architects are advocating

mass-produced industrial systems, a development one cannot help feeling
should have begun in the early thirties and come to its full flowering
in the late fifties.

   *   *   *   *   *

  Was and is this technological time lag inevitable?

   *   *   *   *   *

  One result is that social acceptance cycles have been seriously
retarded. Offering a "responsive environment" to most people today
would be like offering a 20th century sports car to an 18th century
parson whose highest aspiration is a gold-embossed coach-and-four. Lack
of gas stations apart, it is doubtful if he would accept it.

   *   *   *   *   *

  These, then, are the contexts within which we will have to work:
socail and economic, temporal and ecological. Given that traditional
building techniques cannot meet them, at least within reasonable costs;
given that we must meet them somehow, and fairly soon; what are the
options?

  At the moment, mass-produced industrialized building systems are
the favorite. While it is certain that industrialized methods will play
some part, it may be instructive to give the assumptions of present
approaches to industrialized building a close scrutiny . . .

II. "Industrialization as Saviour," A Thumbnail Sketch

Industrialized building can
be characterized as:
"the breakdown of conventional
buildings into a set of standardized building elements
which are centrally manufactured
in series under factory
conditions and subsequently
transported to the site where
they are assembled using a
minimum number of programmed
operations into finished
buildings. Successful
industrialization calls for
large homogeneous markets,
heavy financing, and
advanced management techniques."[1]

1. This statement is the
author's translation and
condensation of a statement
of the Centre du Batiment
Industrielle.

Sixty-five percent of all building in the U.S.S.R. is industrialized. About 30% of West European building is constructed using industrialized methods, but it is expected that by 1980 more than 70% will
be. By contrast, under 15% of American building is constructed using
heavy industrial methods.

   *   *   *   *   *

  Market parameters: The degree of possible industrialization is dependent on the extent to which it is possible to develop and extend
repetitive production. The limit to the repetitive production of a given
set of elements is determined by market demand, which is in turn constrained by the range of functional requirements the system is capable
of satisfying within a fifty kilometer distance from the factory, which
should make for flexibility . . .

  However, the flexibility of any given system is a function of the
number and size of the element types that go to make it up. In general,
the number of element types required to maintain a certain level of
flexibility decreases as element size diminishes. Bricks, for example,
are an universally flexible building material. Unfortunately, as element
size decreases, labor costs and construction time increase.

   *   *   *   *   *

Each stage of the industrial process imposes its own demand.
Thus we have:

1. A market demand for flexibility calling for a large number of element types or small unit size.

2. A conctruction requirement for large element sizes (approaching the limits of transportability and erection equipment capacity) and a small number of highly finished element types to cut down labor costs and construction time.

3. A manufacturing requirement for long runs of as few element types as possible in order to maximize the economies of serial production.

        *   *   *   *   *

In Russia and Central Europe the option chosen has been a steady increase in element size and the strict limitation of the number of element types at the expense of flexibility. This is predicated on a serious housing shortage, the high price of labor, and a centrally-regulated economy.

Even in other countries, though, the trend has been from bricks to blocks, to room-high slabs, to wall-size panels, to assemblies at the scale of a room or even an apartment . . .

        *   *   *   *   *

However, industrial systemization doesn't stop at building components. The manufacturing requirement for a minimum number of standard element types produced in long runs is matched by a demand for assembly techniques utilizing a minimum number of standard operations repeated many times, which induces a "learning curve" that reduces the time needed by a construction gang to assemble any one unit, so increasing productivity. This goal is achieved by limiting building variation to a small number of "typified designs," usually from four to six, in a production run of a thousand dwellings.

Finally, the fact that assembly equipment has to be moved from one part of the site to another, favors certain types of layout (or "extension plans") that accommodate crane lines, moving formwork, etc.

In this way industrialized building in Europe has led from standardized building elements, to typified dwelling units, to work rationalized extension plans, usually on a grid system.

        *   *   *   *   *

While it is important to look at the materials, tools, techniques, and concepts behind industrialization, it is even more important to look at its likely physical and social consequences, the functional and experiential quality of the environment produced, and its ability to absorb and support change.

        *   *   *   *   *

Heavy panel building methods have strict architectural design consequences, since the slabs are often of parameter size in two directions, and can only be jointed along their edges to produce box-like cells that are automatically of parameter size in three directions.

What little flexibility there is, is strictly limited in most building systems to initial design. There is little or no chance for change once the building is completed.

The shift from traditional craft-based building methods to industrialized building has been marked by a corresponding shift in materials used; from brick, timber, and tile--materials which, with work, allow reconstruction and adaption--to steel, concrete, and glass--materials which are hard and immutable. The chances for individual expression and functional adaption are small.

If people cannot affect buildings they can be expected to have very little affection for them.

The possibility for future occupants to have any individual say in the design of their industrialized dwelling is practically non-existant while the formation of the local environment is taken out of the hands of the local community. These functions are abrogated by administrative authorities and large-scale developers as pseudo-clents; inevitably the resulting projects tend to reflect their values and requirements more than they do those of the actual inhabitants. The results can be seen in any modern suburb skirting a large European city.

Despite these observations, European industrial building systems are deemed a success worthy of emulation by many American architects. Yet several large-scale attempts to introduce industrial building systems into the U.S. have failed to take root. Why, and what is the nature of the European success?

The success of European building systems lies not so much in reduced costs: None of the systems is capable of producing industrialized dwelling units at a price that the average worker can afford. Rather the main advantage is a radical reduction in construction time. This, allied to the capacity of the industry to undertake vast projects, makes it extremely attractive to governmental housing authorities. The administrative convenience of only having to deal with one company and the ease of inspection and control also work favorably.

Their wide acceptance in Europe is predicated on large-scale housing shortages that have made European housing a sellers' market since the beginning of the modern movement. Another factor is the relative scarcity of inexpensive building terrain, which favors high-density building. These two features have led to the emergence of centralized housing authorities as large-scale pseudo-clients, backed by government subsidies and capable of providing guaranteed markets. These conditions and the high-volume money flows involved make it financially attractive for large building conglomerates to form and enter the market. Finally, their ability to broach the mass housing problem in terms of sheer quantity, if not of cost-effectiveness, and quality, conjoined with the lack of any demonstrably viable alternative, settles the issue.

It must also be remembered that having made the initial investment, it is economically difficult and expensive to scrap industrialized building systems. It's a one-way street. By contrast few, if any, of these conditions hold to anything like the same degree in America, and from their very nature it is doubtful if we would want to create them!

The cost question raises some interesting points:

1. Building systems are primarily structural in their approach. Yet only from 30 to 40 percent of total costs is attributable to support structure and cladding. The remaining 60 to 70 percent is consumed in equipment and finishes. Thus a 10 to 15 percent reduction in construction costs due to industrialized building works out as only a 3 to 4 percent reduction in overall costs--a price many people would be willing to pay for greater diversity and in-use adaptibility. This gives "package systems" as exemplified by mobile homes a definite advantage over the "catalogue systems" so dear to the architect's heart. (Even greater advantages could accrue to approaches that integrated structure and equipment or, better yet, transformed structure into equipment.)

2. Much of the shortened production time, where the major overall cost gains are made, can be attributed to the use of CPM and PERT techniques combined with better budgetary control, techniques which could just as effectivaly be applied to the construction of neo-gothic follies.

3. What other cost gains there are, are largely due to economies of scale in quantity surveying rather than the efficacy of industrial building systems as such. (The best industrial building systems are about 12% cheaper than traditional construction methods. Some are actually more expensive.)

In the light of the above it is interesting to adk: What are the criteria that building technology would have to take cognizance of?

Mass Production:
Some Characteristics of Conventional Buildings as Industrial Products

a. We live in buildings. With no other product do we have such an intimate relationship.

b. We live among buildings. Together they form the framework of our everyday lives. In aggregate they make a system that is, hopefully, more than the sum of its parts: a neighborhood, a city, a socio-physical environment, a language we live in.

c. In aggregate, buildings are extensive. They articulate the immediate material interface between us and the natural environment.

d. Buildings have long lifetimes. The mean life of most buildings made today extends to the beginning of the third millenium. In making them, we shape the shape of a tomorrow we know very little about.

e. Buildings are large. Only shipyards and the aerospace industries handle products of equivalent size. Being tied to the human scale, buildings cannot be miniaturized--an effective strategem in other industries, e.g., electronics. ~~Being~~ *Since buildings are* bulky, the costs of the materials they are made from necessarily make$ up a larger proportion of total costs than t*h*ey do for other products. The costs of the metals and plastics in a television set ~~is~~ *are* only a few dollars at most, say five percent of total costs, while in the average building component they are in excess of forty percent. Th*e*s*e* cost*s* cannot be essentially reduced by component part industrialization.

f. Most building components are of comparatively simple shape and composition. This being the case, very little machining or processing is required. Hence costs cannot easily be reduced by innovations in production equipment. Increased mechanization yields diminishing marginal returns at an early stage.

g. The cost-density of buildings is low. Having low weight-to-value and volume-to-value ratios, the distance they can be economically transported is strictly limited.

h. Buildings are context dependent. At points of interaction the environment of most products is standardized. The environment of buildings is people. They cannot be standardized, or so we like to think. Cars, by contrast, are not made to fit their garage. They are made to fit highly standardized roads and traffic systems where the operations they have to perform are strictly proscribed. Buildings have to fit sites whose topography, load bearing capacity, climate, points of access, orientation and surroundings, etc., etc., all vary from place to place.

i. Buildings are socially sensitive. For any given building, the occupancy level, status, life-style, living patterns, and functional requirements can change abruptly in ways that are difficult to predict. If buildings cannot adapt then they quickly become obsolete. The urban renewal of industrial system buildings may become a major problem in the not-too-distant future.

j. Buildings are fixed in location. Being tied to a site means that, with the exception of mobile homes, buildings cannot be factory assembled, distributed, stocked, and marketed nationally. We move between them rather than their moving between us. A large proportion of the work has to take place on site.

k. Buildings are expensive. Buying a home is the largest single investment most people make in their lives, equal to two to six times their annual income. In aggregate, the building stock represents the nation's largest capital investment in anything.

Not So Much a Step Further as a Step Different: The Basic Options

The interconnections and consequences of this far-from-exhaustive list of production characteristics are multiple and far-reaching. When they are allied to the environmental and social considerations, it becomes extremely difficult to see how the current mainstream approach to building industrialization could ever be successful. The purely technical

points e through f, above, which are inherent, seem to preclude the
possibility of any major breakthrough being made by a component part
building system.

*   *   *   *   *

Nevertheless, it is tragic that precisely in an era of environmental
concern, and rapid social and technological change, mainstream architects seem to be committed to an industrial approach that will make
buildings both less responsive and longer lasting as immutable objects.

*   *   *   *   *

What is called for is not so much a step further as a step different. Conventional housing does not seem to be a very good subject
for current mass-production techniques and, conversely, current mass-
production techniques do not seem to be well-suited to housing. We
can't take a free ride on the automobile industry. This leaves us with
two basic options:

1.   Abandon "housing." After all, "houses" are a solution to the problem, and not the problem itself, which is the provision of living facilities. (But this runs into the "social acceptance" problem.)

2.   Try to change the technological basis of "industrialization" and
develop a new building technology more truly suited to our needs. (This
runs into a host of problems ranging from research funding to vested
interest groups and building codes.)

3.   Some combination of 1 and 2. This seems attractive, since it would
allow a more evolutionary approach to the problem of retotalization.

That none of these approaches is as hopeless as might appear, is,
I think, being currently proved by the emergence of what we have labeled
"responsive building technology." While no one would claim that it is
an answer to the problems we are beset by, it does serve to highlight
some interesting possibilities--seeds, perhaps, to more powerful solutions.

III. Responsive Building Technology: A Tentative Delineation

Emergent, marginal, dispersed; under-researched and under-financed,
but for all that, stubbornly resilient, responsive building technology
is basically a grass-roots movement issuing from new technology, advocacy planning, intentional communities, the counter-culture, self-help
housing, and experimental construction. Individual approaches and concerns are as diverse as the roots; they run the gamut from high-tech
complexity to low-tech simplification; from computerized environments to
wood shelters. It is impossible to characterize it completely because
it is still inventing itself. Nevertheless, certain broad directions
are emerging, although it is a rash man who would attempt to identify
them to everybody's satisfaction.

*   *   *   *   *

11

From its vague beginnings, responsive building technology has been based implicitly and most often explicitly on technology rather than a "style," be it whether "international style" formal or "Sea Ranch" vernacular. The slick graphics of what has been called "wet-dream architecture" are noticeably absent. What we communicate to each other is information about new techniques, technologies, detials, and systems that work.

Environmental protection, energy conservation, ecological integration, recycling, structural efficiency, ease and malleability of construction and user control, are starting points. The articulation of space comes later, perhaps too much so. The concern is to develop a building technology that is both humanly and environmentally responsive; the beginnings, if you like, of a humanistic biotechnology.

*   *   *   *   *

A distinctive feature of responsive building technology is that it is "function" rahter than "product" oriented; it is concerned with fabricating living facilities rather than houses per se. In this respect it is closer to the communications industry, which is not devoted to a particular product, but rather concerned with developing means of communication, than either the automobile or the industrialized housing sector, both of which are concerned with particular products, cars and houses, rather than with providing means of transport or dwelling systems.

A. The Structural Vocabulary as "Congealed Spirit:" Some Roots and Rationales

In the place of the slab, box, and skeletal constructions of industrialized building systems, which result in discrete orthogonal geometries and component part assemblies, the structural approach of responsive building technology is primarily derived from space structural techniques.

Space structures (probably one of the most important technical breakthroughs in 20th century architecture) simplify and separate structural forces by articulating them in three-dimensional space. The resulting spatial patterns of tension and compression act as indeterminate integrities that cannot be reduced to two-dimensional plan and sections. Bending moments, shear, and torsional forces are eliminated or reduced to second and third order effects.

. . . cable nets, stretched and pressurized membranes, shells, folded plate structures, space frames, and the ubiquitous geodesic dome . . .

. . . radiola and diatoms, firal structure, crystal lattices, analids, reticulated spines, and sea-shells . . .

. . . The result is structurally efficient polyhedral, curvilinear and free-form geometries characterized by minimal potential energy and material useage, large clear spans, and, with some types, great design freedom.

*   *   *   *   *

Space structures open the way for both new production methods and construction techniques, and the use of new materials with low moduli of elasticity and, for that matter, old ones as well, in a plastic and malleable manner.

                    *   *   *   *   *

    Space structures are rarely pursued as an end in themselves. They
are used as a convenience, a technology, useful only insofar as it is
helpful, to be hybridized and bastardized without hesitation. Slab,
post and beam, and skeletal constructions are not outlawed, but are
treated as special cases of space frames, folded plate structures, or
whatever, to be used as it pays. Yet there is an aesthetic, applied and
implicit: a respect for place; a feel for simplicity; a striving for
organismic integration; out of which issues a new space, at once pure
and funky, charged with the sharp pleasure of a tensed cable, the cry-
stalline logic of a space frame, the fat, carveable squishiness of foam,
the billowing responsive mass of an inflatable; softened by growth
and the exigencies of time, place, people, and funds.

B. Materials

Criteria:

*Energy costs of production and assembly

*Resource conservation

*Reuseability, recycling, and biodegradability

*Economic cost

*Appropriateness to context

*Structural efficiency

*Ease of production and assembly

*Environmental control properties

*User adaptability

Foams, fabrics, light alloys, films, plastics, fibers

Light aggregates, laminated wood, composites

Reused wood, stabilized earth, glass and ceramics

Industrial wastes and scrap materials (sulfur and slag, car tops)

    Two approaches to materials are evident, reflecting the roots of
the movement:

    A low-tech, loose-fit Californian approach that avoids "artificial"
    materials such as plastics in favor of local ones such as wood,
    earth, or stone, or scrap and waste materials

    A relatively medium- to high-tech one that constantly experiments
    with new materials and new ways of using old ones.

The conflict between them is more apparent than real (wood is also a
cellulose monomer . . .). The end strived for is the same: environmental integration and user responsiveness. Both approaches see materials as a larger eco-technical time frame of use and reuse and in relation
to user and environmental responsiveness. Only the loop-size, time frame,
and degree of sophistication are different. The spectrum ranges from
attempts to use materials as active interface between form and function,

interior and exterior, almost as biological tissue, and a more passive and meditative static articulation of them that seeks inspiration from their use in indigenous architectures and local traditions.

In practice, the low level of research funding has kept the spectrum closer to low-tech approaches, and this will probably be the case for some time to come. Yet, both approaches can be used effectively in the same building. Antioch College's pneumatic campus is a partial example of this: a wide-span climatic envelope of high-tech materials encloses interior use-structures of wood, stone, canvas, and earth embankments. As yet experience is small; the dialectical synthesis of the local and technological, the simple and the sophisticated, still lies in the future.

Despite their present disadvantages, hopefully curable in time (energy-intensive, polluting, non-recyclable,) many high-tech materials are open to development in a way that wood and stone are not. They allow a quantum leap to new levels of sophisticated simplicity: phototropic glass, optical interference films, superplastic metals, photoelectric surfaces, etc., etc., etc. Thermoplastics are the only material that could conceivably meet all requirements of the building envelope simultaneously: transparency, structural strength, insulation, reflectivity, texture, etc. The variform patterned extrusion of multidensity plastics is a definite future possibility for buildings that has already been realized for furniture.

The Metabolism of Buildings: Homes as Artificial Eco-Systems

If we are to speak of a truly responsive building technology, then we must have one that is equally responsive to environmental as to social and individual needs. One that conserves energy, utilizes scarce natural resources sparingly and efficiently, recycles its materials, grows richer over time, and functions in harmony with nature.

The thermodynamic characteristics of buildings, determined by orientation, insulation, fenestration, shading, albedo, shape, size, thermal capacity, etc., as well as infiltration and ventilation rates, useage patterns, and the efficiencies of installed space-conditioning equipment, begin to lay the basis for an operational integration of building and site--a building ecology. The aesthetic relation of interior to exterior becomes operational rather than sculptural.

The emergent energy crisis is likely to have a major impact on all types of architecture. By forcing us to view buildings as environmental regulating devices that consume energy both in their construction and operation, it confronts us with a set of architectural design criteria that have been ignored for too long by modern architecture. The easy way out is to take a thermos flask approach that simply increases insulation levels and reduces fenestration. The more difficult alternative is the opposite one of ameliorating thermal loads by integrating the building as completely as possible in the site-specific conditions.

Experimental research on solar, wind, and other natural energy systems is the logical complement of work on energy conservation.

Moveable insulation, variable transmission membranes, flat-plate solar collectors, wind turbines, methane generators, and pyrolytic units are all ways of actively interfacing buildings with their contextual and operational environments. Combined with composting, serial water useage, and materials recycling, they lead to the possibility of designing the home as an artificial eco-system.

The catch is cost. Solar energy is not free. The equipment necessary to concentrate, convert, store, and diffuse it costs money. Even when life-cycle costs rather than first costs are used as the accounting base, most of this equipment is still too expensive and inefficient. For a flat-plate solar collector to be competitive with a conventional fossil-fueled mechanical space conditioning system, system costs would have to be halved and fossil fuel prices doubled. Both of these may well happen during the next decade.

In terms of immediate application, low-tech approaches are the only ones that can pay off. A vast amount of work, research, and experimentation remains to be done. This is the area which will probably see the most development during the next five years.. The common goal is the realization of a self-sufficient residential eco-system integrated with and powered by environmental energy fluxes, that, at the very least, meet 70% of the space-conditioning loads, 60% of the electrical power load, and 50% of our nutritional needs.

*   *   *   *   *

The ineffable beauty of a tree . . .

*   *   *   *   *

We are in the process, crudely, slowly, of investing our homes with an integral metabolism. Nor is this necessarily the end of the story, for after the redesign of building physiology, the development of metabolism comes to nervous systems. Cable TV, home computer terminals, the sensitive monitoring and control equipment required to run the energy systems and recycling loops, will, it seems to me, lead in the end to the cybernetic environments envisioned by Nicholas Negroponte, myself, Warren Brodey, Charles Eastman, and many others. Environments with the sophisticated simplicity of an airfoil; the beauty of a tree; truly responsive environments beside which our present efforts will look like the crude beginnings they are.

C. Production

Conventionally, the architect attempts to meet the demand for flexibility through ingenious spatial manipulation rather than through materials, construction methods, and equipment.

Such an approach brings jointing problems, the highest cost item in construction, to the forefront. It is largely the attempt to minimize on-site jointing problems that has, ironically, led to the popularity of box-cell construction systems.

Some recent experiments in building technology turn the problem on its head by, so to say, starting with the joint and its articulation, and making it the basis of the building process. One such process is based on the continuous on=site extrusion of plastic foams or lightweight aggregate concrete, like squeezing toothpaste from a tube in successive layers; another is based on spraying materials onto draped and suspended reinforcement, or onto inflated molds.

Identity of operation is maintained in both cases without requiring the use of geometrically identical building components in repetitive configurations. The need for high precision fit between prefabricated surfaces is avoided altogether since the"joint" is spread throughout the building fabric. Door and window openings are either cut later, or the frames are simply sprayed in place during construction.

Although relatively new in modern building technology, these methods, expecially extrusion, have many parallels in nature, and deep roots in indigenous building technology, specifically in adobe and mud construction methods. The renewed interest in shell structures, developments in materials and, to a lesser extent, in mechanical equipment, have once again brought them to the forefront.

The design freedom is trmendous. Production, assembly, and design become one integral process. The main problems are technical and mechanical. When lightweight aggregates are used in concrete, the slurry tends to separate out in the extension hose. Quality control is difficult to maintain, and cleaning the machinery when work stops is a real bummer. Many technical problems remain to be solved.

Yet the very nature of these problems points up the large returns to be realized by innovative development in articulated extrusion systems and extrudable mixes. Here increased mechanization really pays. Controlled articulation, variable density patterned extrusion, and variform molding, are all possibilities that seem, at least to me, to contain the seeds of a truly architectural building technology that could range from do-it-yourself to highly automated systems capable of great variety and contextual sensitivity. Subsequent additions and alterations should be easy to make. If the extrusion equipment becam cheap enough it would become a normal part of household equipment rather like a saw, or hammer and nails.

Thanks to Domebook, the structure most clearly identified with responsive building technology is the geodesic dome. In many ways this is a pity. Domes are not the most appropriate structure for backwoods

16

do-it-yourself production, being extremely sensitive to precision in
relation to the goodness of seal and the structural strength developed.
Of all structures, the dome is one of the most suitable for machined
mass-production.

A geodesic dome for a responsive environment? Surely nothing could
be more rigid and inconvenient!

While there is a lot of truth in the preceding statement when
domes are small, the real point is that the dome is nothing more than
a clear span climatic envelope that uses a minimum of materials and
allows almost complete interior freedom. The physical separation of the
envelope from support structure and interior partitioning radically
relaxes the production and, for that matter, the design constraints.
Interior structure becomes, effectively, furniture, to be added to and
moved around at will. The range of materials that can be used inside
increases dramatically while the configurative possibilities are far
more loose and fluid. Nor is the dome form obligatory; wood lattice
gravity structures, space-frames, and thin shells can be used to make
envelopes adapted to site conditions and landscape.

While the foam-form is based on advances in production techniques
and materials; the dome form points towards a retotalization of the concept of a house.

*   *   *   *   *

A major limitation of responsive building technology to date is an
almost exclusive concern with the isolated individual dwelling; an ob-
session with the pastoral setting. It has yet to show that it is capable of being used in high-density area, or of being integrated with
existing urban environments.

*   *   *   *   *

The climatic envelope suggests one way of extending the technology
to high density urban environments: Rather than trying to pack and stack
domes, the envelope is extended to enclose whole areas, as Fuller suggested for midtown Manhattan. The interesting thing is that if such a
project was ever implemented it would immediately make all interior
buildings technically obsolete, since they would no longer be subjected
to the wind and snow loads, rain and temperature variations they were
originally designed for. Moreover, activities originally housed only for
climatic reasons could take place in the "open air", so to speak. The
result could be a "Polynesian" architecture using soft, adaptable building
materials, easily assembled, recycled, and industrialized, together with
multi-story support structures. Being unified, sophisticated, adaptive
control systems and new means of transport could be used effectively
eliminate many of the problems of existing urban environments. Noise
and pollution control, total energy systems, and computerized communications could all be integrated into the environment. Avoiding the technological megalomania of a single vast enclosure, with the attendant
problems of thermal stratification, air circulation, condensation,

causing miniature rainstorms, and so on, a collection of smaller, community- and park-sized interconnected envelopes of varying size and shape incorporating both exposed and enclosed spaces and structures, would keep options open and provide a highly varied environment.

Such a development would not only encourage the emergence of diverse and sophisticated urban environments, but would probably have a huge impact on life-styles and social forms. Unfortunately, though, the number of new cities likely to be built in the future is small. Again, Antioch's pneumatic campus is a forerunner of such an approach applied on a miniature scale.

*   *   *   *   *

The grain size of urban growth in the United States is the speculative development project of some 20 to 200 dwelling units. This is perhaps an area we should be directing more attention to in the future. The increased scale, while introducing new problems, would also introduce new possibilities, especially in utility systems, recycling wastes, shared heating and cooling systems, perhaps based on solar energy, community composting, and gas generation. There are almost certainly economies of scale to be gained from integration that could make energy systems feasible that are still impractical for the individual dwelling. Perhaps the most important development could be the emergence of a new sense of community based on a common ecology.

*   *   *   *   *

The building industry has the poorest record of productivity increase of any major industry, approximating 1% a year since 1945.

The reasons usually given are the fragmentation of the industry and its notorious resistance to innovation.

The remedy usually proposed is the industrialized mass-production of housing in central plants. Yet the massively financed attempts to do this have generally failed to date. The causes are undoubtedly complex, but some re-examination of assumptions might be helpful.

The fragmentation of the industry may, to a large extent, be due to the peculiar production characteristics of the product made; that is, it may be ingerent rather than due to poor integration and under-capitalization. Much of the "innovation" being resisted is directed towards mass-produced building systems. That is, the technology being introduced may be directly antagonistic or of limited use to the existing structure of the industry which is, consequently, unable to take advantage of it.

Very few attempts have been made to develop technologies compatible with and supportive of the existing decentralized, dispersed nature of the building industry.

Equipment rental and sub-contracting provide a capitalized pool for production that is probably far more open to and supportive of innovation than is generally recognized. It provides an open production facility, not tied to any particular design or system. Most of the responsive building designs shown in this volume would never have seen the light of day without it.

* * * * *

D. Design: The Evolutionary Building

One of the most interesting features of responsive building technology is its tendency to blur the hard and fast distinctions that have existed until now among design, production and assembly, use, and demolition.

Design becomes a continuous activity that merges with user control; production and assembly shade off into adaption, extension, and upgrading. The system is in a constant state of adaption and evolution in response to changing needs and visions of the occupant, an articulation in time. In fact it becomes difficult to speak of a building's age or lifetime. Like the human body, it is constantly changing its tissue while maintaining a more slowly changing functional pattern. Over a fifty year period, little if any of the original building fabric may remain, and the building's configuration may have altered radically.

The evolutionary dwelling has a number of advantages over the static, finished building. One of the most basic is financial. Houses are expensive. Too expensive. To keep the market price competitive, most houses are built on the basis of lowest possible dollar cost per square foot of finished structure. This initial low cost is usually purchased at the price of high operating costs. Energy utilization is poor and heating and cooling bills are high. Optimized life cycle costs considerably increase first costs. The efficient use of energy calls for increased insulation levels, double glazing, and so on. Further, all natural energy space conditioning systems are capital-expensive. Yet our credit is limited.

* * * * *

One answer is to shift from lowest first cost of finished structure to lowest possible incremental costs and high terminal value, i.e., to "grow" our homes. A "start-up" structure is built and moved into immediately; over time the structure is extended and added to, new systems are incorporated, and the older onew integrated or sold. After some time a relatively stable state is reached, and the mature dwelling enters a long cycle of tuning, upgrading, and adaption. The whole process is analogous to ecological succession from a grassland to a forest climax state. Development is essentially an on-line real-time process guided or initiated by the occupants. Every home has its history. At a community level, the more evolutionary homes there are the easier it becomes to "start up," and the richer the options become. The growth of one building stimulates the growth of another as parts are upgraded and changed, and enter a common spare parts reservoir, for recycling. The homes made from salvaged windows and doors are only the first beginnings.

If such a building approach becomes widespread, it could play a
large part in reducing urban decay and building obsolescence rates.
A rigid construct, run-down, demolish, reconstruct building cycle is
expensive both financially, socially, and environmentally. We need no
longer be trapped in a past that no longer fits us. The future's present can find its own configuration. Although this growth of an evolutionary architecture has been largely unconscious, invented by necessity,
nevertheless its full realization implies a radical reconsideration of
the design and fabrication of homes as we know them. A reconsideration
that involves its own abdications. No longer will we be the Form-Givers
of the Age, the high priests of property, erecting art-objects on sub-
urban pedestals or designing prestige monoliths for the business dis-
trict. Rather than definers we will have to become the designers of
possibilities within which people find their own definitions.

*   *   *   *   *

If we compare the bits and pieces of our not-yet-born technology
with our original list of industrial characteristics of housing, it is
amazing how many of them we have managed to wiggle around. This may be
because I composed the list and then analyzed responsive building technology in terms of it, i.e., answered my own arguments, but the fact
remains that the arguments can be answered, at least potentially,

*   *   *   *   *

A doubtful demonstration of feasibility is, however, not enough.
It's easy to design alternative architectures but unless we can show a
social and economic basis for them, we are merely building castles in
the air.

*   *   *   *   *

It seems unlikely that H.U.D., or anybody else for that matter, is
going to come up to us and say, "Here's a few million; get on with it."
Nor is responsive building technology likely to sweep the suburbs by
popular demand. Where, then, is our social base?

*   *   *   *   *

The adult population can be classified socio-economically in many
ways: by occupation, income level, social class, residential location,
and so on. One of the most interesting calssifications is by age. If
we divide the adult population into fifteen year age brackets--19-34,
35-49, 50-65--and look up their purchasing power over time, then a couple
of interesting things emerge. Up until the 1930's the dominant consumer
group was the 50 to 65 age graoup. From then until the early sixties,
the age of the short-back-and-sides and the economy-sized family package, the 35 to 49 age group was dominant. More recently, largely due
to the post-war "baby boom," economic dominance has shifted to the 19
to 35 age group.

*   *   *   *   *

These are the people who, as soon as they started buying records, turned beat music from a sub-cultural phenomenon to the popular music of today; who, a little older, stood clothing fashions on their head; who dig beanbag chairs, water beds, and inflatable furniture. And we all know what happened when they went to college . . . Here is our constituency: ourselves and our contemporaries.

       *   *   *   *   *

Even so, there is no royal road to realization. Our economic dominance is one of numbers rather than wealth per person. We will have to evolve our own architecture by the sweat of our brows, one step at a time, socially. Wider acceptance will come, if at all, slowly and by the back rather than the front door; through stages that may, only may, go something like this:

*Experimental construction
*Counter-culture
*Owner-built homes
*Recreation (second) homes
* . . . publicity . . .
*Generalized acceptance

The next, at most two, decades, will probably be decisive. We have a long way to go and but little time. But the vision is there: evolutionary domestic eco-systems, responsive environments, a new peace and joy. The people are there: you and I. And we are beginning to get the technology together. Time may be short, but it's on our side.

An Attempt to Derive the Nature of a Human Building System from First Principles

Christopher Alexander

I wrote this paper for a seminar discussion in 1970. There is little in it which is precise enough to convince a person who is in a mood to doubt. However, for all that, I do believe that careful empirical study of the psychology of space, and the laws of efficient structure, will, in the long run show that my conclusions are fundamentally correct. For this reason, I am publishing them in this sketchy form with the idea that they may help some of my colleagues who are already looking in a similar direction.

When a person designs a building, he usually starts with certain known structure types: column and beam, load-bearing walls, stud construction, monolithic reinforced concrete, etc. The building forms which designers have created by this process are very unsatisfactory. To begin with, they fail to meet many important needs. What is far worse, though, these structure types are so sharply distinct, and the choices between them so arbitrary, that one is left with the feeling that none of them are really quite right, and that no one has ever plumbed the problem of defining the class of structures which are actually correct for a human building. For instance, comparison of columns-and-beams with load-bearing-walls, leaves you with the feeling that there

*The psychological arguments for Postulates 3, 6, 10, 11, 20 and 24 are given by the patterns <u>Ceiling height variety</u>, <u>Indoor space</u>, <u>Columns at the corners</u>, <u>Thick walls</u>, and <u>Sheltering roof</u> which will appear in the first edition of <u>The Pattern Language</u>, to be published in 1973. They are summarized, in part, in Alexander and Jacobson, <u>Specifications for a Human Building System</u>, this volume.

are certain pros and cons for each alternative, but that
final choice among these alternatives is more or less
arbitrary.

In this paper, I shall try to overcome this
arbitrariness by arguing from first principles. I shall
start with certain postulates, based on the human needs
which occur in a building, and the laws of nature, and
try to derive, from these postulates, a general description
of the morphology--(i.e., the class of structure) which
is correct for human buildings. As you will see, I believe
we may conclude that any building structure which meets
human needs, and follows the laws of structure, will have
the general character of the room illustrated by the
following drawing:

In order to derive the morphology of such a
building without prejedice, it will be necessary to avoid
assumptions about "types" of structure, like load-bearing-
walls, shells, or column-and-beam--and instead carry on
the discussion at a level of description which could
apply equally well to any of these so-called types, and
also to the very much greater variety of "mixed types"
which lie between them. I begin with the most general
description of a building:

Postulate 1.

From a human standpoint, a building may be viewed as a collection of indoor and outdoor spaces, each one defined by human or social purposes. If you think of each of these spaces as a solid lump, then you can visualize the building as a three dimensional arrangement of these lumps.

Postulate 2.

A Series of Postulates Concerning the Shape of the Lumps

Each space has a horizontal floor. A change of floor level will be treated as a transition from one space to another.

Postulate 3.

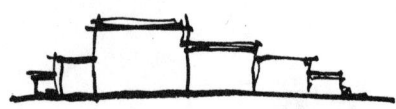

The ceiling heights of spaces vary according to their social functions. Roughly speaking, the ceiling heights vary with floor areas--large spaces have higher ceilings, small ones lower. (Ceiling height variety pattern)

Postulate 4.

The edges of the space are essentially vertical up to head height--i.e., about 6 feet.

Postulate 5.

Above head height, the boundaries of the space come in towards the space. The upper corners between wall and ceiling of a normal room serve no function, and are wasted: it is therefore not useful to consider them an essential part of the space. This does not mean that the structure must have the configuration shown. It does mean that the most general shape for the space has this configuration, and needs no more, so a structure which enclosed more space would be wasteful.

If another floor comes above, it needs to be like this, because the upper floor is flat.

Postulate 6.

Every space is convex in plan. This means that there are no re-entered angles in a space. Wherever such a re-entered angle does occur, it is considered to be the junction of two spaces. (<u>Indoor space</u> pattern)

Postulate 7.

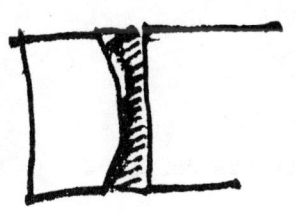

The boundary of any space, seen in plan, is formed by segments which are essentially straight lines-- though they need not be perfectly straight. The reason is this. A curved boundary makes a convex space on one side, but a concave space on the other--which is un- acceptable, by postulate 6.

This means that every space is essentially a polygon in plan--though of course not necessarily rec- tangular. And if a space does have a boundary which is curved in plan, the wall must be thick enough for the next door space to have a straight boundary in plan.

Postulate 8.

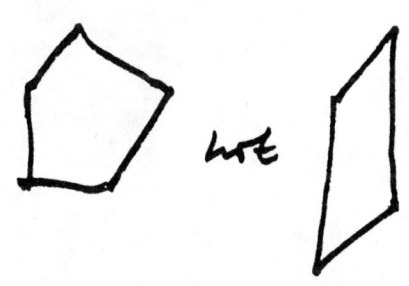

As a general rule, no space has any acute- angled corners. Acute angles are almost always useless: it is almost impossible to make an acute angle in a room, which works. Together with postulates 6 and 7, this means that the corners of the spaces are obtuse angles between 90 and 180 degrees.

Postulate 9.

A building is a packing of polygonal spaces in which each polygon has a beehive cross section, and a height which raises according to its size. This follows from postulates 1 - 8.

Postulate 10.

Postulates Concerning the Enclosure of Spaces

Within the building every space has a floor and a ceiling. It may or may not have windows, doors, partitions, etc. Each space must be partly defined by the material which forms its vertical boundary. However, it is by no means necessary for all this material to be there--a large part of it can be missing, and the space will still be defined and felt. (Indoor space pattern)

We assume, therefore, that the material in the boundary of a space need be there only to the extent that it is psychologically necessary to create the virtual space in people's minds.

Postulate 11.

A space with a polygonal plan is defined psychologically at least, almost entirely by its corners. There are exceptions, but as a general rule, this means there will need to be material in those parts of the boundary which form the corners of a space. (Columns at the corners pattern)

Postulate 12.

Postulates on Flexibility

The statistical distribution of spaces of different size (small, middle, large, etc.) is largely fixed by the nature of man and society--and does not need to change during the life of a building.

Attempts to change large spaces into small, or vice versa, always fail, and are not worth including in the concept of flexibility. This is because the three most important characteristics of any room--its height, acoustics, and natural light--are all critically related to room size, and will always be wrong after any change in which the size of spaces is itself changed.

From postulate 12 we may derive:

Postulate 13.　　　　　　　The need for so-called flexibility in a building can be completely taken care of by changing the amount
of enclosure round different spaces. There is never any
sense in trying to change the basic spaces themselves.

Postulate 14.　　　　　## Postulate on Design

At some stage in the design process, it is
possible to specify a building as a three dimensional
arrangement of spaces, in which all the spaces have the
characteristics defined above. This is the stage which
immediately precedes the design of the load bearing structure.

Postulate 15.　　　　　## Postulates on Structure

To visualize the problem of defining the structure,
in the most general sense, imagine the following process.
Make a lump of wax, for each of the spaces which appears
in the building, and construct a three dimensional array
of these lumps of wax, leaving gaps between all adjacent
lumps.

Now, take a generalized structure fluid, and
pour it all over this arrangement of lumps, so that it
completely covers the whole thing, and fills all the gaps.
Let this fluid harden. Now dissolve out the wax lumps
that represent spaces. The stuff which remains is the
most generalized building structure.

This general structure is homogeneous. There
are no distinctions, yet, between parts of it that work
in compression, or in tension, and no distinction, yet,
between columns, beams, arches, vaults, walls, etc.

Postulate 16.　　　　　The problem of defining an ideal structure,
is the problem of defining rules which will tell us how
to take this homogeneous, ideal structure and make it
into a real structure.

To make this imaginary ideal homogeneous
"structure" into a real structure, three steps have to
be taken.

a. We may move the original spaces around a little, in order to improve the global stability of the structure.

b. We have to define the positions of doors, windows and openings between spaces, and remove the material from the ideal structure, wherever these openings occur.

c. We have to define the distribution of thickness in the remaining ideal structure to optimize its resistance against actual loads-- and specify the tension-compression characteristics of its different portions.

When these three steps are done, we shall have a complete, workable structure, with its geometry and thickness at every point specified in detail, and the compression and tension characteristics known at every point. At that stage, it will be possible to choose actual materials which have the geometry and stress characteristics of each piece within the whole.

Consider first the arrangement of spaces, in plan.

Postulate 17.

It is natural to expect that the corners of spaces, where the edges of different spaces meet, will have to carry the greatest load--since it is at these points that loads will change direction, and it is these parts of the vertical structure which will be subject to the greatest shear, bending and torsion. For this reason, it is natural to expect thickening at the corners. Although we need not think of the thickening as columns, it is congruent with the intuition already expressed in postulate 11, that a polygonal space is defined by its corners.

Each corner will either be a three or four way corner. This follows directly from postulate 8. A five way corner would create at least one acute angled space. If the connection is four way, it has to be right angled, to meet postulate 11. This means there are essentially three kinds of valid interior corners, in plan.

The T-junction is inherently less stable than the Y-junction, but makes sense if there is another horizontal load coming in from an arch, or other horizontal member, spanning the larger space.

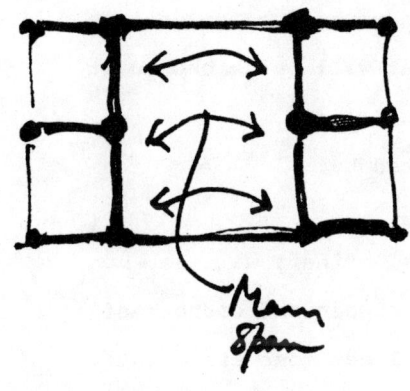

Vertical continuity. If the corners of spaces are most critical, and thickest, then these corners must be vertically continuous just as columns are in a conventional structure.

There are two ways of guaranteeing this.

Either   a.   Each corner, at a given level, must have corners below it at all lower levels. This is rather restrictive. It implies either that the floor plan at the second level is the same as that at the first, or that it is the same, with certain corners left out, which makes the spaces upstairs larger than those downstairs, and is unlikely to make sense in social terms, since larger spaces are usually more public, and need to be closer to the ground.

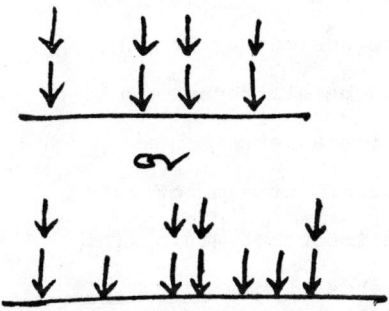

Or   b.   To get around this difficulty, we can say that the corners of spaces at the second floor, must at least fall above <u>walls</u> (i.e., boundaries between spaces), on the floor below, but not necessarily above columns.

This means that the lower walls will contain extra thickening at certain points which are <u>not</u> corners, to carry loads from upper floors, and the lowest floors will have the largest number of these extra "columns."

It is clear, therefore, that the wall between the corners of a space, will not in general be a continuous homogeneous non-loaded membrane, but will be thickened at certain points, to carry concentrated loads.

This principle is elaborated by:

Postulate 20.

Struts and ribs in walls. Define a wall as any part of the vertical boundary between the corners of a space. In general, this wall will act most efficiently if it is non-homogeneous, and braced and stiffened by thicker parts, which come out from its surface, at right angles to the surface.

There are several reasons. A homogeneous surface is never most efficient unless it is acting in pure tension. The walls of a space will rarely, if ever, be acting in pure tension. If they are subject to bending, compression, and shear, they need to be stiffened. This will need a wall which looks more like a leaf than a flat wall--the ribs may be all vertical, or some of them diagonal and horizontal. Even more important, this kind of wall is required to meet the demands of the <u>Thick wall</u> pattern.

Postulate 21.

Rounding corners. Right angled openings in a structure are the weak points in a structure, liable to cracking or rupture. All openings should, if possible, be rounded or angled off, so as to lead the forces gently round the opening.

This means that all windows, doors, and corners between columns and beams, should be chamfered at the corners.

Postulate 22

Stiffening of all open edges. There are extra load concentrations at any free edge, so that all the openings round doors, windows, etc., must be stiffened by ribs or thickening, of the type already described under postulate 20. As a comment on existing ways of building, this suggests that window frame and door frames should play more of a part in the overall structure than they do today.

Postulate 23.

Horizontal continuity. In traditional column and beam structures, it is important to keep beams as long as possible, running through several columns, to reduce effective bending length. This is made possible, in part, by a grid of columns. However, it does not require a grid of equal spacing. It can be guaranteed, equally well, by a grid in which adjacent grid lines are unequally spaced: Thus, for example, this column grid allows perfectly adequate horizontal continuity in beam members.

Limited lining up of spaces of this kind will allow the beams to run continuously from one space to the next. Since, by postulate 5, the space for the beam is triangular in cross section, this will effectively create a system of intersecting triangulated barrel vaults.

Postulate 24.

Roof postulate. Floors need to be flat, and must therefore use a slightly inefficient structure, to create a flat upper surface. The roof does not need to be flat on top, except where there are roof gardens. For this reason, we should expect roofs to have the most efficient structural form - some type of dome or vault, either a single one, or a multitude of them covering individual spaces. This is perfectly consistent with postulate 5, which defines a similar shape for the inside of the spaces. In short, top storey spaces, will have their ceilings formed directly by the roof, which will be a sloping, dome or vaulted structure,

31

except where there are roof gardens. This is also consistent with the psychological demands imposed by the
<u>Sheltering roof</u> pattern.

Any building which satisfies these 24 postulates,
will be made up of interior spaces which look more or
less like this, but loosely packed, both horizontally and
vertically, with some reasonable degree of continuity,
but without being on any exact grid, either horizontally
or vertically. <u>As you can see, the structure is an
archetypal one, and contains echos of many many different
traditional building types. It is slightly reminiscent
of a barn interior</u>.

Christopher Alexander is the founder of the Center for
Environmental Structure in Berkeley, California

Specifications for an Organic and Human Building System

Christopher Alexander and Max Jacobson

Introduction.

    For the past several years, we at the Center For Environmental Structure have been trying to create a process of building which leads, once again, to environments that can be described as <u>organically whole</u>, or <u>alive</u>.

    What do we mean by such terms? We do <u>not</u> mean that the environment will literally <u>look like</u> a biological organism, any more than a lovely mountain cabin looks like a tree, a stream, or a rock. Nor do we mean that the environment will <u>act like</u> a biological organism, any more than the same cabin literally speaks to us or has an intentional animate life of its own. We mean that the environment - the social and physical environment <u>together</u> - form a living system;[1] and that like any living system, this one can be more alive or less alive, more whole or less whole. This idea is more fully described elsewhere,[2] but we are already sure that objective indices of the state of health of environmental settings can be specified.[3] When we use the terms organically whole, or alive, we are referring to the state of health of the social and physical systems which together make up the environment.

    We have found that a building process which is able to make buildings which are organically alive needs four elements.

    1) <u>A common pattern language</u>: A process for building which can improve and evolve through public debate. Our first attempt at such a pattern language

is almost complete and will be published at the end of this year.[4)] The patterns in the language are physical solutions which resolve conflicting human tendencies, conflicts which frequently recur in the existing environment.

2) <u>User design</u>. Users must once again design their own environment, using this common pattern language to inform their work. Only then can they feel competent, responsible and mature. They know their own needs and the particulars of their own problems better than any professional designer.[5)] Our experimental work has shown that laymen can make competent house designs using an explicit pattern language.

3) <u>Repair and piecemeal growth</u>. Every act of building must be thought of as repair of the existing environment. The idea of building anything once and for all, or building all of a new facility prior to occupancy, is directly opposed to the way of nature. Nature is continually involved in ongoing, piecemeal growth and repair. It is not enough to have users initially designing their environments; it must also be possible for people to improve their existing environments gradually through repair.[6)]

4) After several years of work with these ideas, it has now become clear that the organic process of building we envisage requires a fourth element: <u>a human building system</u> specifically designed to support the pattern language, user-design, and piecemeal growth and repair. After trying to work with currently available building systems, we have been forced to the conclusion that no building system now available is compatible with these three concepts. Yet we know that many traditional building systems used to be compatible with them; so it must, in principle, be possible to have such a system. In this paper, we will discuss the task of developing such a building system, and will give a list of specifications which it will have to meet.

1. Compatibility with the pattern language.

Several of the patterns in the current pattern language are particularly sensitive to the nature of the building system, and are difficult, if not impossible, to realize with presently available systems. Here are some examples.

Light on Two Sides (1969).[7)] This pattern discusses the fact that any social space or room should have daylight on two sides to prevent glare and a general atmosphere of "gloominess." If the arguments and data presented in this pattern withstand the test of professional criticism and debate, far-reaching changes in our building practice will be needed. To let every room get light from two sides, buildings will have to be much thinner; and they will need much more wall surface per unit area.

It is clear that many current building systems do not allow this to happen. The most obvious examples are the large office and commercial buildings which have very large spans and minimum surface area per unit floor area. In such a building, only the spaces at the four corners of each floor can be lit from two sides (About 15% of the total interior space in a medium sized building). Even in tract home construction, the trend toward use of 25-30 foot wooden trusses leads to box-like plans in which a third of the spaces are lit from one side only.

Sheltering Roof (1972). This pattern sets forth the argument that man has a deep psychological need to feel the presence of a building's roof as a protective element. Specifically, it specifies that it must be possible to see the roof, to reach out and touch the overhanging eaves. In the case of flat roofs, the users should be able to walk out onto it and use it as a balcony, walkway, porch, or garden.[8)9)] The roof's presence must be felt throughout the entire building - on the ground floor one can see the eaves serving as the windows' and porches' "eyebrows," while in the attic the whole form of the room takes the shape of the roof overhead and dormer windows stick out beyond the roof line.

Most current building systems ignore this need. Big office buildings hardly ever step back from floor to floor to allow contact with the roof surfaces. In conventional house building the pitch of the roof has either disappeared altogether (with no compensating use of the flat roof), or been given a shallow token slope with eaves that are far too high to touch.

Alcoves. A series of patterns deal with the need for various types of alcoves - small intimate spaces which open into larger, more public spaces. For example, the pattern "Family Room Alcoves" (1968) discusses the conflict that exists between the desire for a family to remain in close contact and the individual members' desires to carry on separate activities during the evening hours. The pattern specifies low-ceilinged alcoves around the family room, with partial views between them, each being between one and two meters deep. Other patterns which call for alcoves are "Window Place" (1972), "Master Bedroom Alcoves" (1969), "Corridors Which Live" (1970), "Activity Pockets" (1968), and "Reception Alcoves" 1969).

Since alcoves are typically at the outside edges of larger spaces, it is often the outside wall of the building which has to form them. This puts considerable demand upon the building system. Curtain-walls don't allow it. Most pure wall-bearing structures (like concrete block) don't either, since wall crenelations on the upper storeys would have to be carried straight down, putting too much constraint on the lower floor plans.

Columns at the Corners of Social Spaces (1971). Social spaces range all the way from nooks and alcoves (around 15 square feet), up to large meeting rooms (10,000 square feet or larger). In traditional systems, each one of these spaces was defined by structural members, and we believe that this connection between space and structure fills crucial psychological needs.

Yet many modern systems make this either impossible or structurally superfluous and irrelevant. For example, most modular systems (and all fixed-span modular systems) are incapable of allowing the structure to follow the social spaces. The same goes for typical concrete column-and-slab and steel frame construction.

Thick Walls (1967). This pattern discusses the problems caused by hard, smooth wall surfaces that are difficult to personalize, and specifies that wall surfaces need to be "thick" and "carvable" so that as time goes on each wall will begin to receive shelves, niches, and nooks according to the users' needs. The thickness of the walls is needed too, for sound and heat insulation, for storage space, as well as providing subtle transitions between one space and another as one passes through the thickness of the separating wall.

But current building practice treats the wall as a skin whose main function is to seal the inside from the outside. Again, curtain-walls are obvious offenders. Stud walls and concrete block walls are hardly better - the first because of its thinness and the second because of its hardness. The new use of wood panel systems for home construction is completely incompatible with this need for thick walls. So are the molded plastic and dome technologies. And imagine the fate of the poor man who tries to hang up a picture in a pneumatic house!

Ceiling Height Variety (1968). Every social situation has an appropriate ceiling height. If the ceiling height is wrong, the situation gets disturbed. Roughly speaking, the ceiling height over a given social group should be proportional to the horizontal diameter of the social group.

Many traditional building systems had this pattern
in them. Vaulted systems had it, for example, since
the height of a vault is proportional to its span.
Trussed systems also had this feature (when no ceiling
was superimposed) since the depth of the truss varied
with the span. But modern concrete slab buildings have
a uniform floor-to-floor depth regardless of the social
spaces within them. In these buildings, it is only
possible to get the pattern by suspending ceilings at
different heights; and this violates the psychological
need for structure and social spaces to be congruent.
The same trouble occurs in houses, when uniform long-
span timber trusses span all the way across the house,
and interior spaces are formed by partitions.

This small number of examples should show that
patterns often have profound implications for the choice
of a building system. We have found that no existing
building system is compatible with the patterns we know
to be important.

We turn next to the kinds of requirements that must
be put on a building system to enable users to design
their own buildings.

2.   Compatibility with user design.

The main feature which a building system must have,
to allow user-design, is an <u>explicit, conceptually simple
set of rules which tell a person exactly how to turn
a schematic sketch into a functional working drawing</u>.
We say "functional" since we do not mean an elaborate
conventional working drawing, but any drawing which the
builder can build from.

This was commonplace in the traditional Japanese
house design process. The family fixed the arrangement
of rooms which they wanted. Only three additional rules
were then needed, to turn the room plan into a working
drawing.[10)]   Such a drawing is shown below (Fig. 1).

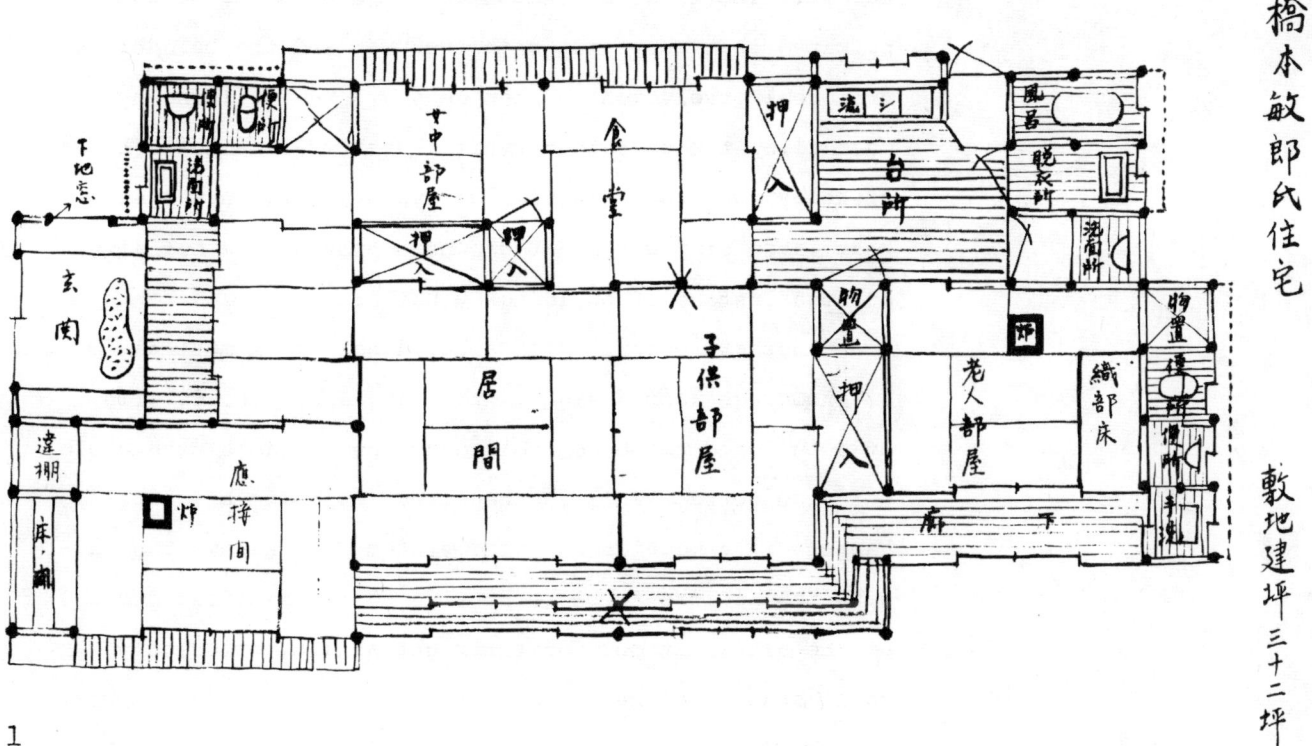

1

There is no need for an architect in such a building
process since the user-designer can specify all the
relevant details that the builder needs. And since the
functional working drawing he makes is far less detailed
than today's typical working drawings, the builder will
be able to express his own creative capacities as he
carries the plan through to completion. He will determine
the exact sizes of structural members, grades of materials,
and levels of workmanship. Just as user-design gives
more wholesome and fitting plans, so true builder-
participation will stimulate the builder to the same
levels of responsibility and expressive power which were
common in traditional societies.

Current building practice doesn't normally allow
the user to make a simple drawing and have it built
directly, because buildings are too complicated. Govern-
ments take on the responsibility for judging the worthiness
of designs before construction, and this means that drawings showing every detail must be submitted for a building

permit. The legal and technical apparatus are so complicated that they shut both the user and the builder out
from real involvement. When we say a new building system
must allow the user to get his sketch of a building built
directly, we are asking for a revolutionary new system
that is so simple that it can be approved as a building
process instead of building by building.

Another feature which a building system must have
to encourage user-design is that it allows the details
of a building to be controlled by the design of the whole,
not vice versa. This is the direct opposite of what
happens in a modular building system. It allows the layout of buildings to be responsive to the minutest demands
of the site, without perfect right angles, or exactly
equal spacing of bays. And it allows the builder to carry
out the design without "regularization" since he can be
confident that the details can be fitted into the larger
decisions about entrances, room corners, and so on.

The straight-jacket of modularization can perhaps
be made more clear by analogy with painting or biology.
In a painting, for example, the life of the whole stems
directly from the fact that the thousands of daubs of
different color, size, and shape are all laid down in
response to the overall image as it develops. In the
growth of an organism, each tiny cell takes on a form
that is subject to the overall form of the surrounding
cells. This is the source of "aliveness" in organic
forms. If all the cells were exactly alike it would be
impossible for the organism to be alive; the same goes
for a painting.

Very few of our current building systems allow for
the details to be executed in response to higher order
actions. In any strict modular system, the global design
is fixed by the conditions which the details impose on
it. The overall shape of a geodesic dome does not come
from the site and the client but from the system of
struts and connectors. In a curtain-wall office building,

the positions of the main columns are fixed by the available panel dimensions. Even tension and shell structures take on shapes that are more due to the hyper-critical needs of structural integrity, pre-fabrication, and assembly, than to overall planning decisions.

We seek a building system in which the details can be adapted to the needs of the whole. We want the builder to be able to lay out the overall building according to the client's instincts, with the confidence that he can later place the 2X4's, joists, or whatever else to fill in this general layout. The knowledge that the details of construction can always be fitted smoothly into the larger planning decisions is essential if the user is to take part in design. So long as the user feels that every planning decision hinges upon detailed dimensions of panels, windows, and door knobs, he will continue to rely upon professional designers.

Another point. Some users will want to help in the actual work of building, and many of them will want to repair their buildings after they move in. It will be best then, if the building system allows the user many different degrees of participation. A few will build for themselves; more will contract out the difficult initial stages and finish the building themselves; others will help the contractor all the way along; many will want to supervise the initial layout phase; most will do repair and modification work themselves.

This range of participation implies that the construction process must be radically simplified, to include more non- and low-skilled labor. Carpentry and brick-laying, for example, require labor from highly skilled trades, and are thus beyond the average user's capacity. The same is true of concrete formwork and steel welding. The building system we seek will allow low-skilled, machine-intensive participation in about the same way that a rented chain-saw lets an average suburbanite cut a year's firewood in a day.

Another question concerns level of finish. Available building technologies demand hi-skilled labor even to obtain a moderate level of good workmanship. Formica work, cabinetry, and drywalling simply cannot be done by low-skilled users without the result looking very amateurish. If users are to take part in the building process itself, it is crucial that the building system be capable of wearing different levels of finish with equal dignity. For example, traditional Japanese mud-and-wattle wall panels could be left in their raw and rustic initial form, could be roughly leveled, or could be further refined to create a perfectly plane and uniform surface. Each level of finish had its own character and integrity. Contrast this kind of option with today's emphasis upon "optional parts." One is an optional process - the other an optional product.

User-design would also be encouraged if the design process were more integral with the construction process. In today's building systems, the spaces only appear when the building is finished. Concrete column and slab construction is one of the best examples. The spaces don't appear until the very end of the building process, when the non-structural interior and exterior panels are put in place. But user-design is much easier when the spaces precede the structure in the building process. This means that the user can make a full-scale mock-up towards the end of his design process; and then use the mock-up as the beginning of the actual construction.

The need for full-scale mock-ups is not a crutch for the lay designer. In some European cities, for example, builders are required to erect full-scale bamboo framework mock-ups of projected office buildings so that the townspeople can see their impact on the city, before the building itself has gone too far. We have found, over and over again, that people cannot visualize spaces accurately unless they put up some sort of rudimentary structure - bamboos,

sheets, string, or 2X4's as a mock-up. If the nature of
the building system somehow lends itself to this kind of
mock-up, so much the better.

We have discussed some of the requirements which a
building system must meet if it is going to allow users
to design their own buildings. But can laymen actually
use a pattern language with construction patterns to
design their own buildings? Our preliminary experiments
suggest that it is feasible, and that it does create the
organic architecture we are looking for.

For the sake of experiment, we developed a post-and-
beam system, abstracted the rules which governed its
use, and introduced these rules into the pattern language
in the form of five construction patterns. We found that
laymen could indeed use this language to produce a
functional "working drawing."

In this experiment we were more interested in testing
the capability of lay designers than in proposing a
final building system. Yet, the simple wood post-and-
beam building system we used includes many of the principles we have discussed in this paper. The usual post-
and-beam system was modified to eliminate the need for
modularity, and to make spaces appear at the very beginning of construction. It was designed so that Thick Walls,
Columns at the Corners, and Low Ceilings were generated
automatically. Other patterns like Family Room Alcoves,
Window Place, and Sheltering Roof became cheap and easy
to build.

Below we show a picture of a model built with this
system and some examples of house designs done by laymen
using the pattern language (Figs. 2 - 6). Having seen
that laymen can, in fact, use a pattern language to produce
functional working drawings, we now turn to the third and
final group of demands which we are putting on the building
system we seek.

2

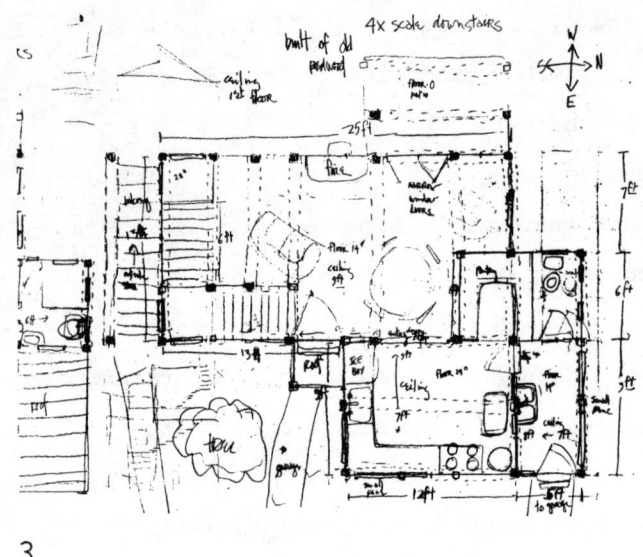

3

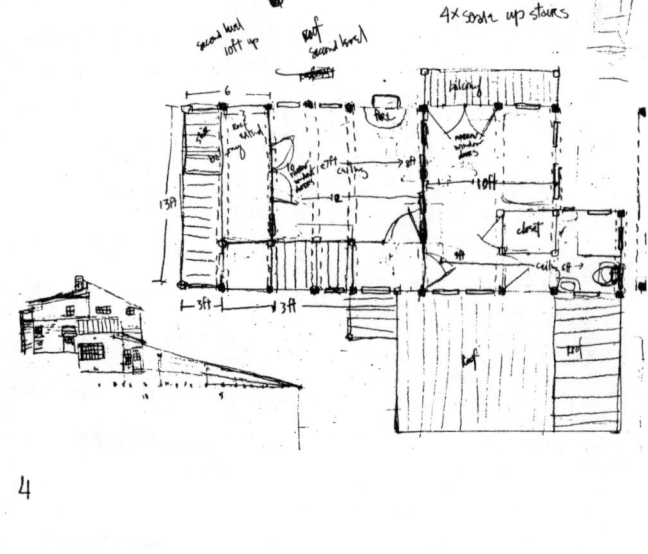

4

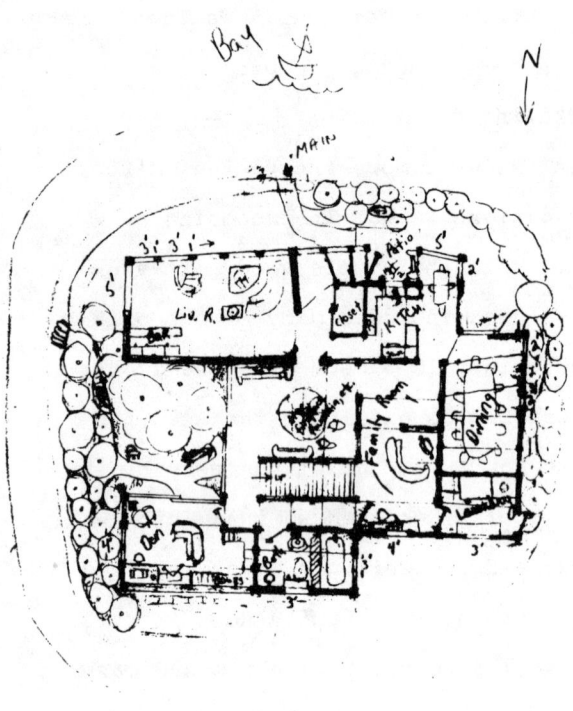

Bay

Floor 1

5

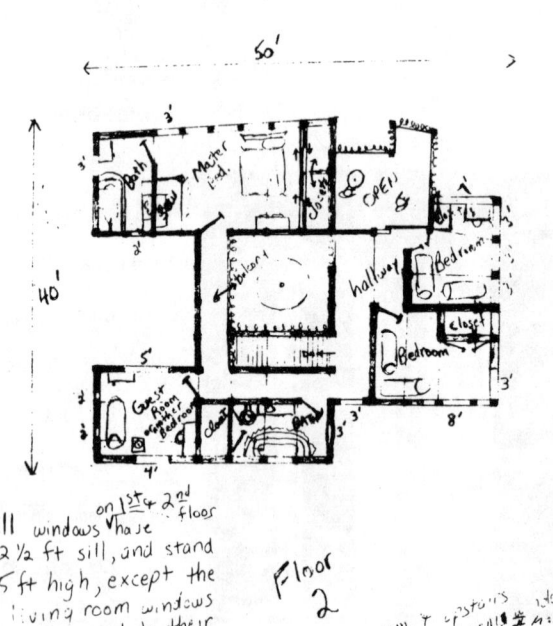

All windows have
2½ ft sill, and stand
5 ft high, except the
living room windows
which extend to their
respective ceilings, as do the dining room & upstairs bedroom windows.

on 1st & 2nd floor

Floor
2

6

3. Compatibility with repair and piecemeal growth.

A building cannot be organic or alive unless it is built <u>gradually</u> and <u>repaired constantly</u> during its lifetime.

7

It is useful to remember that traditional buildings last for centuries not because they are so sound, but because they are continually repaired. These buildings are built with a <u>relatively low</u> initial capital and labor investment - low relative to the total expenditure over their lifetime. The physical fabric of the buildings disintegrates slowly all the time and is constantly being countered by the users' reconstruction and repair (Fig. 7). And this never-ending dialogue between growth and decay means that the actual form of an individual building keeps changing over time and becomes more and more finely adapted to the particulars of the site and to the users' needs.

In contrast, current building technology is based on the notion that participation in the natural process is to be avoided at all costs. Buildings are either designed to last only a few years with the idea that they can be torn down then; or they are designed to last forever, and made of materials which never have to be repaired. In the first case (ticky-tack construction) the buildings decay too rapidly via normal use to be repaired. In the latter case (concrete and steel) the buildings are made of such terribly hard, permanent, and monolithic materials that slow modification is again out of the question. In both cases the outward forms are rigidly immutable, and never have a chance to get better, or more subtly adapted to peoples' needs.

In both cases, living and working in buildings are seen as <u>destructive</u> activities - buildings are only thought good when they are new, fresh out of their wrapper. It is all part of the consumption society. Buy it new,

sell it or throw it away the moment it is used. <u>The
environment cannot become healthy, or alive, until we
begin to conceive the process of using a building as a
creative, reparative activity.</u>

If our buildings are made of monolithic reinforced
concrete then repair and modification are out of the
question. Yet when the same material is used in the form
of human-scaled bricks or blocks, repair can happen.
Would you rather repair a patio made of individual bricks,
or one made of a continuous sheet of concrete? (Fig. 8)

8

One objective question to ask of any building system
is "Do the normal everyday marks of use spoil it or give
it a <u>patina</u>?" Compare plate glass and leaded windows.
A crack ruins a large sheet of glass; a new pane enriches
a window made of many small panes. Or compare formica
and wood table tops. Burns spoil formica and make wood
more beautiful.

The fact that so many of our modern materials tend
to be spoiled by use, instead of taking on a patina, is
caused by our tendency to "technologize" our building
materials, i.e., to create materials which do a very good
job of solving a very narrow range of problems. While
natural materials possess a wide range of moderately
adequate properties, technology is used to create materials
with superior properties, but in a very narrow range.
Thus, formica is superior only in terms of "wipeability,"
and this is obtained at the expense of many many other
useful characteristics.

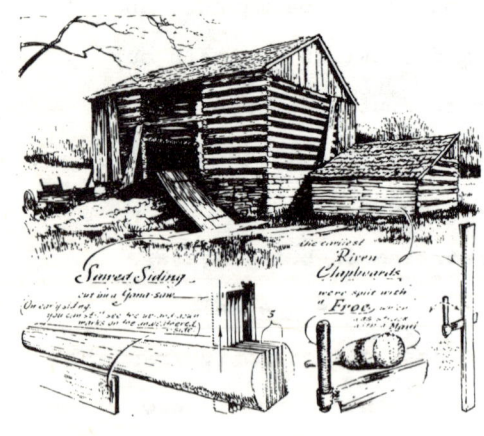

9

The tendency for modern materials to be spoiled
easily is made worse by the fact that the materials are
not allowed to speak their own language. Compare the
processes of making wood shingles with a froe, and with a
saw. The froe chips away shingles. The chipped shingle
is as good as a sawn shingle, yet the role of the machining
is kept to its bare minimum (Fig. 9).

We see then that one of the most obnoxious qualities of our present building technology is its demand for an almost neurotic level of non-productive perfection. We have recently heard that the frieze-work of Greek temples was designed so that the natural accumulation of dirt would fall in the shadows, accentuating the relief rather than defacing it.

The environment will not be continually repaired by the users unless it is <u>transparent</u>, an observation first made by Paul Goodman.[11] This means that when a man looks carefully at a building, he can understand precisely how to reproduce it somewhere else. Compare stud construction with post-and-beam construction on this score. It is virtually impossible to understand stud construction without taking the building apart destructively. But post-and-beam is completely clear and "transparent," even to a casual observer. A Canadian student at the Center once remarked that in all the really good buildings he could think of, one could look at the foundation and know how to complete the building (Fig. 10). It is obviously impossible for users to repair their own buildings unless they are transparent in this sense.

Finally, ongoing repair will be impossible unless the components of the building system are easy to handle. We must be careful not to propose building systems which depend upon super-human machines like cranes and bulldozers (which most current building technologies rely upon). They require highly-skilled labor to operate, and are so expensive that they make user-repair and modification almost impossible.

To begin thinking of the entire construction process as repair, requires reorientation of our current ideas about the economics of construction. The idea of continual repair and piecemeal growth clearly implies much smaller chunks of building at one time. At first glance, this might seem to fly in the face of well-established economics of scale. Yet we recently learned that, in Sweden at least,

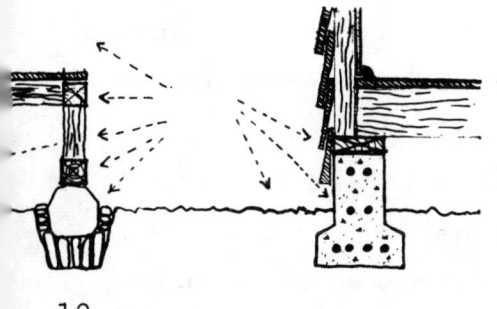

10

the cost of administration represents fully two thirds of the building costs in a house, because of the very large size of the firms which build the homes. Large organizations make things in the total more expensive, not cheaper.

And we suggest reconsidering the bald assumption that buildings should be as cheap as possible. If the users are continually involved in the creative repair of their buildings, then the buildings' final monetary value in terms of materials and labor may be enormous. We need a building system which enables generations of users to create a very expensive building indeed, over a very long period. This in turn means that we need new methods of financing which replace lump sum loans by incremental loans.

4. Specifications For an Organic and Human Building System.

We have now discussed the requirements placed upon a building system in order that critical functional patterns can be realized, that the users can again design their own buildings, and that the resulting environment can grow slowly, be continually repaired, and modified. We have seen that our present building systems are not capable of meeting these requirements. We now summarize our discussion, with a list of specifications that any new building system must meet in order to produce an organic and human environment.

Pattern language.

1. It must be possible to build narrow buildings, with a high wall to area ratio, so that rooms can all have daylight on two sides. (Light on two sides)

2. It must be easy to build steeply pitched roofs, with dormer windows and roof gardens set into them, without expensive flashings. (Sheltering roof)

3. Ceiling heights must be able to vary throughout the building, in a way which is integral with the structure, not "fake." (Ceiling height variety)

4. It must be easy to form alcoves at the edge of larger spaces, again in a way that is integral with the structure. (Alcoves)

5. Structural columns must occur at the corners of all social spaces. (Columns at the corners)

6. Walls must be thick and "carvable" to let people make them "their own." (Thick walls)

User-Design.

7. It must be possible to derive an explicit, conceptually simple set of rules which tell how to turn a schematic design into a functional working drawing.

8. More expressive control must be handed over to the builder, so that he takes creative charge of details, and doesn't simply work like an automaton, from machine-like drawings.

9. The building process must be so simple and reliable that it can be approved as a process by Local authorities, so that detailed drawings of individual buildings no longer need to be submitted for approval.

10. At each stage of the building process it must be possible to place and shape structural elements in response to the positions of those larger elements which have already been put in place.

11. The user must be able to take part designing the building to any extent he wants: helping to build it, helping to finish it, taking full responsibility for building it, or repairing it occasionally.

12. The process must require a minimum amount of hi-skilled labor.

13. The building system must be able to wear different levels of finish with equal dignity.

14. It must be possible to build full-scale mock-ups in the last phase of design, and then possible for them to become part of the building's final fabric.

Repair and piecemeal growth.

15. All building elements are light enough to be carried by one or two men, without help.

16. Building elements are not blemished by use, but take on a patina instead.

17. The fabric of the building is structurally redundant so that parts can be added and taken away without endangering its stability.

18. Materials must display their own color and texture, and machining kept to a minimum.

19. The building must be transparent, in the sense that anyone who looks at it can see at once how it is made.

20. The building process must not rely on the use of complex or expensive machines on site.

21. The building process must not require a large managerial and technological organization.

22. Budget and financing must provide for long-term piecemeal construction and repair of buildings, not merely for an initial capital budget.

A tree grows under the dual influence of inner generic patterns and external particulars, and by continuous growth and repair. That is what we mean by organic growth. A building process which will allow towns and buildings to grow organically in the same way as the tree, must at least meet these 22 specifications - and probably many others too.

Max Jacobson is a researcher at the Center for Environmental Structure in Berkeley, California

Christopher Alexander is the founder of the Center for Environmental Structure in Berkeley, California

Footnotes

1) Barker, R.C., <u>Ecological Psychology</u>: Stanford University Press, 1968. Barker introduced the term "behavior setting" for such a system.

2) Alexander's forthcoming book <u>The Timeless Way of Building</u> (in press).

3) Jacobson, et al. Center For Environmental Structure, "A New Method For Assessing the Health of Environmental Settings," paper in preparation.

4) Center For Environmental Structure, <u>A Pattern Language - First Edition</u>: to be published by Harvard University Press in December, 1972.

5) Alexander, <u>op</u>. <u>cit</u>.

6) The idea of repair and piecemeal growth is discussed both in Alexander's book and in <u>A Planning Process for the University of Oregon</u>, C.E.S., (in press).

7) This pattern, along with all the others that are mentioned in this paper will be included in <u>A Pattern Language</u>. The dates in parentheses refer to the writing date of the individual pattern.

8) "Building Stepped Back" (1968) also discusses this use of flat roofs.

9) Rand, G., "Children's Images of Houses," in <u>Env. Des.: Research and Practice</u>, Proc. of EDRA 3/AR 8, U. of Calif. L. A., January, 1972. Rand shows that the roof is a primal element of the archtypal house, even for children raised in hi-rise apartments.

10) Engle, H., <u>The Japanese House</u>, Chas. E. Tuttle Co. Rutland, Vt., 1964.

11) Goodman, Paul and Percival, <u>Communitas</u>.

Graphics

Engle, H., <u>The Japanese House</u>, Chas. Tuttle, Vermont, 1964.

Sloane, E., <u>An Age of Barns</u>, Funk and Wagnalls, N.Y., 1967.

Shoemaker, A. L. (ed.), <u>The Pennsylvania Barn</u>, Penn. Folklore Society, no date.

"I went around my house and took pictures of things I can't fix . . .
Here's a doorknob. I just can't find how to get <u>into</u> this thing
to fix it or replace it.

"I'd really like to get inside a telephone to see how it works.
But when you turn it over, here's what it looks like. It's designed
to be indestructible by the curious.

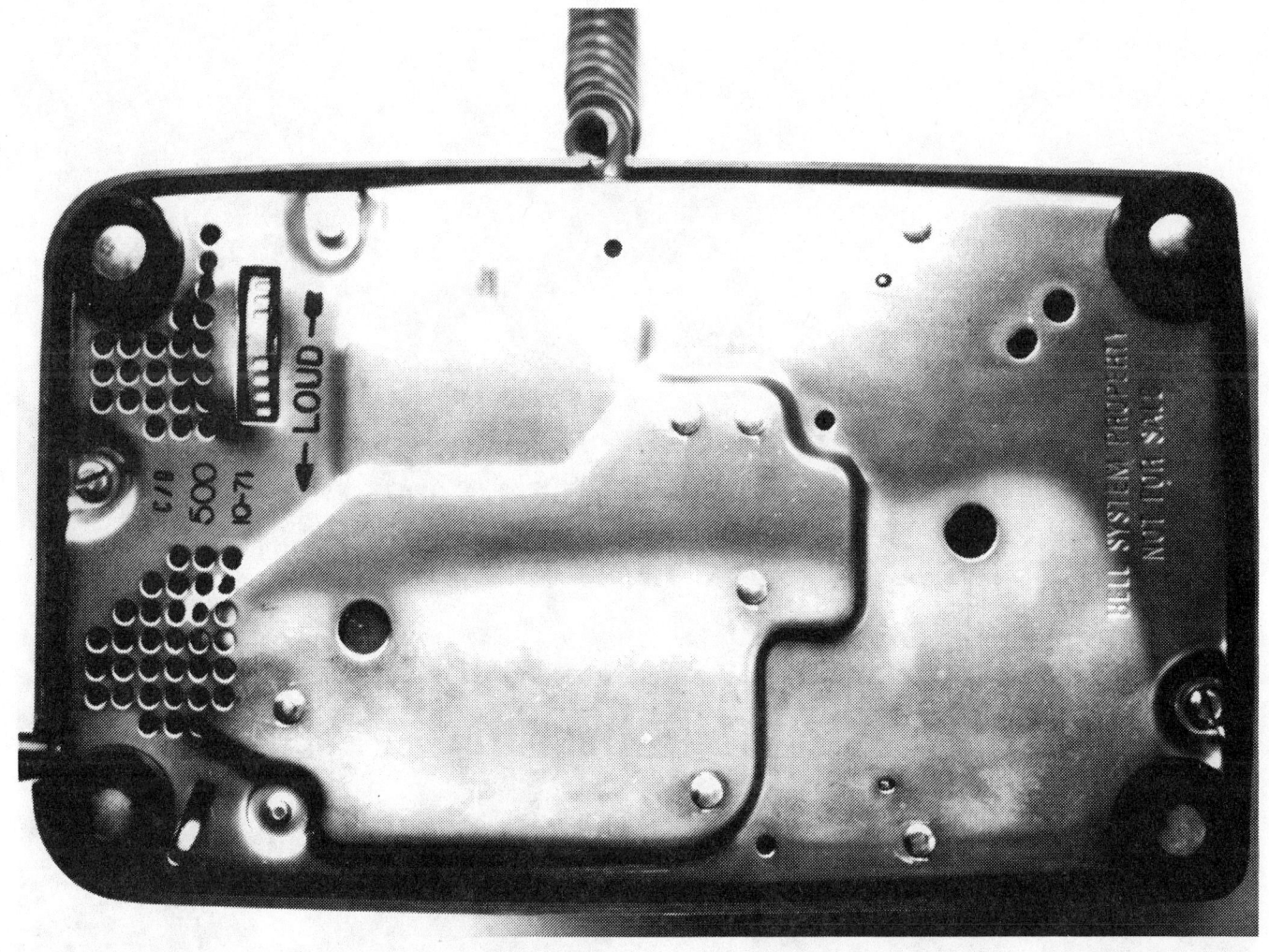

"Here's a complicated mechanical device--a flute--which is nevertheless totally transparent and potentially repairable.

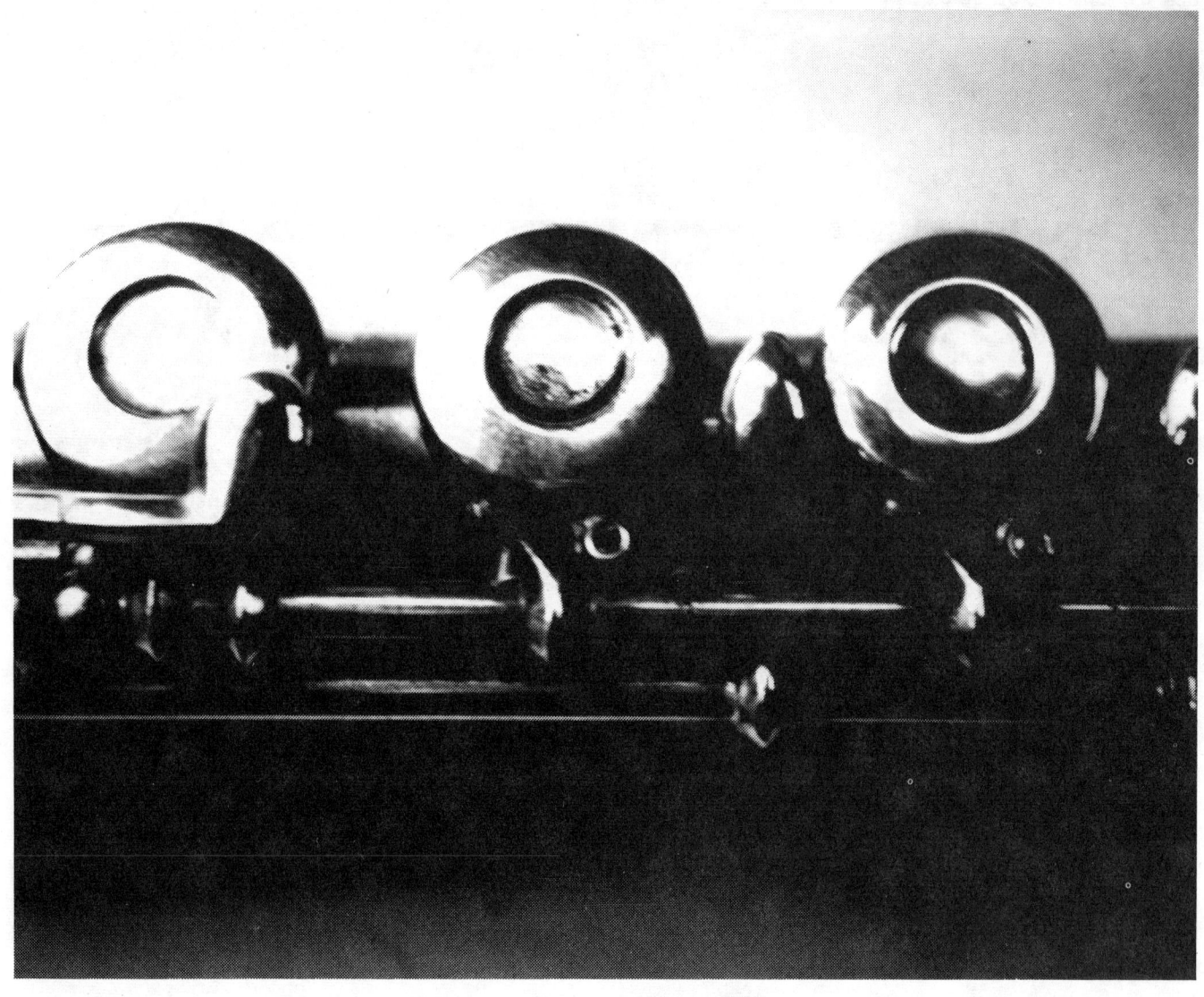

"This is a house design done by a fellow who's a carpenter, and he's a pretty competent person. He wanted to design a house for himself, so he did. He drew this in plan, and in perspective.

"He began to get just a little bit nervous about technical details, and he wanted some advice, so he went to an architect who looked at his design and said, 'Look, this is like the first quarter of Architecture 1. Give it to me and I'll fix it for you and make it a little more competent.' He produced this design, and I swear to God, he's building this house. He got talked out of his own design."

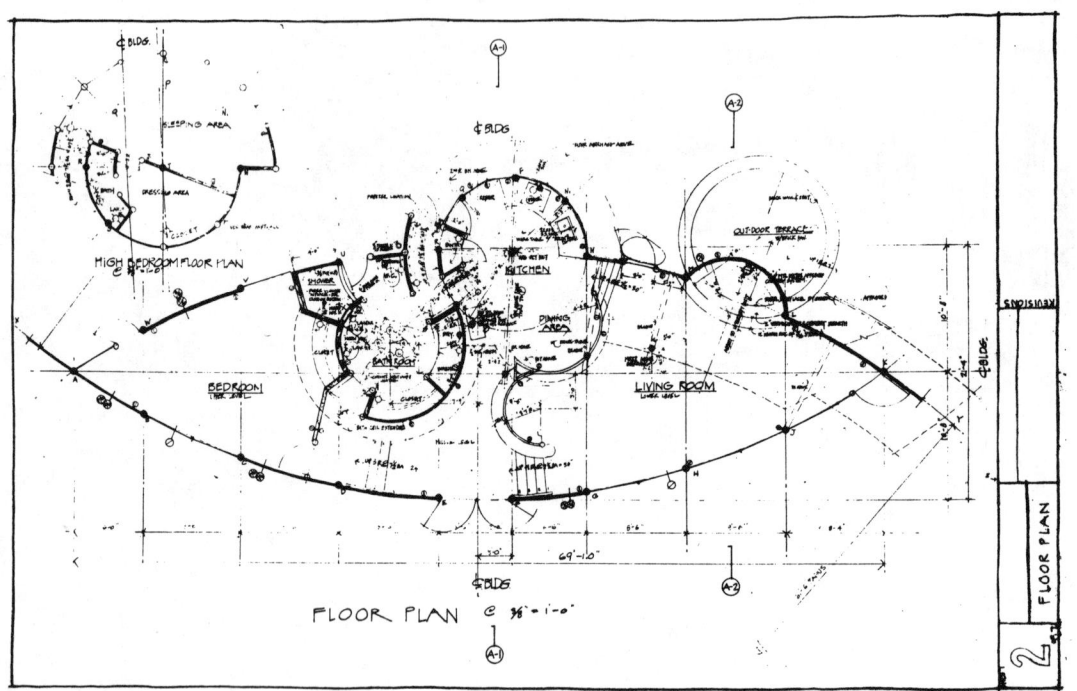

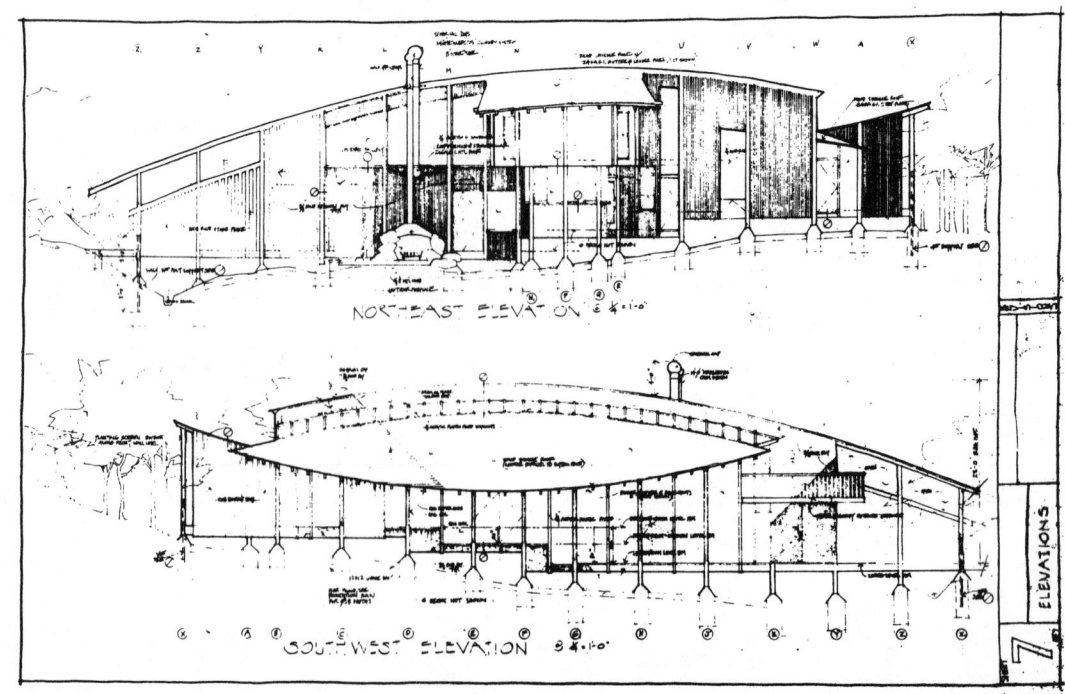

Essay Edward Allen

It is the primary task of a work of architecture to respond to human needs, physical and emotional. It is not the task of the occupant to respond to the configuration taken most conveniently by a piece of building material, nor to a structural span that is especially economical. We must demand at least as much compliance and compromise from the materials of a building as we do from its occupant.

A great number of buildings have been planned on modules ranging between two and ten feet. Such modules are efficient in terms of materials, but for human use they are usually wasteful and inefficient. A modular office building or factory may show little distortion when compared to a modular house or theater, but all are warped in some degree by being forced onto a grid. A truth being learned very slowly is that the only viable modules range upward from twenty-four feet or downward from eight inches. A man can subdivide, modify, and occupy a space twenty-four feet or more on a side; and with pieces eight inches or less in dimension he can assemble enclosures of space as he requires. The ideal modules for general use are infinitely large or infinitesimally small—an unlimited clear span, or a molecule.

A material small in three dimensions, as a grain of sand, is the most responsive and least demanding piece from which a building can be built. Such pieces can be joined in an infinite number of plastic ways. A material small in two dimensions, such as a steel wire, a cotton thread, or a two-by-four wood stud, is reasonably responsive if it can be cut to any required length. A material small in one dimension, a sheet material, is responsive only if it can be cut to any desired shape, and then bent, stretched, or molded. A material large in all dimensions, as a large block of stone, or an unmodified piece of space, is responsive only if it can be hollowed out or carved in a plastic way to meet human requirements; or if it is composed in turn of very small particles which can be molded, joined and manipulated at will, as in a ball of wet clay.

The practice of shop fabrication of building components has grown at a rate much slower than was predicted in the early postwar years, and will probably never reach its projected peak, because prefabricated components are closely tied to unresponsive modules by the nature of the machinery which produces them. In the near future we will see a renaissance of site fabrication, as buildings are erected by machines such as the spray gun which are capable of rapid on-site assembly of very small particles of material.

At the same time we will see considerably less structural expressionism. A wall whose functions of load bearing, weather resisting, insulating, and energy-carrying are carefully articulated into distinct elements is not as responsive to human use as to its own internal problems. A material or assembly which can gracefully assimilate all such functions and still let itself be molded into spatial enclosures and light-gatherers, without undue regard for its own nature or dimension, offers infinitely more exciting possibilities.

(Reprinted from
Arts and Architecture,
November 1965)

Edward Allen is an Associate Professor of Architecture at M.I.T.

An Address
to the Shirt-Sleeve Session
on Responsive Housebuilding Technologies

Steve Baer

If we were a gathering of people concerned with the production and
use of milling machines, or a gathering of paint application
specialists, perfume salesmen, Christmas Card printers, or producers
of plastic brief cases, it would be out of place to attempt any
lengthy philosophical discussion about what we were doing. Why?
We might soon decide to forget the entire enterprise; our operation
itself might not stand the inquiry. This is not the case. We are
builders and our product will always be needed by our fellow men.
Let us in our inquiry follow as far as we can. We will find no barriers
since we, ourselves, are the users of our product. I think it is
important to consider this very carefully and, indeed, to ask if it
could be any other way. What we do must benefit ourselves or we are
slaves and our activities are controlled by fear.

We all know that architects and builders receive a great deal of
abuse from our society - the housing crisis, the crumbling cities.
In fact, the final course in Architecture School might best be taught.
by a cocker spaniel to show how in the face of overpowering criticism
one looks guilty and shows one's eagerness to make amends and do
better in the future. We are told of the automobile manufacturers
and their accomplishments, the boat builders, the airplane manufacturers, and NASA and its space craft. Now, who has paid for all
these marvels? We have. So the citizen after being bled almost to
death for such marvels questions the architect and builder about
the miserable hovel in which he lives. We must discuss this -- what
is wrong? Are the people who make bricks, who pour concrete, who
install windows, who pour tar on roofs, who negotiate loans for
such structures, who draw the plans -- are we a special cult of
incompetent fools slouching and bumbling through our days while all
around another breed of man goes on with great courage, skill, ingenuity
producing an enormous gamut of sophisticated products -- X-ray
machines that can look into your very body, tiny electronic calculators
that fit in your pocket and give you the power of a methematical
prodigy, airplanes that fly from Chicago to New York faster than
you walk across town.

Are we, the civil designers and builders, and idiot sub-breed who
become steadily more inefficient, slow witted and timid? The automobiles, airplanes and now rocket ships -- roaring, careening,
zooming over and between us -- as we, rooted to foundations, fidget
with our old tools. A look at our cities and houses makes this seem
probable. What are we to do? Can we copy them, those whose skills
produce everything from snow mobiles to guided missiles? Why not
simply blame everything on them? I would guess we can no more reach
out for salvation to these devices, the men who make them, the
organizations who sell them, the logic, style and technique used
than could a sick dog find salvation by metamorphosing into one of
the ticks it sees growing fat on its own body.

In a sense the constraints of the economy and popular taste on the architect and builder are the same sort imposed upon the designer of a bus station, airport, or railroad depot. The point is not to build the most comfortable and attractive station, but, rather, to hold people before and between the purchase of tickets. The ticket, the ticket, get the ticket. Anyone who has ever tried to sleep in a bus or train station knows that the arm rests on the benches are placed precisely to prevent anyone from sleeping on the benches. Like the bus station bench the house and the city, itself, is then not seen as an end, but is rather the staging grounds for other activities and, of course, even more, its own short life adds to the ill ease and discomfort.

What is it when people compare ocean liners, airplanes, automobiles and rocket ships to houses? Of course, the house is lacking the kind of excitement these other devices have. Listening to such comparisons between houses and machines to what finally becomes, virtually, comparisons between conduct of life and the production and sale of equipment where invariably the house or culture is seen the beggar for the power and glory of machinery. One thinks of other absurdities such as a pathetic child, ignorant of gravity, on the down side of a teeter totter envious of the one in the air and pleading with it to toss him whatever it is he has that makes him rise. Houses could be machines for living, marriages could be romances, children could learn not to make the mistakes their elders have.

It is a problem we all share. Even the astronauts have to go home. We take the scientist, the systems analyst, the industrialist, the pilot, the astronaut. Take them to the town, the house. This is where I live; this is where I sleep. I don't want to fight. I don't need to fly; I don't need to go to the moon. Here is where I sit. Here is where my friends sit. Here is where we eat. Here is where we drink. Here is where we talk. Here is where we embrace, make love. Here is where our children grow up. Here is where I'll be healthy and sick; here is where I will die.

Then the house should be beautiful. It should be honest. It should not belong to the bank. It should be complex; it should be simple; it should be calm; it should be active. One should know of those who built it - did they find the work rewarding?

This does not lead to a rejection of glass, steel, aluminum, concrete, plastic. It does lead to questioning both the properties and histories of such materials. Our own house, which we are just completing, includes about three tons of aluminum. Thus, in terms of energy, the structure has already consumed enough energy to heat the building for from three to five years. This does not lead one to snap rulers and tape measures, to smash thermometers and clocks. Nor does it make one long for some gigantic mutant Chevrolet, Boeing 707, or NASA space capsule.

Man's interests in objects and equipment has taken a new turn as he views himself as the potential creator of someting different than his offspring and less demanding than his culture. We quite clearly feel ourselves equipped with new power to assess ourselves and all that is around us. We have left the earth, we have drilled holes in it -- at all times there are thousands of people skimming tens of thousand of feet above its surface. Yet I assure you all of this would quickly fade if we possessed a few different talents. Imagine the outcry if,

Steve Baer's own house is a zonohedral structure of stressed-
skin aluminum-and-honeycomb panels, designed by him and his wife.
A windmill pumps water from a well. A solar collector (right fore-
ground) heats water for household use. Wall panels are dropped by
means of hand-operated winches to expose water-filled barrels as a
means of collecting solar heat for space heating (left).

Steve Baer's house, interior.

Steve Baer's house, outdoor courtyard.

let us say, General Motors, The Rand Corporation, or NASA labs
discovered how to build a rabbit. A new product, developed by teams
of scientists. The outcome of years of research. Of course, some
malcontent somewhere would mutter, "It's just a God damned old rabbit.",
but this bitter comment would be audible to only a few. Once this was
done in a few generations you could perhaps convince people that any
agency in Washington produced different species of animals that were stocked
in different parts of the country.
of animals & birds

People are eager for new power and new possessions, yet most of the
things people strive for seem meaningless if shown alongside what they
have already done. Those who lead in these silly struggles are, of
course, only too happy to confirm that, yes, this prize which they _heart, eyes_
have won most certainly was the answer. Man possessing his organs, _nose_
inheriting a language -- whether he has or does not have an automobile
or $10,000 -- in one sense is the difference between two three bedroom
houses one of which has carpet on the stairs to the cellar and one
of which doesen't.

Our civilization seems a giant parasitical machine harnessing various
leaks and vibrations in men's enterprises with themselves. Anything
that is addicting or has a long way to fall soon sets up shop within
the constellation of activities and flourishes, solemnly rolls, seeps,

chatters, blinks on its way, and is noticed by itself, set free from
its source -- as free as a phonograph blaring in a tract house in
San Diego 300 miles from the power plant and 20 years and three
pressings away from the recording studio. But where do things come
from? What is the point? The questions emerge, re-emerge, and can
not be shut off.

The citizen looking out past the flimsy sheet rock, fiberglass and
celotex - to the chromed Chevrolets, Fords, Cadillacs, Lincolns,
Buicks, Datsuns, Saabs, Volkswagons. Or, peering up through the smog
between immense illuminated signs eighty feet tall made of thirty inch
steel tubing bolted to enormous concrete footings with special high
tensile strength bolts, strands of double 0 copper wire the only
other roots it needs, masts soaring into the sky, a geyser of strength,
efficiency, electricity to proclaim ARCO, TEXACO, STANDARD, MOBIL,
CHEVRON. Great decorative blossoms attracting motorists to their
bases where, unlike the coy flower from which the visitor must suck
the nectar, there are electric machines which actually pump the
gasoline into the car. What accounts for these gigantic erruptions
of disorder where clearly both the blossom and the bird are doomed?
The same oil companies prominently advertise ecology slogans. Are
the executives(wealthy, well educated, respected)-- really pimps,
cheaters, and thieves, or are they merely pawns at the mercy of the some
vegetable gangsters. Has the oil,rotted vegetable matter, ancient
flowers from the mezozoic age staged its own clean up, anti-polution
campaign? For, after all, oxygen, animal life's prime resource, is no
help to the plants. Researchers in Sweden have found our present
oxygen rich atmosphere to be slowing down plant growth. Oil, lying
for millions of years buried, conscious of its polluting sin to not
burn, has it found a means to get out and give back what it took from
its descendents millions of years ago, employing a distinguished
succession of inventors and scientists - Sadie Carnot, WL Wright,
JJE Lenoir, Nikolaus August Otto, Sir Dugald Clerk, Joseph Day?
Should we not recognize the technique of the plant and flower, unable
to restrain itself to mere television and magazine adds it reveals
its signature, the mast and blossom.

The builder is a bag of dreams holding tools with which he constructs
houses, towns, cities. If he ends building the props for his own
nightmares he has betrayed himself. This happens; it is awful. If the
economy provides for no other sort of activity let us quit - why should
we have the shitty end of the stick? Why should we build the shitty
end of the stick?

Today, a new culture, a new architecture must emerge. Let us investigate
nature, the properties of materials, the symmetries of our space
and see if we can forge objects, buildings, towns which are satisfying.

Steve Baer is an inventor, author, and the proprietor of Zomeworks
in Albuquerque, New Mexico.

Section II

"Shall we forever resign the pleasure of construction to the carpenter? What does architecture amount to in the experience of the mass of men? I never in all my walks came across a man engaged in so simple and natural an occupation as building his house."
--Henry David Thoreau

Seven Axioms for the Owner-Built Home

Ken Kern

1. *BUILD ACCORDING TO YOUR OWN BEST JUDGMENT.* At the apex of the poor building hierarchy—and perhaps the greatest single impediment to good housing—is convention. Building convention takes two forms: first, there is convention which is socially instilled (commonly called *style*), which can be altered through education. The second type is more vicious, and politically enforced. Building codes, zoning restrictions, and ordinances all fall into this class. In urban jurisdictions, politically controlled convention calls the shots for practically every segment of the building industry. Ordinance approval or disapproval makes the difference between having a house or having none at all. Or it may make a difference of $1,000 (average)—wasted because of stupid, antiquated building laws.

So if we are to be at liberty to build our own home at less cost, we must necessarily be free from all building code jurisdiction. This means we must locate outside of urban control—in the country or small township districts.

2. *IN BUILDING YOUR HOME, PAY AS YOU GO.* A building loan is another type of legalized robbery, added to that of the building codes. More than any other agency, banks have been successful in reducing would-be democratic man to a state of perpetual serfdom. The bankers have supported and helped to determine social and political convention, and have amassed phenomenal fortunes through unearned increment. As "friends" of the homeowner they have made it possible for him to take immediate possession of his new home—and pay for it monthly for 20 to 30 years. Most people who fall into this trap fail to realize that the accumulating interest on their 30-year mortgage comes to more than double the market value of their home! If one expects any success at all in keeping the costs of his new home down to a reasonable price, he must keep entirely free from interest rates.

3. *ASSUME RESPONSIBILITY FOR YOUR BUILDING CONSTRUCTION.* The general contractor has become such a key functionary in practically every building operation that one soon loses sight of the fact that he is a relative newcomer to the housing scene. Not many years ago the contractor's job was performed by a supervising carpenter—a so-called master builder who had control of the whole project. Once people realize how little is involved in implementing a set of house plans, they will better appreciate the fact that the contractor is the most expendable element on any job.

Excessive profits are made by the general contractor for coordinating the work stages and assuming the responsibility for a satisfactory completion at a specified cost. For this service he receives 10% of the total cost of your house. Besides, he receives an even greater percentage on all materials which go into the structure. The contractor is an expensive and non-essential luxury for the low-income home builder.

4. *USE NATIVE MATERIALS WHENEVER POSSIBLE.* Much of an architect's time is spent in keeping abreast of the new "improved" building materials which manufacturers make each month. Many of the products are really worthwhile: but more often than not they are entirely beyond the reach of the average home builder. Basic materials, like common cement and structural 2 x 4's, have not appreciably advanced in price over the past dozen years. But some of the newer surfacing materials and interior fixtures have sky-rocketed in price during this same period.

By not using these high-cost materials, one of course nips the problem in the bud. Instead, emphasis should be placed on readily available natural resources—materials that come directly from the site or from a convenient hauling distance. Rock and earth and concrete and timber and all such materials have excellent structural and heat regulating qualities when properly used.

5. *SUPPLY YOUR OWN LABOR.* Building Trades Unions have received—and not unjustly so—a notorious reputation as wasters of speed and efficiency in building work. We all know that painters are restricted to the 4-inch brush and that carpenters are limited to the 14-oz. hammer (upon threat of penalty from union officials). Apparently more width and weight might conceivably speed up a project to the point where some union man would prove expendable.

The disinterest that the average journeyman has in his work, despite his high union pay rate, is appalling. The lack of joy-interest work or acceptance of responsibility among average workmen can be accounted for partly by the de-humanizing effect of the whole wage system. So long as the "master-and-slave" type of employer-employee relationship continues to exist in our society, one can expect only the worst performance from his hired "help." So until the dawn of the New Era approaches, one would do well from an economic, as well as from a self-satisfying standpoint, to supply his own labor for his own home insofar as he can.

6 *DESIGN AND PLAN YOUR OWN HOME.* Another ten-percenter with whom we can well afford to dispense in building a low-cost home is the architect-designer-craftsman-supervisor. Experience in this branch of home building has led me to the conclusion that *anyone* can and everyone *should* design his own home. There is only one possible drawback here: the owner-builder must know what he wants in a home and must be familiar with the building site and regional climatic conditions. Without close acquaintance with the site and a clear understanding of family living needs, the project is doomed to failure no matter who designs the house. An architect—even a good architect—cannot interpret a client's building needs better than the client himself. Anyway, most contemporary architects design houses for themselves, not their clients. They work at satisfying some esthetic whim, and fail really to understand the character of the site and the personal requirements of the client.

7. *USE MINIMUM BUT QUALITY GRADE HAND TOOLS.* If the house design is kept simple, and the work program well organized, an expensive outlay in specialized construction equipment can be saved. The building industry has been mechanized to absurd dimensions. And even with more and better power tools, labor costs rise. Or at times where labor savings occur, the difference is taken up in the depreciation and maintenance of the equipment which saved the time in the first place. Whatever way you look at it, a certain amount of work must go into building a home. If a prospective home owner is unprepared to accept the challenge of building his own home—and falls into the power tool trap—then he must be prepared to spend greater sums for a product which could very well prove inferior.

Now that I have presented the *ideal* program for the owner-built home, I should retrace my steps and face the sheer realities of the situation. Obviously, not all people can locate their home site out of building code jurisdiction. Nor can many people expect to finance their home from their weekly pay check. Very few people have the native ability to design an inexpensive and attractive home—one that truly fits their needs and site conditions. Even more rare is the person who can carry through all phases of building construction, or who even has the necessary free time to devote to a house building effort. And how many people do you know who could take the raw material resources and process them into building materials for wall, roof, and floor? One has only to observe current owner-built home flops to appreciate the fact that we are dealing with a disturbingly complex problem—a problem that demands a comprehensive solution.

It is unquestionably our drive toward specialization (stemming from a basic failure on the part of our whole educational system) which is primarily responsible for modern man's inability to provide directly for his own shelter needs. Despite this drift, I sincerely believe that the owner-built home can be an economic as well as esthetic success. It *has* been so for centuries, for thousands of families—if not millions—and continues to be so today. Furthermore, the process of building one's home can become one of the most meaningful and satisfying experiences in one's life—as indeed it should. Owing to the physical limitations of the owner-builder, and those impositions fostered by society in the form of restrictions and general mis-education, one can expect only to *approach* the completely self-tailored home. On one or more scores compromises are in order, but to the extent that the owner retains full control over his design and his work, he is successfully participating in creative building.

From The Owner-Built Home, by Ken Kern, Oakhurst, California, 93644.
Reprinted by permission of the author.

including

History   Tools   Design
Permits   Codes   Materials
Foundation   Plumbing
Floor   Framing   Electrical
Enclosing   Heating
Ventilation   Cabinets
Finish Work ...... and ...

...... an occasional sidelight ......

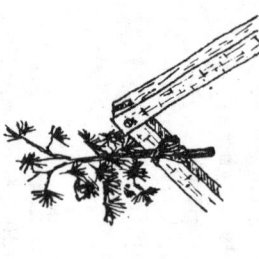

used to be ... when a rafter
was first put in place, folks
would say a prayer of thanks—
giving and lash a small tree
to it

The Stud House Council could use your information. If you know
of sources of free or cheap materials, vacant structures that could be
made liveable, architects or architecture students that could help
people design, carpentry tricks or anything else you think we might
be able to use; please let us know. We'll try to let you know.

······ punch out these black dots ·····

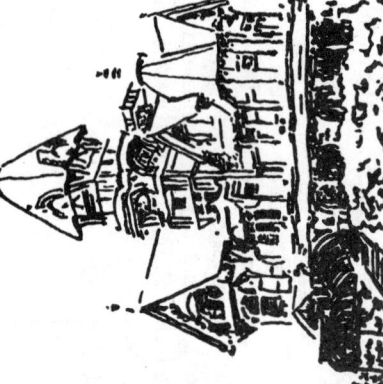
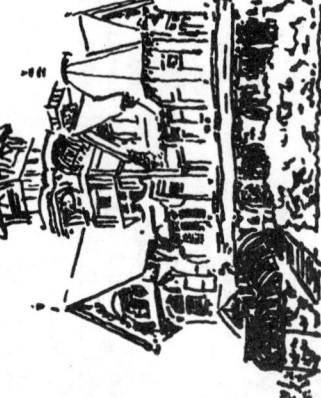
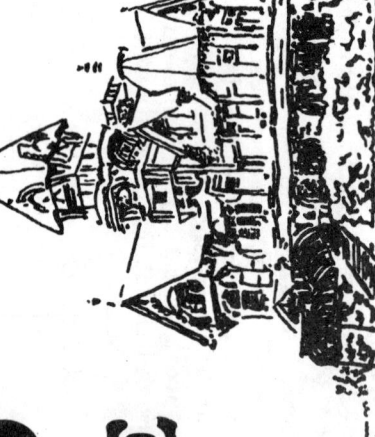

STUD
HOUSE

BEGINNING A SERIES

THAT WILL HELP YOU

BUILD OR REPAIR A HOUSE

We are a group of people recognizing the need for spreading some
news about building stud frame houses. About 1,500 new stud houses
were started in Lane County last year (about 95% of all new houses),
and about 4,500 houses were added to. Most of this work was done
by contractors.

You don't need to be a contractor to build a stud frame house.
Thousands of owner-builders have been proving this for years. Yet,
while it is easy to build a stud frame house these days, no concise
explanation of this extremely simple but most versatile of building
systems exists. The layman has difficulty finding out what his
ancestors have known how to do for centuries——build his own house.
STUDHOUSE is a series of articles appearing in the Augur that
will give instructions and helpful hints to anyone planning to build or
repair a house.

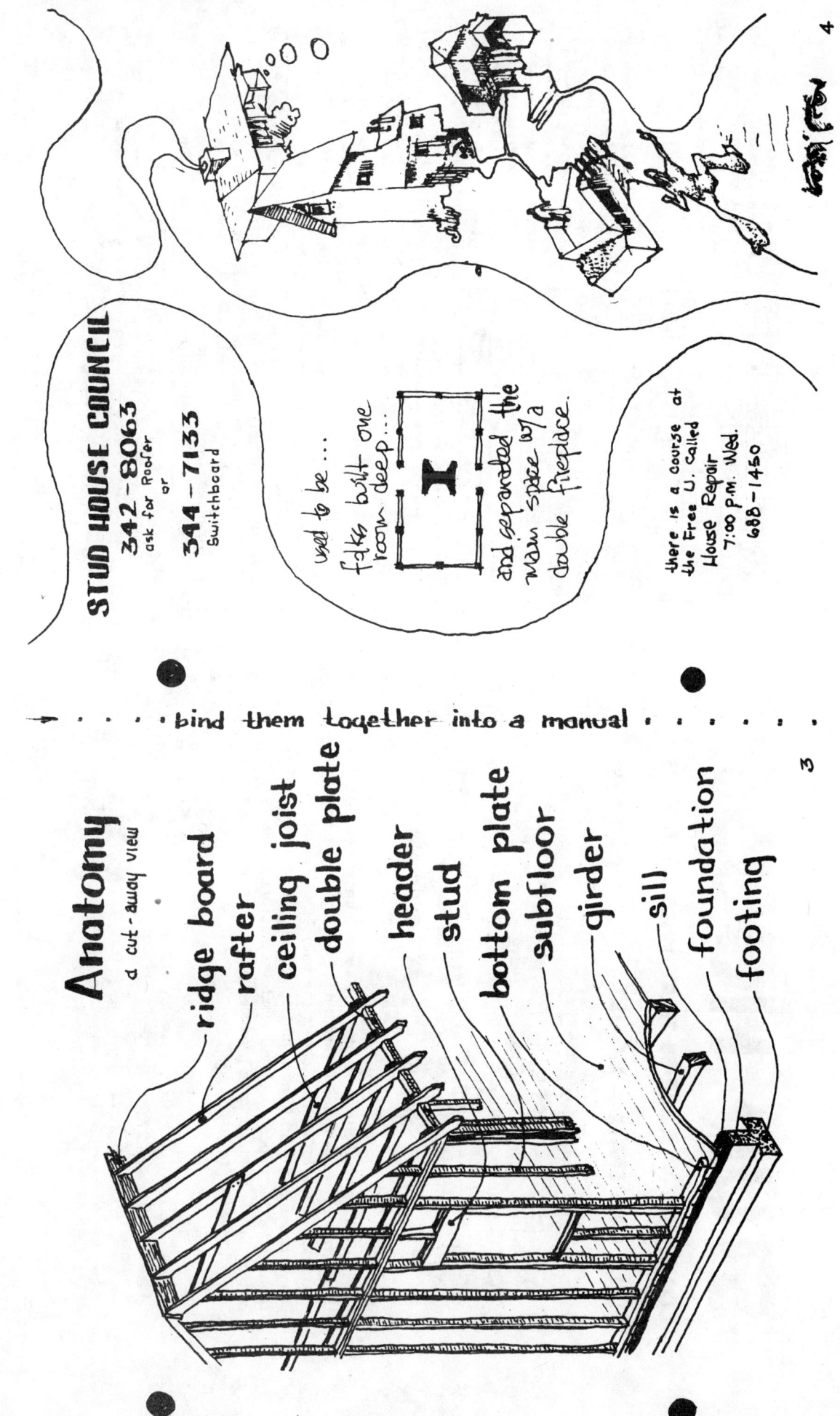

STUD HOUSE COUNCIL
342-8063 ask for Roofer
or
344-7133 Switchboard

used to be...
folks built one room deep...

and separated the main space by a double fireplace.

there is a course at the Free U. called House Repair 7:00 p.m. Wed.
688-1450

→ · · · · · bind them together into a manual · · · · · · ·

Anatomy
a cut-away view

ridge board
rafter
ceiling joist
double plate
header
stud
bottom plate
subfloor
girder
sill
foundation
footing

4

3

STUD HOUSE

CONTINUING A SERIES THAT WILL HELP YOU   part two

BUILD OR REPAIR A HOUSE

↑ Hi HOUSE-BUILDERS! Let's talk about....
... WOOD... it's outasight.

Wood is <u>warm</u>. Leave some unfinished where people can rub it with their hands. It will contour & shine according to how it's touched, where, and by whom........ Think about it ... be seeing you

Thoreau said:

What of architectural beauty I now see, I know has gradually grown from within outward, out of the necessities and character of the indweller, *who is the only builder*—out of some unconscious truthfulness, and nobleness, without ever a thought for the appearance, and whatever additional beauty of this kind is destined to be produced will be preceded by a like unconscious beauty of life.

a friend just put his rafters in peace. He laughed a small tree to one of them and gave thanks. Then sat down with his friends and fellow builders and celebrated it with jugs of wine & fine country food

——— cut here   save it for a friend

The building code of Hammurabi, founder of the babylonian empire, the earliest known code of law.

CODES (cuneiform)

"If a builder build a house for a man and his work is not strong, and if the house he has built falls in and kills the house holder, that builder shall be slain."

Nowadays if a builder has built a house for a man and the house falls in and kills the man, that builder will be sued. But his building has been inspected by the state building inspector. Before he started, the builder had his plans checked by the inspector and was issued a building permit. The inspector made sure the man followed the rules set forth in the Uniform Building Code which purpose it is:

Purpose

Sec. 102. The purpose of this Code is to provide minimum standards to safeguard life or limb, health, property, and public welfare by regulating and controlling the design, construction, quality of materials, use and occupancy, location and maintenance of all buildings and structures within the city and certain equipment specifically regulated herein.

☞ Now only the most populated counties in Oregon (Lane included) require a building permit. All counties however must comply with statewide regulations governing electricity and plumbing.

▸ Building inspectors tend to be very helpful - in both helping you get a permit and helping you meet the code.

▸ This code is a very thick book which you're required to own for every type of building. We will try to pick out those sections that you will need to build a stud house.

There is a bill before the state legislature that would enforce a revised version of the U.B.C. in all counties throughout the state above.

design a...

Layout

If you're in a hurry to build a house, hire a contractor. It will cost you at least twice as much, but it will be done fast. Since time is money to a contractor, contracted houses are designed for speed of construction, not for the delight of the people who will live in them. Contractors have no time to design each house individually. This is why you can see houses with identical facades across the street from each other, one facing North, the other South.

If you're not in a hurry, you have many advantages:

1. You will have time to fit your house to the special conditions of the building site. Every site is unique and presents special problems and opportunities. The person who designs his own house can relate his house to its unique site in a unique way.

2. You will have time to design a house that includes the things you want instead of the things a contractor thinks you want. The owner-designer is potentially the best designer since he is the one who knows best what he wants.

3. You can make a design that includes materials which are less expensive but take a longer time to find and a longer time to install.

4. You can do the work yourself saving unbelievable quantities of money.

There are lots of people who are familiar with the problems of designing houses and who are willing to help. Some of the most helpful;

The Owner-Built Home and preliminary house design (send sketch, etc.)
Ken Kern
1961; 300 pp.

$10.00 postpaid

from:
Ken Kern Drafting
Sierra Route
Oakhurst, Ca. 93644

Build Inc.

an alliance of local students & businessmen who are helping people design & build their own house.

LAURAN DAVIDSON
342-4893

STUD HOUSE
watch this column for design tips

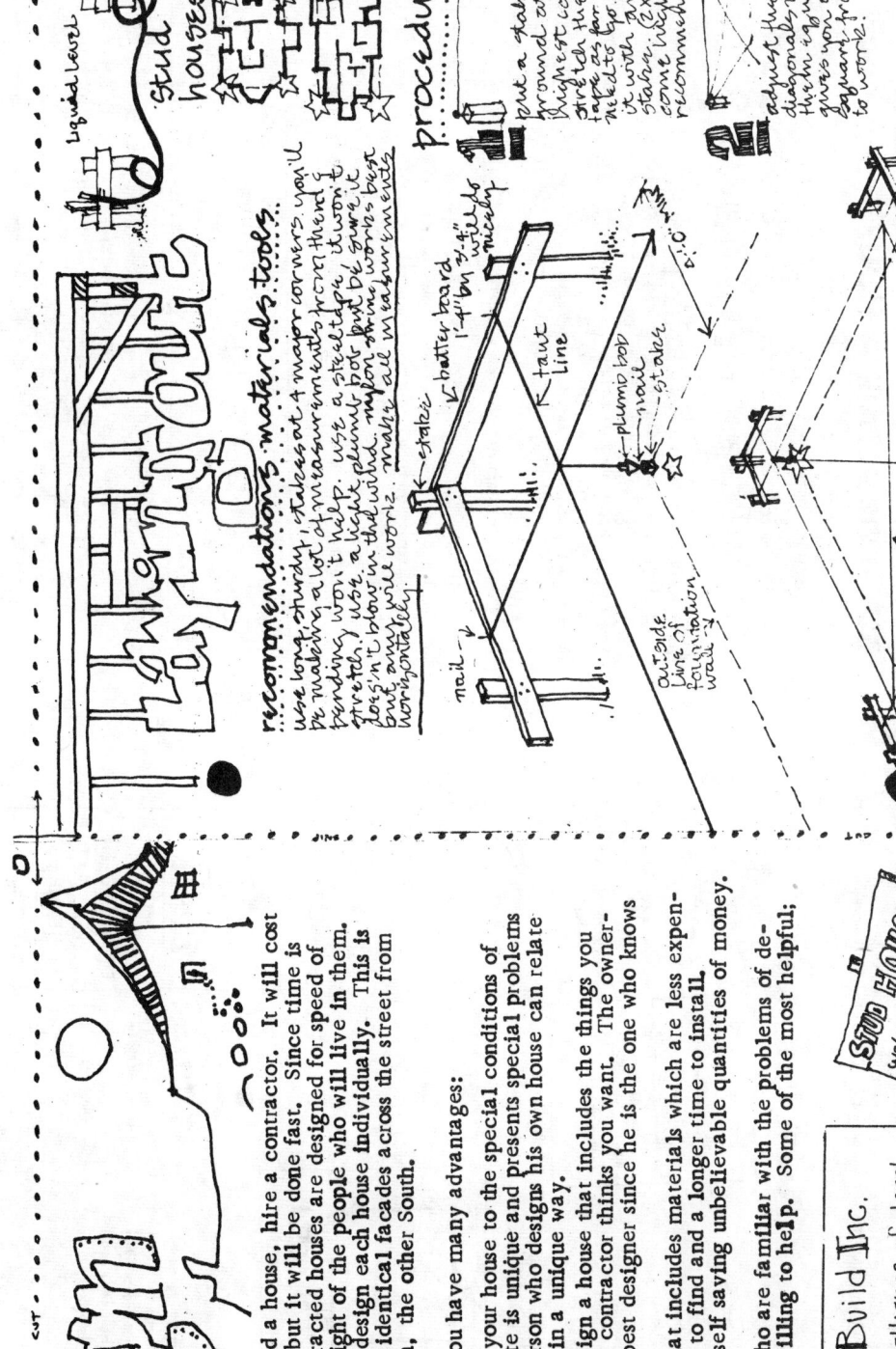

recommendations: materials, tools

use long, sturdy, straight 4 major corners you'll be making a lot of measurements from them & trending won't help. use a steel tape. it won't stretch. use a light plumb bob but be sure it doesn't blow in the wind. nylon twine works best but any wire works. make all measurements horizontally.

procedure

put a stake in the ground at your highest corner. stretch the steel tape as you need to for nailing it onto another stake. (2x4") stakes come slightly recommended.

adjust these two diagonals making them equal & making sure you square layout you want to work.

you can now string out your foundation lines as illustrated to the east. its then a matter of foundations dig it.

(the owner-built home by Ken Kern is good buy) 8

stud houses

7

STUD HOUSE

CONTINUING A SERIES THAT WILL HELP YOU

BUILD OR REPAIR A HOUSE   part three

BUILDING PERMITS

WHO NEED THEM?

1. People planning a building larger than 8ft x 8ft square — unless the building is to be used only for agricultural purposes.
2. People planning on setting-up a trailer house or house trailer or mobil home or mobil house.
3. People planning on making an addition or alteration to an existing building that would alter any structural part of the building.
4. People planning on changing any plumbing.

If there is any doubt the best thing to do is call the building department to find out. If you get caught building without a permit they can stop you from proceeding by giving you a "work-stop order". In order to start work again, you must get a permit for which you pay double fees. About 180 people get caught each year in Lane Co.

If you live in a city or town check first to see if your town has a bldg dept. If not call the county.

If you live in the county:
Lane Co Bldg & Sanitation
135 6th E. Eugene
342-1311

Septic approval — If you want to build a house where there is no sewer system, you need to apply for septic approval. This is a permit to install a septic tank. Septic approval is granted on the basis of whether the ground will support a septic tank and whether any water supply will be endangered. Fee is $5

WHAT DO YOU NEED FOR A PERMIT?

1. TAX-LOT-NUMBER ... the number that legally describes the location of the building site.
2. MONEY ... the building department will charge you a fee based on floor area. a permit for a 1000 ft² house is about $65 nowadays. There is a $5 minimum.
3. SEPTIC APPROVAL ... for any building or trailer that will be occupied where there is no sewer hook-up.
4. TWO SETS OF DRAWINGS ... are set for you ... are for them ... including ...

plot plan
- property lines
- proposed building
- setback distance

foundation plan
- dimensions
- footing (shown dotted)
- foundation wall
- piers
- distance between girder centers
- chimney footing
- vents

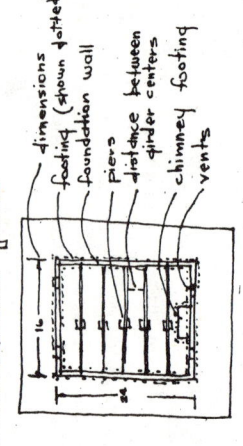

floor plan - a horizontal slice 4' above floor level
- dimensions
- window placement & size
- walls
- kitchen fixtures
- size of rooms
- door location and size
- steps
- bathroom fixtures

two elevations
- windows
- distance to ground level
- front elevation
- general shape

section
- size, grade, span of structural members and how they go together

these drawings don't have to be done by a professional draftsman. The main thing is that the building official be able to understand how you are going to build your house. They will help you if you need it.

foundations

heres some criteria about foundations from the building code (sect 2905). ■ min. dimensions → you must also have 1½ sq. ft. of ventilation space for every 25 linear ft. of exterior wall.
- there must be a 18"x24" min. crawl hole.
- foundation plates (or sills) must be bolted 6' on center with ½" bolts sunk 7" into foundation wall with a minimum of 12" from the end of the piece. ■ the plate itself has to be either treated or redwood. ■ the foundation must be stepped where the ground slopes more than 1" in every 10'.

NO. stories	foundation wall thickness	footing width	footing thickness
1	6"	12"	6"
2	8"	15"	7"
3	10"	18"	8"

these are some minimum foundation dimensions for stud walls. TABLE 29-A (UBC)

PROCEDURE

1 excavation actually begins the work on your house. begin digging (outside line of foundation wall) dig footings regarding code dimensions. then place 2x4's as shown in drawing.

2 after the footings get built the forms for your foundation wall and attach them to the 2x4's on the footings. use wire to hold the forms together and diagonal bracing perpendicular to the wall isn't a bad idea either. suspend anchor bolts from the top of the forms and your ready to pour the concrete once again. to have ready mix'd concrete delivered and poured saves mountains and insures a quality concrete. the price varies depending on how far they must truck. but will run about $15.00 a yard.

a good book for foundation construction is Nelson Burbanks "house carpentry simplified" tis in the city library.

11

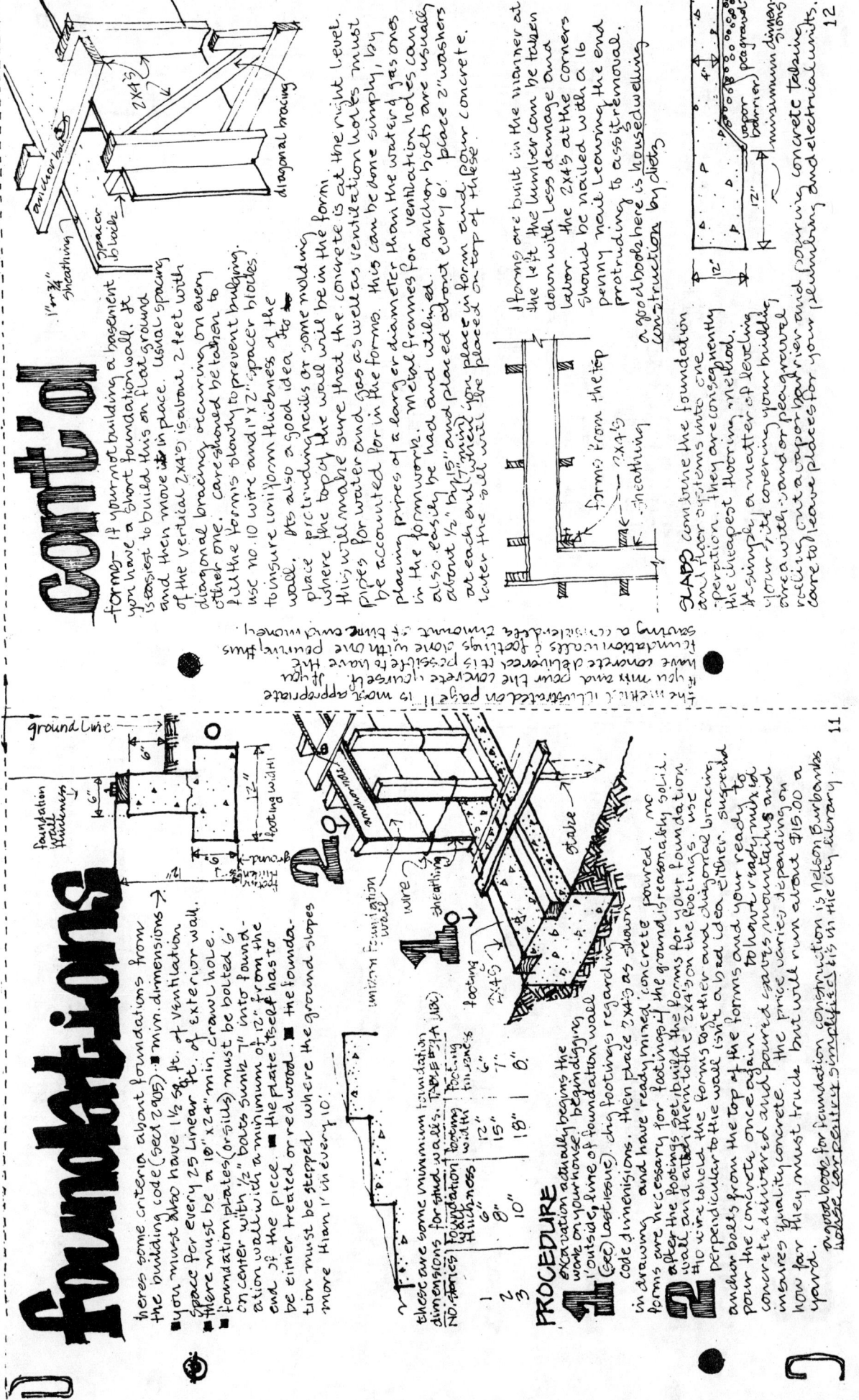

the sketch illustrated on page 11 is most appropriate if you mix and pour the concrete yourself. if you have concrete delivered it is possible to have the foundation walls & footings done with one pouring thus starting a considerable amount of time and money.

cont'd

forms— if your not building a basement you have a short foundation wall. it is easiest to build this on flat ground and then move it in place. usual spacing of the vertical 2x4's is about 2 feet with diagonal bracing occurring on even or odd. fill the forms slowly to prevent bulging. use no.10 wire and 1"x2" spacer blocks to insure uniform thickness of the wall. ■ its also a good idea to place protruding nails or some molding where the top of the wall will be in the form. this will make sure that the concrete is at the right level.

pipes for water and gas as well as ventilation holes must be accounted for in the forms. this can be done simply by placing pipes of a larger diameter than the water/gas ones in the formwork. metal frames for ventilation holes can also easily be had and utilized. anchor bolts are usually about ½"x15" and placed about every 6'. place in form and pour concrete at each end (12"min) you place in form and pour concrete. later the steel until the ... placed on top of these.

if forms are built in the winter at the left the lumber can be taken down with less damage and labor. the 2x4's at the corners should be nailed with a 16 penny nail leaving the end protruding to assist removal.

a good book here is house dwelling construction by dietz

SLABS combine the foundation and floor systems into one operation. they are consequently the cheapest flooring method. it simply a matter of leveling your site, covering your buildin area with .. and/or pea gravel rolling out a vapor barrier and pouring concrete taking care to leave places for your plumbing and electrical units.

12

AUGUR/20

temporary electricity

Permit — the power company will hook up a temporary service only after the service has passed the inspection of a government electrical inspector. This inspection is paid for in the form of a permit — permits cost $39 from the Bureau of Labor (in county) or from the city building department. **The National Electrical Code** is the standard followed by the inspector. It requires...

service head must be 9' above ground.

mast must be of at least 3/4" conduit.

meter base

meter supplied by the power co.

post must be sturdy — must project at least ten feet above the ground. you can use a tree.

service panel must be at least 4' above ground — it must be of an approved weather proof type or it must be protected from a driving rain. (this way you can use the panel you're going to use in your house)

ground wire must be no. 6 bare copper connected to the neutral bar in the service panel & to the ground electrode with a clamp.

ground electrode must be driven 8' into ground. can be an approved type or a 3/4" water pipe.

have the person you buy all this from show you how to permit it —
total expenses will run from $25 to $45.

If your thing doesn't pass the first inspection, the inspector will leave a note explaining why. after you fix it he will come again to inspect — no extra charge

- - - - - - - - - - - - - — snip - - - - - - - - - - -

STUD HOUSE

CONTINUING A SERIES THAT WILL HELP YOU

BUILD OR REPAIR A HOUSE   part four

— foundations — a summary —

1. LAYOUT — the process of transferring your plan to the ground. this was explained way back on page 8.

heavy string
batter boards
diagonals
outside line of foundation wall

2. FOOTINGS — its usually easier to use forms instead of pouring right into a trench. this way you'll have something to keep the concrete level. Stake some 2x6 about every four feet but don't nail the boards to the stakes until you've made your level marks to do this, pick a corner stake and mark it. mark the other corner stakes at the same level (about $4/day rental) or a hose with water stretch a string to mark and nail the forms to the in-between stakes and water stakes. Pour the footing and rest!

batter boards
2x6 forms
stake cut below form top
use a transit on a tripod or clear tubing in the...

3. FOUNDATION WALL — the easiest forms to make are

16"
96"
front
back

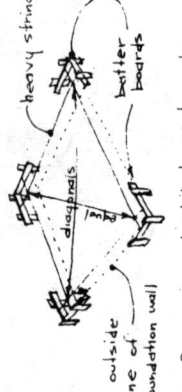
spacer
forms
strap
footing
side brace

put your forms together with a metal strap under the bottom every 2', a spacer across the top every 4 and a side brace every 8'. don't forget to provide for vents, girder insets and anchor bolts.

13
14

SUBFLOOR SYSTEM

The following is what is easy, economical and simple, also conventional. It is the most widely used system at this time. It goes like this...

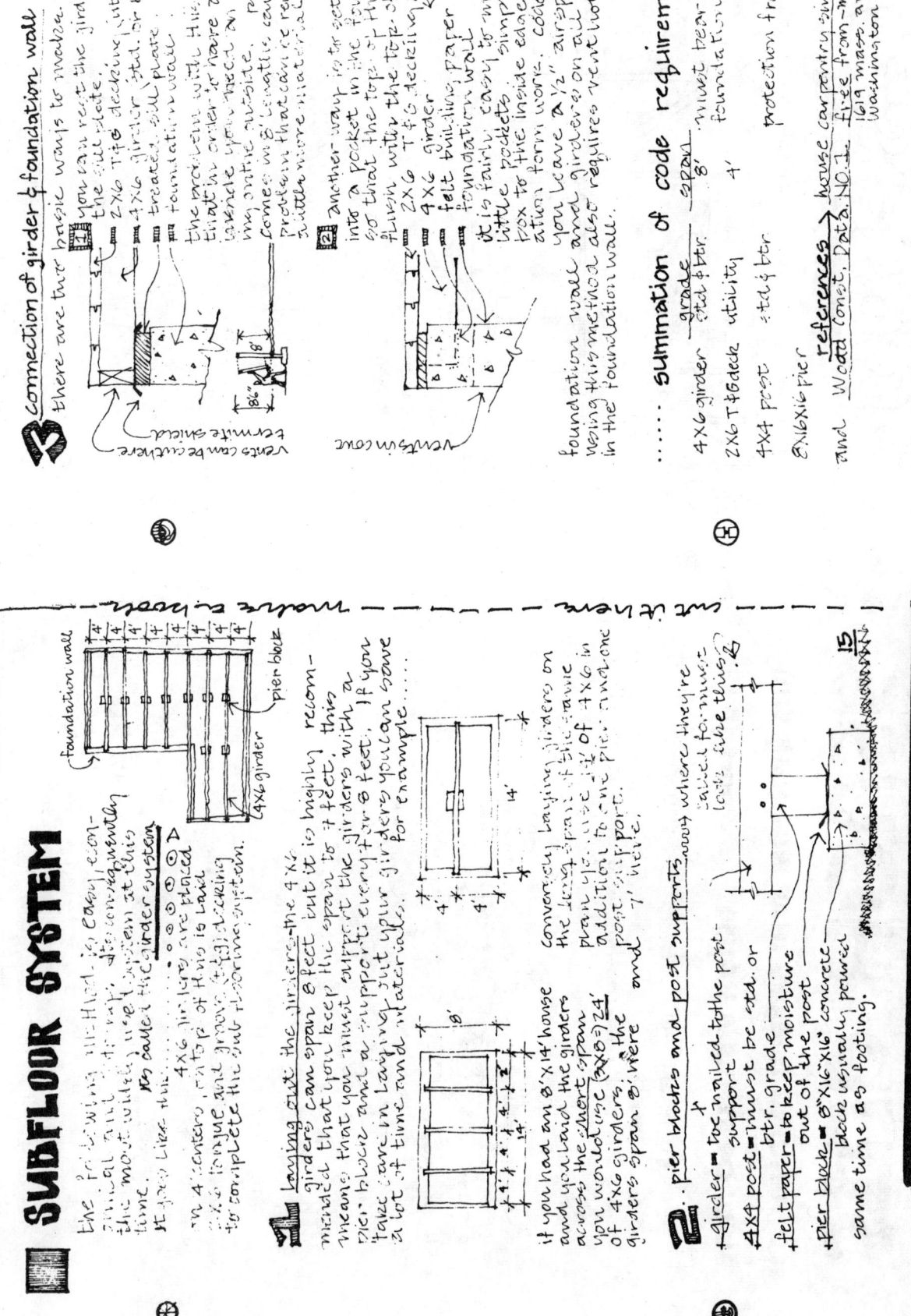

4×6 girders are placed on 4' centers, on top of this is laid 2×6 T&G decking & 2×4 decking to complete the subfloor system.

foundation wall

pier block

4×6 girder

1. Laying out the girders—the 4×6 girders can span 8 feet, but it is highly recommended that you keep the span to 7 feet. this means that you must support the girders with a pier block and a support every 7 or 8 feet. If you save care in laying out your girders you can save a lot of time and materials. for example...

if you had an 8'×14' house and you laid the girders across the short span you would use (2×8=7)24' of 4×6 girders span 8' here

conversely laying girders on the long span, if the same plan (4'=3) use 3' of 4×6 in addition to the pier and one post support here and 7' here.

2. pier blocks and post supports

→ **girder** = toe nailed to the post support

→ **4×4 post** = must be std. or btr. grade

→ **felt paper** = to keep moisture out of the post

→ **pier block** = 8"×16"×16" concrete block usually poured same time as footing.

↗ connection of girder & foundation wall

there are two basic ways to make this connection

1. you can rest the girder directly on the sill plate.
2×6 T&G decking
4×6 joist (utility grade)
4×6 girder (std. or btr. grade)
treated sill plate
foundation wall

the problem with this method is that in order to have an 8" wall (which you need in an 18" wall) over the outside... plywood siding comes in 8' lengths, considering a 6" problem that can be remedied with a little more material and work.

vents can be either here
termite shield

2. another way is to set the girder into a pocket in the foundation wall so that the top of the girder is flush with the top of the sill.
2×6 T&G decking → same grades
4×6 girder
felt building paper
foundation wall

it is fairly easy to make these little pockets simply nail a box to the inside edge of your foundation form work. code requires that you leave a 1/2" airspace between your wall and girder on all sides.

foundation wall also requires vent holes to be left in the foundation wall.

using this method

vents in some

..... summation of code requirements ....

	grade	span
4×6 girder	std & btr	8' → must bear on 4" of foundation wall
2×6 T&G deck	utility	4'
4×4 post	std & btr	→ protection from water
8"×16"×16" pier		

references → house carpentry simplified ; .16
and Wood Const. Data No. 1 free from wood assoc.
1619 Mass. ave NW.
Washington D.C. 20036

cut it here

STUD HOUSE

CONTINUING A SERIES THAT WILL HELP YOU

BUILD OR REPAIR A HOUSE   part five

Howdy folks! I'm back again to talk about WOOD....

did you know that wood shrinks 5 - 7 % from the time it's cut from the tree till the time it's dry. Sure does! Shrinks mostly along the line of the annular rings and perpendicular to them. Shrinks very little in length.

SHRINKS: SOME / LITTLE / MOST

If your wood is too wet it needs to be seasoned. Stack it on a level support and put lath between each row to allow air to circulate. Protect the pile from rain and ground moisture & wait a couple months. Or you can buy kiln-dried lumber. Wood gets stronger as it gets drier!

Shouldn't build with wet wood! Cut a piece exactly 6" wide and dry it in a slow oven 48 hours. If it shrinks 1/4" its too wet!

When buying wood for a house I would recommend making a master list and then call every yard in the yellow pages for an estimate. Never know which one will be cheapest on any given day. They will want to know:

1) QUANTITY—how much in lineal ft. or board feet (1"x12"x12")
2) GRADE—CONSTRUCTION, STANDARD & BETTER, or UTILITY. the code requires certain grades for certain jobs.

You can't save much by buying a little bit at one yard, a little at another... they start charging you for delivery. If you can buy your lumber in mid-winter the prices are likely to be at their yearly lowest.

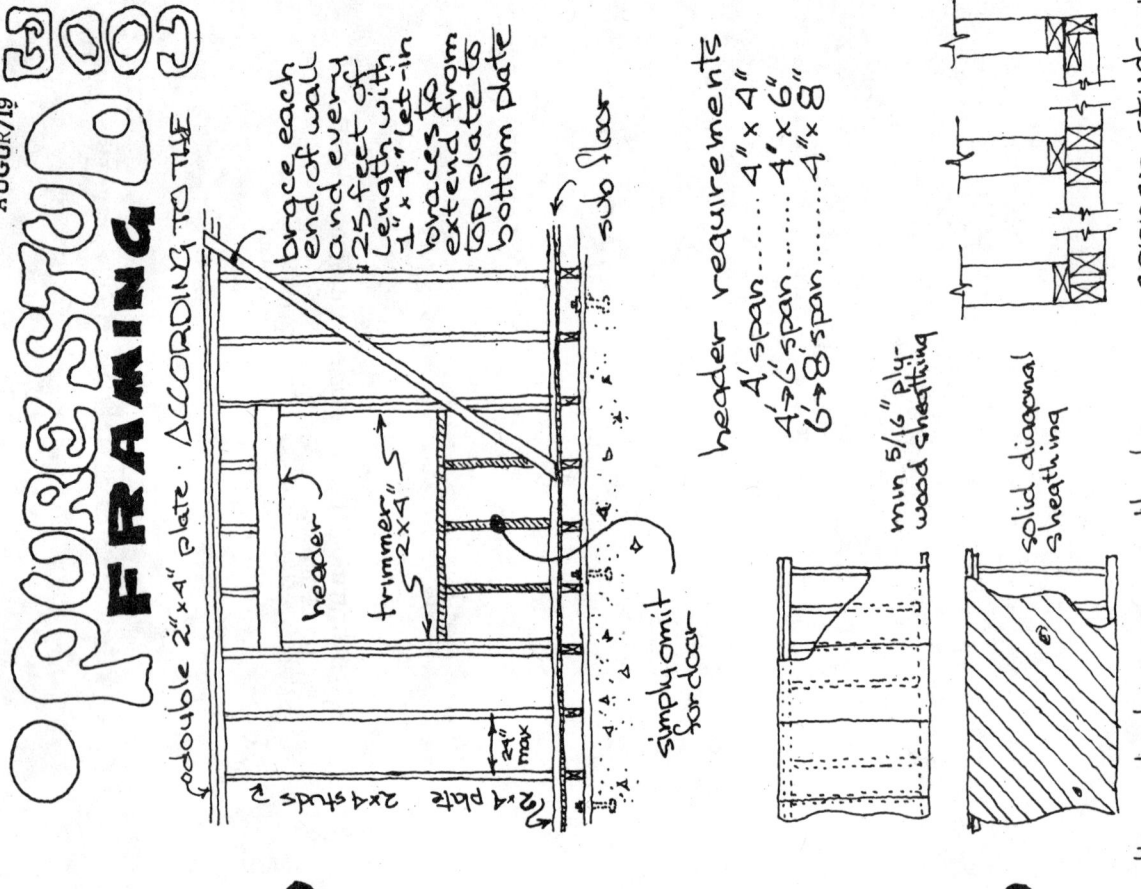

PURE STUD FRAMING   AUGUR/19

double 2"x4" plate · ACCORDING TO THE

header
trimmer 2x4"
24" max
simply omit for door
sub floor
2x4 studs
2x4 plate

brace each end of wall and every 25 feet of length with 1"x4" let-in braces extend from top plate to bottom plate

header requirements
4' span.....4"x4"
4'>6' span.....4"x6"
6'>8' span.....4"x8"

min 5/6" plywood sheathing

solid diagonal sheathing

alternate bracing methods

corner studs 18

ROUGH PLUMBING

BEFORE THE FLOOR GOES IN, YOU'VE GOT TO PUT IN WHAT'S CALLED ROUGH PLUMBING. THIS INCLUDES EVERYTHING THAT'S BASIC EXCEPT THE FIXTURES (SINKS, TOILETS, ETC.)

THERE ARE BASICALLY THREE SYSTEMS INVOLVED IN PLUMBING: HOT WATER, COLD WATER, & WASTE AND VENT.

THE COLD WATER SYSTEM STARTS AT THE WATER MAIN OR WELL OR SPRING AND ENTERS THE HOUSE. THE ORIGINAL PIPE MUST BE AT LEAST 3/4" ALL THE WAY TO THE HOT WATER HEATER:

RELIEF VALVE — WATER HEATER — UNION — 3/4" — TO COLD SYSTEM — SHUT OFF VALVE — UNION SO YOU CAN DISCONNECT IT.

THE UNIONS - BOTH COLD WATER INLET AND HOT WATER OUTLET MUST BE AT LEAST 12" BEFORE THE PIPE GOES INTO THE HEATER. ON THE INLET SIDE, BEFORE THE UNION, IS THE PLACE FOR THE MAIN SHUT-OFF VALVE FOR THE WHOLE HOUSE. ON THE OUTLET SIDE, THERE HAS TO BE A RELIEF VALVE IN CASE SOMETHING GOES WRONG. THE RELIEF VALVE HAS TO BE 3/4", HAVE "ASME" STAMPED ON IT, AND HAVE A HAND TEST LEVER ON IT.

In this county you can't use plastic pipe inside the house for cold or hot system.

WHERE THE COLD WATER PIPE TAKES OFF AFTER THE SHUT-OFF VALVE, IT CAN BE REDUCED TO 1/2" BUT THIS WILL REDUCE PRESSURE.

THE HOT-WATER SYSTEM, WHICH STARTS AT THE HOT WATER HEATER MAY BE REDUCED TO 1/2" ALSO BUT AFTER THE FIRST TEE IN THE LINE.

GARDEN — TOILET — SHOWER — SINK — BASIN — INLET — HOT WATER HEATER — HOT WATER RELIEF

THE WASTE AND VENT SYSTEM IS EASIER — HERE'S A LIST:

A BASIN HAS A 1 1/2" DRAIN AND VENT.

A SHOWER OR TUB HAS A 1 1/2" UNLESS THE HOUSE IS FHA, THEN IT HAS TO BE 2" WITH THE SAME RESPECTIVE VENT.

THE KITCHEN SINK HAS A 2" DRAIN WITH A 1 1/2" VENT AND IT HAS TO HAVE A CLEANOUT.

THE TOILET HAS A 3" DRAIN AND VENT WITH A CLEANOUT.

THE SEWER IS 4" FROM THE FIRST TEE.

VENT — VENT — VENT — SINK DRAIN — TOILET DRAIN — CLEANOUT — CLEANOUT — TO SEWER — 4" — 3" — CLEAN-OUT — 20

IF YOU'VE GOT ANY PROBLEMS AT ALL, GO DOWN AND SEE THE COUNTY BUILDING DEPT. THEY'RE REALLY NICE.

plastic pipe is approved its cheapest + easiest

19

STUDHOUSE is a series of columns which appeared originally in Augur. Its author is Robert Thallon.

Self-Help vs. Helplessness

David Conover

For the past year and a half I have been connected with a program in Cambridge, Massachusetts, which, in spite of the federal bureaucracy, uses federal monies to support self-help dwelling rehabilitation. I work predominantly with lower-income families, but the provision of decent housing for low-income families is not at all what the Government has in mind.

Consider the case of an individual who wants to upgrade his house, and consider the people around him who say they want to help: planners, developers, local building departments, local zoning and planning boards, relocation specialists, property managers, and architects. Each says, "I want to help."

What does the individual say? "I want to do it myself."

The dichotomy is that the ring of professionals, do-gooders, experts, agencies, and institutions, who want to help, makes the individual so helpless that he has absolutely nowhere to turn.

It happens in other areas of life as well. Take the man who says, "I'm broke, I want a job." All around him are people who say they want to help: job counselors, psychologists -- to figure out why he hasn't been able to make it, maybe even a parole officer.

All the man says is, "I can get the job. Just give me a chance." But the professionals, instead of teaching him to get his own job, prefer to get jobs for him. The problem is one of helplessness.

I started working in the inner city primarily because I am a resident of the inner city. I've lived in a renewal area, and now I am working in a model cities area. It has become increasingly clear to me that the ring of experts perpetuates helplessness and contributes to inner-city decay.

Renewal agencies, for example, work in terms of identifying renewal territory, studying, planning, and identifying structures in substandard condition. They look at the physical facilities, without looking at (or consulting with) any of the people therein.

Banks create another sort of helplessness. In Boston, we have a particularly difficult problem in what is called BBURG, Boston Bank and Urban Renewal Group, which is willing to loan money to Blacks for home ownership, but only in certain sections of town. They are building a new ghetto in terms of financing. If a family wants a mortgage to buy a house, the bank says, "Sure, we'll lend you money -- so long as you live over here."

Many of these families have had the capital to buy a certain house, but have found that the FHA has classified the house as substandard, and has insisted on putting the house in standard condition before guaranteeing the mortgage: a new plumbing system for three thousand dollars, plus perhaps new electrical and heating systems. This often puts a family on the brink of losing the house -- and the deposit they put down to purchase it. In Boston, federal rehabilitation money under section 312 often takes over six months to obtain.

Speaking of federal money, the Cambridge Model Cities Agency, for which I work, rips-off over 80 per cent of the money that comes into Cambridge. Over 80 per cent is drained off before it ever gets out into the community! Where does that money go? It goes to the salaries: the director and executive director (both with master's degrees or higher), project directors, department heads, communications specialists, in-house evaluators and researchers. We look at ourselves as a ring around each individual saying, "We want to help."

Nowhere in this metropolitan area have I found tax assessors who are willing to give understanding to low-income families in search of shelter. Yet recently a large developer was able to work out a half-million-dollar-per-year tax understanding with the assessors of Cambridge.

So we see that these agencies, plus others -- building departments, code enforcement agencies, even private institutions-- all tend to inhibit any kind of direct action the self-helper is likely to want to take. This leaves the tenant or homeowner, who would like to paint his walls, or change some of the plumbing, or improve his electrical system, unable to stand the high cost of deteriorating housing.

We have got to get beyond the local institutions and their representatives to allow center-city dwellers to do things for themselves. Outside the central cities, there are nearly 200,000 houses built in this country by do-it-yourselfers each year, without any governmental interference, but also without governmental assistance. In the inner cities, where there are many agencies trying to help homeowners, it is virtually impossible for a do-it-yourselfer to do very much. I've found that here in the East, people are willing to work inside their homes, with moonlighters doing the work, or with their own labor, evenings and weekends, but they just don't touch the exterior at all, because

they're so afraid of upward reassessment, and so afraid that the
building inspector will come by and condemn their work.

If a family decides to do things through approved channels
and hire a general contractor, other problems often appear. The
contractor says, "Look, I'll fix your front stairs, rebuild your
bathroom, build a new kitchen, and I'll do a fine job. I can
get started right away. You'll be happy with it." But it's
seldom that simple. Often the homeowner finds that the tradesmen who show up to do the work never were made to understand
exactly what they had to do. And to find the general contractor
and get him to carry out proper superintendence of the work
becomes one of the largest single problems. The general contractor,
who supposedly coordinate the various tradesmen to get a job
done, is entitled to a margin of profit, about 15 percent in this
area. He often finds it so difficult to locate qualified tradesmen who are willing to work for him, that he has little time
to be on the job. Instead, to make financial ends meet, he
spends perhaps two days a week hustling other jobs, takes time
to find the least expensive tradesmen and puts together the least
expensive materials for the maximum amount of money. This
basically means that the general contractor cannot do decent
work on the existing housing stock, unless it is in the field of
really major modernization.

Personally, I've found that the best rehab work comes out
of moonlighters. Most of them don't have to depend on a moon-
lighting job for their main income, and if necessary, they can
spend twice the amount of time they were originally contemplating
to get the job done. Often there are neighbors who are working
for the local utility company, or who happen to be steamfitters
and can do some plumbing. They don't have to worry about a
permit, because the building department representative will never
show up. In short, oftentimes if you do the work illegally, you
can get it done well, you can get it done for a fair price, and
you just don't have the hassles with the agencies and the other
folks who want to "help out."

In the program that I was connected with, we were able in
one year to work on fifty-two structures. More than half the
families took on the role of general contractor -- they found
subcontractors, made contractual arrangements, coordinated the
work, and made all the decisions. And they also made sure that if
they didn't like what they were getting, that they held onto the
purse strings as leverage.

Twenty-two of the fifty-two families that same year did at least 25 per cent of the work themselves. Many of them had never lifted a hammer before, but felt pretty good after they had done it and succeeded. Of the fifty-two families, seven hired general contractors, and in every case the product was poorer. In most of these instances, unfortunately, the tradesmen who did the work could not speak English, causing communications problems between owner and worker.

There is a high level of paranoia that exists, not only among homeowners in looking after their own property, but among tenants as well. The level of the art of home improvement and rehabilitation is so low that they would rather not do anything at all, than become involved in a process that takes their money and leaves them with an unsatisfactory product.

This brings me to what I want to talk about: One of the most viable ways of being able to improve our housing is to get people to do more things on their own. It is to start to attack the kinds of hopelessness that exist, to build self-confidence, some level of skills, some real joy in being able to get things done on one's own. It is to have someone to back up the individual -- whether that happens to be another individual like a neighbor, or an institution that ways, "We will lend some form of stability," whether it's actual or imaginary. It is to teach people to do things for themselves. Who wants to hire a plumber to fix a faucet for ten dollars, when, if you've been taught to do it once, you can do it for yourself every time after that? The level of independence that's generated in just that one act is a real boost to people who traditionally have been denied the chance to try things on their own. The secret is provide maximum information to people who need to know, while maintaining an advocacy position, as opposed to a paternal position of "We'll do it _for_ you; just tell us what you want."

Our next-door neighbor last year decided that he was going to build a patio of brick. I had done the same thing, and I mentioned to him, "Well, you know, all bricks look about the same, but they vary in composition and firing, so make sure that those you pick out are suitable for exterior use." And he thought he had, I guess, but his brick patio, as of this spring, is about two-thirds dust. With all the variables involved, he did a fine job, and worked hard for two weeks. But he selected a brick that could not withstand wetting and freezing action, and his patio started spalling away immediately in the New England winter. Take this experience of a middle-class, middle-income homeowner, transfer it to a low-income person, and superimpose this kind of failure on his outlook. If there were agencies around that could advise families about what materials to use, where to get them, and what tools are required to get the job done, then these agencies would be working not only in support of better housing, but in direct support of our entire environment. And as we go down the economic ladder, the environment plays an increasingly important part in our mental health and our attitudes.

What I'm suggesting is that groups of people who are know-ledgeable about building should work to improve the existing housing stock by supporting people in working things out for themselves. The process of housing is really the product. If we constantly bypass the process of involving each person in his own environment, we are going to find ourselves more and more locked into unworkable state and federal programs. We will be more entrenched in our attitude that we are going to do it for them, and we will do nothing but perpetuate the level of help-lessness that already exists.

David Conover works for the Cambridge Corporation in Cambridge, Massachusetts.

Obstacles to Owner-Building: A Quest for Responsive Housebuilding Technologies

Ian Donald Terner

The impounded funds and domestic budget cuts of the second Nixon Administration focus special attention on the one housing program that wasn't cut-- indeed, couldn't have been cut-- because, in fact, the federal establishment has barely recognized its existance, much less funded or supported it. That program, or perhaps more appropriately, "non-program," encompasses the broad and significant array of self-help housing activities in the United States.

The sheer magnitudes of this effort are impressive. In 1970, approximately 160,000 homes were built by owner-builders-- accounting for approximately 17% of all U. S. housing starts and nearly one out of five of all single family starts.[1] Owner-builder output equalled more than three times the combined totals of such giant builders as National Homes, Levitt, Boise Cascade, and Kaufman and Broad. In addition, it was calculated that "home repairs and remodeling in which homeowners bought the building materials and provided all the labor" rose 36% between 1967 and 1969, and fully 100% between 1967 and 1971.[2] At present more than half of the remodeling or additions of kitchens, bathrooms, attic or basement rooms, carports, patios and breezeways are "self-help" projects.[3] Moreover, this activity is expected to continue to grow, since public opinion surveys disclose continued pent-up demand for these items. According to these polls, home improvements now constitute "the first spending preference of most Americans", superceding "such traditional family favorites as air travel and a new car."[4]

Yet what is perhaps most surprising about these figures is not that they are so large and significant, but that they exist at all in the face of severely formidable obstacles. In particular, four are prominent:

1) lack of government support

2) lack of financing

3) lack of information

4) lack of responsive housebuilding technologies

Government

In many areas, it is actually illegal to owner-build. Administrative procedure requires that building permits be issued only to licensed contractors, thus excluding potential owner-builders. This barrier is not fatal, however, as many owner-builders have learned that for a "fee," some contractors will take out a permit for them. Nonetheless, this view typifies governmental attitudes toward owner-builders, which range from a view that they are an outright safety menace, to the condescending attitude that owner-builders constitute merely a handful of rabid do-it-yourselfers, nostalgically looking backward in time toward "the good old days when people really did build their own houses."

Financing

In many cases, commercial banks balk at the idea of extending a construction loan to lay persons who want to build their own house. Thus many owner-builders are required to seek initial financing internally-- that is, either from savings, family, or friends in order to get started. The irony is that once construction is finished, banks are delighted to offer mortgages to owner-builders, once the bankers can see how well built the home is, and the degree of commitment and pride to maintenance and upkeep that has been engendered by the household's participation in the actual construction process.

Information

One of the greatest obstacles to owner-building is the problem of inadequate information. The question is: Precisely how does one build a house? What type of construction should one use? How much self-help? What kind of self-help? How to assemble plans? Land? Materials? What are the legal requirements? The alternative financing methods? The time phasing, etc.?[5] All of these present tremendous barriers to the novice builder, and very few of these questions, if any, are addressed in a typical American schooling experience. Tragically, after some twelve years of public education, the typical high school graduate has little insight into how to face one of his most awesome and important adult responsibilities-- that of providing shelter.

This lack of information relates back to the lack of government interest in the owner-builder, not only by the schools, but also by the U.S. Department of Housing and Urban Development (HUD) which for years has steadfastly ignored various expert proposals to mitigate this information gap by the provision of federally sponsored "housing advisory services." [6]

Responsive Housebuilding Technologies

This final, and perhaps most important point for this discussion, focuses on technology-- the actual tools and methods for building. While HUD has ignored pleas for housing information services, it has focused on housebuilding technology, although of precisely the wrong kind. Its banner technology program, "Operation Breakthrough" not only ignored the individual owner-builder in favor of America's corporate giants, but it even carefully eschewed the goal of lower-cost housing. [7] The widely reported failures of the program are well-known,[8] and indeed typify the growing gap between the highly-mechanized, labor saving trends in the construction industry, and the labor-intensive, easy-to-use technologies which can contribute to the needs of self-help builders and low income households. A dramatic example of this gap may be found in the painting subsector of the construction industry, as embodied by two widely different technologies: on the one hand, mechanized spray equipment; on the other, paint rollers, water-based paint, etc. [9] In diagramatic form this technological "fork in the road" is represented in Figure One.

ONE

TECHNOLOGICAL "FORK IN THE ROAD"-- The painting sector of the construction
industry exemplifies the two directions which new technologies may take.
On the one hand, conventional brush technology has yielded to mechanized
spray equipment which implies added capital investment and fewer, more
highly-skilled workers. On the other hand, rollers, water-based paints,
etc., require relatively large amounts of labor, but contribute by reducing
the necessary skill level and start-up costs. The former are typical of
most technological innovations in construction, and are generally unsuitable
for owner-builders; while the latter represent the kinds of tools and methods
which are truly responsive to self-help needs.
(Source: Massdesign, Architects and Planners, Inc. [10])

Indeed, much technology in the painting industry has followed the general
trend toward capital-intensive mechanization-- common throughout the construction
sector. New and costly spray equipment allows fewer, more highly-skilled
painters to cover more area than ever before possible. However, the other
cluster of technology has yielded an entirely different result, and in fact,
is representative of the main thrust of this discussion. Innovations in the
form of rollers, instead of brushes, and quick-drying latex and water-based
paints which can be cleaned up with soap and water-- coupled with small aerosol
spray applicators, inexpensive polyethelene dropcloths, long handles instead
of ladders, etc.-- have reduced the equipment and skill-level necessary for
house-painting to the point where virtually anyone who can afford the paint
itself can have an acceptable, even professional-looking paint job.

These relatively simple innovations and applications of modern technology
have liberated the dweller from reliance on the skilled journeyman painter.
The home-owner or renter must still purchase materials; although the market
serves him well, since it is extremely rich in competing brands, prices,
varieties, colors, and performance aspects of the paint.

The technological reduction of the skill-level needed to paint is significant for another reason as well, for while it has increased the dweller's
freedom to paint at his own initiative-- without reliance on others-- it has
also increased his freedom <u>not</u> to paint. The dweller can now hire non-professionals (students, unemployed, etc.) to do his painting for him-- at greatly
reduced costs.[11] Furthermore, there is evidence that some journeyman painters
have also reduced prices, perhaps partly because they find themselves more
productive using rollers (although painters' unions at first refused to permit
them)[12] and partly from the pressures of increased competition by self-helpers
and non-professionals. Yet it seems that painting is the exception, and that
the vitality of owner-builders and self-help rehabbers (who are not even included

In the earlier statistics) exists in spite of technology, and not because of
it. The fact remains that it is still very difficult for most people to undertake construction on their own--even if they have somehow managed to solve parts
of the government, finance, and information problems on their own.

The question then emerges: "What can be done?" If "heavy-type" Operation
Breakthrough technologies have been irrelevant, what contributions might science
and technology make to the field of responsive house-building tools and methods?

From the painting example, one answer is clear: some forms of product
technology can significantly contribute to reducing the skill and experience
requirements for participation in the building process. Technology can increase
the individual's self-reliance by making construction easier. It can liberate
the individual dweller from his present reliance on masons, carpenters, plumbers,
and electricians in the same way that it has liberated him from the painter. And
in some extreme cases, it may even mitigate the dweller's presently crucial
dependence on the mortgage banker, land developer, or municipal utility network,
all of which have been known to make highly discriminatory service judgments
based on a dweller's race, social status, or life style-- as opposed to his
economic abilities and potentials, or their own public service responsibilities
and obligations.

An Innovations Agenda --
It is possible to envision a technological, or "innovations agenda" designed
to produce improvements in those areas of construction which are still uncommonly
difficult to undertake without special training or prior experience.

The innovations agenda may be organized under four broad headings:
1) structural/envelope systems; 2) water/waste systems; 3) power/energy
systems; and 4) foundation/footing systems.

THE STRUCTURAL/ENVELOPE SYSTEM-- Clearly one of the most difficult construction
tasks is to make a building stand up-- straight, safely and securely. The
easiest thing to build is a lean-to, a structure leaning against an object
already straight, erect, and secure, be it a tree, a wall, an adjacent
building, or whatever. A view of early-stage squatter areas in developing
countries reveals an abundance of these appendage-like structures, simply
because they are easier to build than free-standing structures.

In light of this, a persistent technological "dream" has been to create
for adults what toymakers have succeeded in creating for children-- simple
sets of construction components with easy, virtually fool-proof joining systems
such as Erector Sets, Tinkertoys, Lincoln Logs, Lego, etc. Indeed, there have
been a number of attempts at "erector set technologies" [13] with commercial
claims that two adults, with no special equipment nor prior experience or training

in construction techniques, following a simple set of instructions, could build a wide variety of single and multi-story structures-- in much the same manner that a child builds with his toys. The structures were claimed to be safe, fire-proof, wind and earthquake resistant, very inexpensive, and absolutely simple to build from a small number of standardized, light-weight components. [14]

Yet to date this technological dream has not been fully realized-- although there is little doubt in this author's mind that it can be done. Several attempts have come very close, but there has yet to be a cost breakthrough. However, technological progress in this area of cheaper, lighter, and stronger materials brings this dream closer to fruition. The implications of such a technology are potentially enormous, for they would enable dwellers to solve one of the most difficult aspects of self-construction-- that of achieving structural integrity.

The task of enclosing or providing a weather-envelope around such a structure is a related problem. In some cases, when the building system incorporates structural panels, the weather envelope is combined with the structural solution. On the other hand, if the structure is only a frame, then various types of interior and exterior walls become a separate consideration, although their fabrication then becomes a function of various environmental requirements-- but not structural requirements. Self-built walls, which need not be structural, are much simpler to build-- as evidenced by the abundance of woven reed, and thatch, mud block, and thin sheet material in use in developing areas. Such walls can also be replaced or upgraded at the dweller's initiative without jeopardizing the structural safety of the house when used in the context of a permanent and secure structural frame.

WATER/WASTE SYSTEMS-- The second general item on the innovations agenda centers on water-waste systems. Conceivably, through technology, the skills required for installing plumbing systems and sanitary fixtures could be made no more difficult than hooking up a garden hose. The question must be asked: "why should pipe be rigid?" If it were flexible, many costly fittings and joints could be eliminated entirely, and those which remained could be designed for lay installation. The special and costly skills of the journeyman plumber may someday be thus challenged in the same way that the skills of the journeyman painter are challenged at present.

Furthermore, not only in the U.S. but in all countries of the world, the dwellers' presently critical reliance on public water and sewage systems could be reduced through technological advances. It is not impossible to imagine a self-contained, recirculating, water/waste system, whereby a generally finite amount of water could be used, purified, and then re-used. Further, it is conceivable that such an appliance might not be larger nor more costly than,

for example, a modern washing machine. The implications of such an applicane
for developing areas could be enormous-- particularly when one considers the
millions of squatter and owner-built houses in those countries which <u>are</u>
equipped with major electrical appliances, such as televisions and refrigera-
tors, yet most of which do not have running water.

In the United States, a recycling water/waste appliance could have an
equally important impact.[15] As municipal utility systems are forced to expand
in order to accomodate population growth, and as the dangers of pollution mount,
the public sector faces major new investments in water treatment facilities.
Thus an individual water/waste appliance could potentially reduce the need for
such investment, and could also significantly reduce the costs of developing new
land for settlement. Further, such an appliance would reduce the now critical
dependence of a dweller upon a developer or a municipality to provide such
services at the place where he wishes to live. Such technology might also
imply potentially more intense utilization of land already under settlement
by replacing, in some areas, the need for individual land parcels to accomodate
both a pure water well and a septic tank.

POWER/ENERGY SYSTEMS-- Perhaps the closest analogy to the gradual shift
that was observed in painting-- i.e. the shift from journeymen to non-professional
labor, exists in the wiring and electrical phases of construction. Vast numbers
of squatter and owner-built houses in developing areas are electrified-- often
through illegal and clandestine "tapping in" to government or private power
lines. Even in the U.S., where it is sometimes illegal or perilously unsafe,
many non-professionals and home-owners wire houses in attempts to bypass the high
costs of the licensed electrician.

Modern technology and innovation have played some role in liberating the
dweller from a dependence on the electrician, but it is patently obvious that
technology can go much further in making the electrification aspects of home
building far safer and far easier. For example, recent studies sponsored by
the U.S. National Aeronautics and Space Administration (NASA) have led to the
development of "scotch-tape wiring," as shown in Figure Two, i.e., a wire with
a profile as thin as a human hair which can be manufactured in an adhesive-
backed strip much like a standard piece of cellophane tape. [17] This wiring
can even be used with adhesive-backed switches, resembling ordinary switches, but
which in no way penetrate the wall surface. Such "stick-on" wires and switches,
and perhpas someday, wall and ceiling fixtures as well, can be completely
safe, simple to install, and can have virtually the same, or even improved
appearance over conventional circuitry which must now be laboriously buried
inside walls, floors, and ceilings by highly skilled, licensed laborers. It
is entirely conceivable that in the future, technology may make "plugging in"

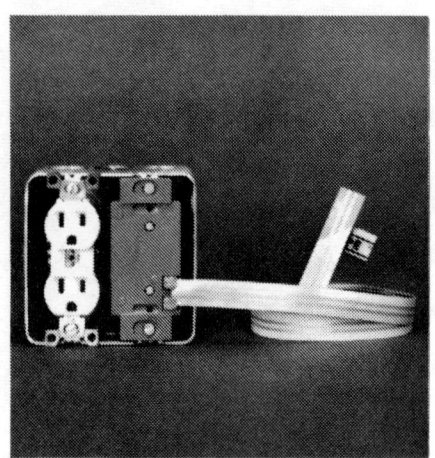

TWO

SCOTCH TAPE WIRING-- For the layman, one of the most difficult aspects
of wiring or rewiring a home involves the wires being "buried" inside walls,
floors, and ceilings. Often the wiring is difficult or impossible to reach
without considerable experience and unsightly cuts into wall surfaces which
are time consuming and difficult to repair. Surface-mounted wiring, with
a profile as thin as a human hair, can now be manufactured in adhesive-
backed strips much like a standard piece of cellophane tape. The wiring can
be used with "stick-on" switches which do not penetrate the wall surface,
or with conventional electric boxes as shown in the photograph. The
system can be installed or moved easily and safely by self-helpers without
reliance on skilled electricians.

(Source: Switchpack Systems Inc. [16])

---

the entire circuit system of a house as commonplace, easy, and safe as plugging
in an electric clock or toaster today. [18]

Technology may also contribute to power and energy generation, as well as
its internal distribution. Today, for example, deep sea buoys, and equipment
for use in outer space commonly generate their own electric power from solar
energy cells. Whether it will ever be economical for individual dwellings to
generate their own power (perhaps through solar cell roofing materials, fuel cells,[19]
or tiny fission or fusion reactors) is highly uncertain, but the question
remains open.[20] It is yet another area wherein technical product innovation
embodies the potential to further support an individual's autonomy.

FOUNDATION/FOOTING SYSTEMS-- Finally, it may be well to touch briefly upon
a fourth potential area of technical contribution to dweller autonomy. Another
one of the most difficult tasks for the inexperienced builder (and not at all
unrelated to his problem of structure, discussed earlier) is the problem of
joining the house to the ground. The problem of laying out and joining a regular
and generally predictable building to the irregular and unpredictable earth
is a major one for the owner-builder and a major expense even for the conventional,
professional builder. Variables such as slope, compaction, drainage, surface
characteristics, and bearing capacity of the soil all can affect the success
and safety of a building's "roots." In addition, the alignment and stability
of foundations often affect how well and how easily the superstructure of the
dwelling fits together. Various innovations yielding inexpensive, simple and
improved jacking and leveling devices, raft foundations, "screw-in" footings,
structural pads, etc, would all be helpful. In addition, technology might
also be applied to improved "air-rights" platforms to utilize space above water,
swamps, highways, etc. Increasing effective supplies of "buildable land"
through technology, can, of course, potentially benefit all dwellers-- not
only those of lowest income.

A Common Theme--

Unfortunately, this is only a most skeletal and tentative outline of an innovations agenda. However, even in such a preliminary effort, a strong common theme runs throughout: all the innovations mentioned form parts of a "componentized" strategy of housing. None are dependent upon a "unitized" or packaged dwelling module. Moreover, like painting, much of the technology called for in the innovations agenda is <u>labor intensive</u>, and runs counter to the nearly universal trend to make construction more capital intensive, less reliant on labor, and more nearly akin to the high productivity format of other manufacturing industries. The component strategy advocated here <u>accepts</u> a high labor input, and in fact attempts wherever possible to trade <u>more</u> labor for reduced material costs-- <u>so long as technological innovation can assure that labor need not be experienced or skilled</u>.

Kits --

The concept of complete and understandable information designed explicitly for inexperienced builders complements the requirement that technological innovation assure that labor inputs need not be skilled. Kits, which include some or all needed materials packaged with step-by-step instructions as shown in Figures Three and Four, represent an aid to self-builders which help to alleviate five kinds of chronic problems: 1) the kit helps to alleviate the problems of design, often by offering a series of tried and proven alternatives; 2) it alleviates the problems of determining what kinds and quantities of materials are needed; 3) the kit reduces time consumed in seeking and purchasing materials; 4) it compensates for a builder's lack of experience by providing instructions, which often not only describe necessary techniques and processes, but also clearly and completely describe task sequencing and time phasing; and 5) finally, the kit can reduce skill levels needed for construction by performing in advance some special processing or some of the more difficult building tasks, such as accurate measuring, cutting, and drilling.

Although kits are usually more expensive than the cost of materials alone, kit manufacturers can buy materials in bulk, and economies of scale may partially offset processing costs, so that the consumer, who might have to purchase retail materials in small quantities, may still find the kit competitive.

The kit concept is well developed in the toy industry. For example, a child selecting a model airplane has a wide array of choices, ranging from completely packaged and finished models, through a large variety of kits, to relatively simple raw materials. The child can choose rather precisely at which level he wishes to participate in the construction. The easy kits, with many parts pre-finished and subassemblies already built up, generally cost more. Yet he

can save money and often gain increased satisfaction by choosing to participate more actively. Indeed, he can choose the exact point of entry that fits
his needs, abilities, and desires. Creation of analogous opportunities in
housing through redirected public policies and expenditures is a goal of this
discussion. At present, the entry options in housing are far fewer than in
the example of the toy industry described above. Many people who are extremely
anxious to trade personal effort for cost savings in the provision of shelter
are continuously frustrated by their lack of self-confidence or their perceived
lack of ability to undertake a construction task directly, or to even know how
to find out information. [23]

Indeed, it is a fundamental assumption on the part of this author that
if good houses could by gotten by trying-- if the quality of a dwelling
environment responded directly to effort-- then the poor would be far better
housed than they are now. This assumption is supported by the overwhelming
eivdence of ingenuity, resourcefulness, and perseverance inherent in the significant numbers of owner-built houses around the world-- houses built despite
incredible odds and obstacles. Indeed these houses reflect no lack of trying.
Hence the present strategy looks in part toward technological innovation to
mass produce simple and inexpensive components which can be assembled by dwellers
in a wide variety of compositions and configurations, and can be directly responsive to needs not only at the time of construction, but as needs change
over time.

THREE

KIT HOUSES-- One of the techniques which can be employed to reduce the
skill levels required for construction is to combine prepared building
materials and components with complete sets of assembly instructions.
In this manner, construction kits can save the inexperienced builder
considerable effort, including design, phasing, sequencing, measurement,
cutting, drilling, and selecting and purchasing materials. Kits are not
a new idea; the advertisement shown above is from the 1918 Sears Roebuck
Catalogue, and promises a "ready made building... in sections ready to bolt
together; wrench and screwdriver are the only tools needed... Doors and
windows are hung in their frames. All necessary hardware furnished."
Over 110,000 such kits were sold by Sears.
(Source: Sears Roebuck and Company [21])

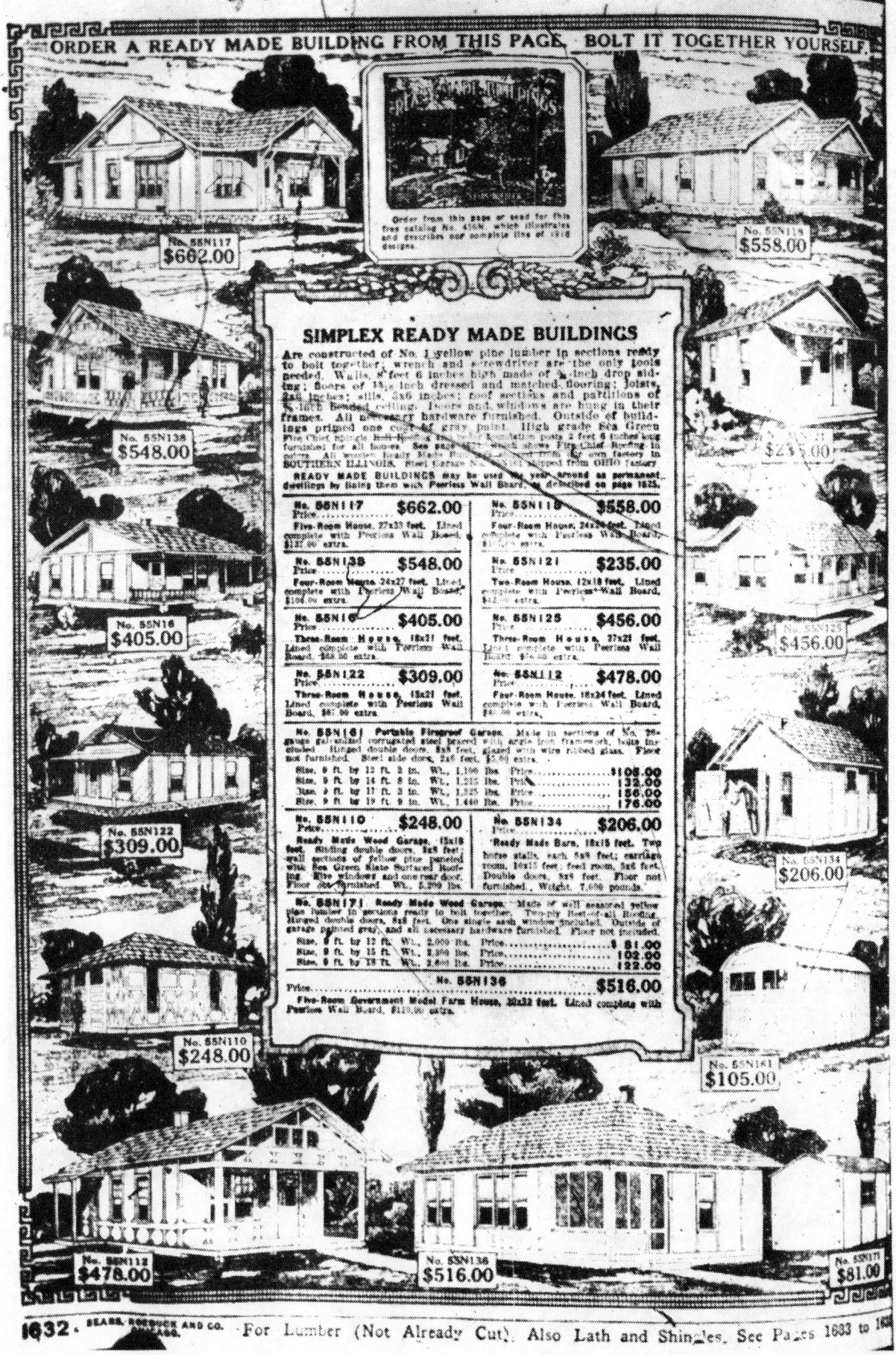

93

FOUR

ERECTOR SET TECHNOLOGY-- The building kit shown utilizes a precut
and drilled wooden frame with bolted joints. In concept, it provides
for adults what toymakers seek to offer children-- a set of coordinated,
simple-to-use, and easy-to-join components which can be assembled by following clear and easy directions. The vacation cabin shown, includes
step-by-step assembly instructions and is designed for use by people with
no prior construction experience, and no access to power tools or equipment.
(Source: Shelter-Kit, Inc. [22])

Financing--

The componentized strategy described above has additional implications for
another aspect of the housing problem-- the area of finance and credit. This
strategy allows a dweller to pay as he builds; a packaged strategy demands
an immediate lump sum-- implying substantial ready cash or an ability to
qualify for credit so that payments can be "smoothed out" and extended over
time. In the case of housing, this is a crucial consideration, since a home
is the largest single purchase that most families will make in their lifetimes,
and accumulated savings are almost never sufficient to pay for a completed
dwelling.

On the other hand, the vast majority of owner- and squatter-built dwellings
in developing areas proceed without access to credit and without a dependence
relationship with a mortgage banker. Family savings are typically used to construct
a habitable but unfinished house which is then incrementally expanded and
improved as the needs of the dwellers-- and their ability to pay-- dictate.
Often over the course of years these houses evolve into fully-finished and
occasionally even elaborate multi-story dwellings.

The fact that these dwellers can and will pay substantial sums for housing
is underscored by the extremely high rents that many are forced to pay simply
because they have no access to land and/or accumulated savings with which merely
to start the process of owner-building. In the developed nations, the process
of owner-construction without mortgage credit is much more rare-- mainly because
land costs are much higher, and squatting as a means to avoid this initial cost
is vastly more difficult. However, owner-building an incomplete yet habitable
house as another means for reducing first costs is still possible, although more
limited because of much stricter construction, inspection and occupancy regulations.
Technology, of course, is limited with regard to increasing the effective supply
of buildable land, although the innovations on the agenda described above might
well help in reducing the need for construction credit. By spreading the
construction process itself into small increments over extended periods (which
in fact is what credit is designed to do for construction cash outlays),
owner-builders can better afford to pay as they build.

Technology can also affect credit and financing in another way. It is well known that families who might never qualify for a home mortgage, can often receive financing for televisions, automobiles, and even mobile homes. Why? Because when a product can be _moved_, it can be repossessed. In the case of mobile homes, this is significant, for even though they are rarely moved after their journey from factory to first site,[24] the fact that they _can_ be moved is enormously important, for it makes them a much more liquid asset. Hence a credit agency will qualify more low income people as acceptable risks for mobile homes than for conventional houses which are bound to the land, to a section of the city, and to a bundle of municipal services (or a lack thereof, as is often the case for low income households). In this context, again, a componentized technology-- one where discreet building components can be returned, replaced, and even repossessed-- can go far toward being as accessible to low income families as major appliances or automobiles.

Oddly, this discussion on intermediate product technologies has spanned, to a degree, both Third World and fully-industrialized nations. This is because in both areas, short of radical change in government or private sector views of housing, the concept of dweller autonomy-- served and strengthened by new technology-- appears to offer one of the few hopes for truly broad-based housing improvement. The challenge to technology is enormous. The job to be done is nothing less than to transform housebuilding-- which is now costly and difficult-- into a process which is relatively inexpensive and easy.

Ian Donald Turner is an Associate Professor in the M.I.T. Department of Urban Studies and Planning

[1] William Grindley, "Owner-Builders: Survivors with a Future," in Freedom To Build, Ed. J.F.C. Turner and R. Fichter, The MacMillan Co, New York, 1972, p. 4.

[2] Kenneth McKenna, "Money Pinch Helps Make Old Homes Look Like New," Sunday News, New York City, New York, September 12, 1971, p. C-24. The author cites Seymour Kroll studying U.S. Bureau of Census figures.

[3] Ibid., Citing the Bureau of Building Market Research.

[4] Ibid., Citing U.S. Public Opinion Polls.

[5] William Grindley, "A Proposal for a Housing Advisory Service in Support of Owner-Built Housing," M.I.T., Cambridge, Mass., Unpublished, 1971, p. 5.

[6] Ibid.

[7] Operation Breakthrough is a U.S. program, begun in May, 1969, by the Department of Housing and Urban Development with the stated goal "to develop, test, and promote the best in technologically advanced systems for producing housing." See: Operation Breakthrough: Questions and Answers, HUD-186-RT (2), U.S. Government Printing Office, Washington, D.C., March, 1971.

[8] See, for example, "Housing: Move out of Modules," Time Magazine, Feb, 19, 1973.

[9] Portions of the following discussion are adapted from I.D. Terner, "Technology and Autonomy," in Freedom To Build, Op. Cit., pp. 199 - 237.

[10] Massdesign, Architects and Planners, Inc., Cambridge, Mass., Illustration from "Self-Help Technology in the U.S." (forthcoming, 1973).

11   In an informal experiment conducted in conjunction with preparation of this chapter, estimates were solicited for painting the exterior of a single - family house in the area of Boston, Mass. Estimates for labor showed keen competition-- and ranged from $50 to $600, with the higher bids relating generally to greater experience and professionalism on the part of the painter.

12   Interview with Wayne Horvitz, labor mediator and negotiator, Washington, D.C., September, 1971. See also, William Haber and Harold M. Levinson, Labor Relations and Productivity in the Building Trades, Ann Arbor, University of Michigan, Bureau of Industrial Relations, 1956.

13   A small sample of four system descriptions of this sort may be found as follows: 1) Rita Reief, "Building Your Own House: 'Simple as an Erector Set'," New York Times, June 28, 1971, p. 26. The article describes a complete kit of parts and tools, including detailed instructions, designed so that two inexperienced people can assemble a small house in four days. 2) John Peter, "Tree House For Grownups," Look Magazine, August 24, 1971, pp. 62-M to 63-M. The house is typical in that it is described as designed "for step-by-step- assembly by amateurs with ordinary hand tools" and with each part being "light enough to carry by hand." 3) I. Terner and R. Herz, "Squatter Inspired," Architectural Design Magazine, London, England, August, 1968, Vol. 38, pp. 367-370. 4) John Hoge, A.F.S. Group Brochure, Cincinnati, Ohio, mimeo, pp. 5.1 to 5.8.

14   N. Mitchell and I. Terner, Squatter Housing: Criteria for Development, Directions for Policy, The United Nations Seminar of Prefabrication of Houses for Latin America, Information Document No. 19, Copenhagen, Denmark, August - September, 1967, pp. 13 - 22.

15   Several systems approaching the performance attributes described above are becoming commercially available in the U.S. For example, Advanced Waste Treatment Systems, Inc., (AWT), of Wilmington, Delaware, a subsidiary of Hercules, Inc., presently markets a "turnkey waste treatment and disposal plant" that can serve between 200 and 4,000 dwelling units. The cost per dwelling unit is between $250 and $375; and the cost per plant ranges from $50,000 to $1.5 million. The company states "that the not-too-distant future will see the introduction of many plants-- with (lower) price tags to match-- that could serve single family homes or even boats." See: "Here's a Totally New Method of Sewage Treatment," House and Home, McGraw Hill Publishers, New York City, Volume 41, no. 2, February 1972 pp. 74 - 75.

16   Switchpack Systems Inc., Del Mar, California, shown in Figure Two is "Surfacepack," surface mounted low voltage switching system.

17   Anthony J. Yudis, "Cambridge Researchers Accept NASA Challenge, Space Age Technicians Envision Nail-Free, Fireproof Housing," Boston Sunday Globe, September 26, 1971, p. B - 39.

18   The Lectri-Pac Corporation of Saginaw, Michigan, packages a self-help electrification kit.

19   Business Week, "The Fuel Cell Goes to the Drug Store," McGraw-Hill Publishers, New York, February 26, 1972, pp. 20 - 21. The article describes experimental fuel cells operating in the U.S. noting that the electric power produced is "one-third more efficient than a typical central power station," cheaper ("about 1/4¢ less per kwh") and noise pollution-free because there is "... no combustion of the natural gas fuel."

20   See, for example, Marshall F. Merriam "Decentralized Power Sources for Developing Countries," International Development Review, Vol. XIV, No. 4, 1972/4, pp. 13-18.

21   Sears Roebuck and Company, Chicago, Illinois, Fall catalogue, 1918, p. 1,632.

22   Shelter Kit, Inc., Franklin, New Hampshire. Shown in Photo: "Unit One."

23   "Owner-built Housing in the U.S.," Report No. 8, of Self - Help in Housing, prepared by the Organization for Social and Technical Innovation (OSTI), Cambridge, Mass., for U.S. Department of Housing and Urban Development, contract H - 1057 - A, Washington, D.C., June 1970.

24   "The Report of the National Commission on Urban Problems" ("The Douglas Commission"), Part V, Chapter 2, cited in Industrialized Housing, Joint Economic Committee of the Congress of the U.S., op. cit., p. 53. The report notes: "The 'trailer' became a permanent abode; and with the change came a new name-- 'mobile home'. Even this name is a misnomer, for it has been estimated that more than 60% of all mobile home owners have never moved the unit they currently occupy. The Mobile Home Manufacturers Association reports that the average stay in one location by mobile home owners is fifty-eight months, which is about the same as for owners of conventional housing. About 70% of the more than 2 million homes produced since World War II have been used as permanent dwellings."

97

Self-help Housing Construction: The State of the Art in Terms of Technical Innovation

Donald Watson

I   <u>Introduction</u> The focus taken in this paper on building
technology should not imply that the key to low cost
housing problems lies simply in technological solutions
or in industrialized building systems. The current self-
help housing programs in the U.S. are successful to the
extent that other factors are properly accounted for,
such as project management, financing, transportation,
personnel, and technical training and support. Several
organizations already exist which address the information
and technical support needs of self-help programs, notably
the International Self-Help Housing Associates and the
Rural Housing Alliance (1). Richard Margolis's <u>Something
To Build On</u> (2) presents an excellent introduction to
many examples of current self-help efforts in the U.S.
One organization, Self Help Enterprises of Visalia, California,
recently completed its 1000th unit of housing, using conventional building methods together with partial shop assembly
of panel components.

Nonetheless, the construction method is critical to many
self-help efforts which tend to become bogged down because
of the long term effort and time required to complete the
conventional self-help house. As evaluated by Robert Troy
in <u>South Today</u>:

"...building time must be cut sharply from
the 12 to 18 months it now takes, by using more
prefabricated units and less unskilled labor
(less, in short, "self-help").

....Self-Help housing can have a significant
impact on rural housing if it is carried out
much more quickly and efficiently. For this
to happen, those involved must be willing to
compromise some on the ideology of self-help
in order to increase speed and cost-effectiveness."(3)

This conclusion, that time and ease of construction are critical, argues for improved construction technology-- but always in view of the basic self-help equation, that the larger the construction involvement by the self-helper, the greater the sweat-equity contribution can be. This question was investigated in an extended series of research reports funded by H.U.D. in 1968 and carried out by OSTI of Cambridge, Mass. (Organization for Social and Technical Innovation) and BSD of Washington, D.C. (Building Systems Development, Inc.). In their June 1969 report Self-Help Housing in the U.S., one of four principal recommendations reads:

"One of the major conclusions of this study is that the use of industrialized building components is, in most settings, the most appropriate building technology for self-help housing. The advantages of component technology were found to be: simplification of the field construction process, reduction of field supervision, simplification of material logistics, increased speed of field construction, and improved building design options." (4)

The specific research in view of this conclusion is elaborated in three BSD Reports and summarized in Report #4:

"The best building method for a given situation depends mainly upon the amount of time the self-helper intends to contribute.....

Highly Prefabricated Methods. A typical self-help builder working only two to three hundred hours can gain the lowest total construction cost by using a highly prefabricated technology, such as the two-skin panel system with the service core.....

Partially Prefabricated and Site Fabricated Component Methods. If a self-help builder wants to work between 300-900 hours, then more on-site labor intensive methods than the two-skin panel system must be selected in order to minimize the total construction cost. We found the one-skin panel with a service core to be the cheapest for this range of self-help hours.....

In-Place Fabricated Methods. Material cost becomes the main determinant of the total construction cost at great self-help labor levels. Above roughly 1000 hours of self-help input, the service core is too expensive, as much of its cost is due to labor performed off-site. Thus conventional... ...technologies are competitive, as they have a higher labor/material cost ratio than the service core, permitting an even lower total construction cost to the self-helper..."(5)

The OSTI/BSD Research ends with a proposal for large-scale marketing of self-help building packages, to be tested in a demonstration project, and awaits governmental funding as such (6). In addition, H.U.D. has also received a number of proposals related to self-help housing as part of the 1969 Operation Breakthrough "Part B" Call for Proposals, but has apparently placed its limited development funds into large-volume industrialized building systems (7). Most recently O.E.O. announced a $4.7 million research and demonstration program to be undertaken by Battelle Memorial Institute of Columbus, Ohio, to investigate self-help housing methods for rural housing (8).

The implication to be drawn from the aforementioned studies is that prefabrication of compatible building components for easy on-site assembly is a viable solution to self-help construction, as long as careful attention is addressed to cost ratios between highly prefabricated, partial prefabricated, and on-site assembled components. This jibes exactly with conclusions made by H. Clarke Wells in a 1966 evaluation of contractor-built housing---which needs to be brought up to date for contractor-built operations in view of recent modular innovations--but which remains valid for field-assembled components:

"Highly mechanized assembly lines that were made to mass produce low-cost houses...don't pay in a market that demands design flexibility... Simpler and less expensive machinery (truss assemblers and conveyor jigs) offer adequate efficiency, plus almost unlimited flexibility... Shipping costs limit the size of the package, the degree to which it's finished, and the distance it can go .... Big components are so expensive to handle that less-finished panels usually make more sense..a still smaller and less-finished package (precut and sub-nailed framing parts) may prove the most practical of all..." (9)

II Examples. With this overview, it is now useful to review a number of building systems currently in advanced prototype or production stages that could be used in self-help housing. The examples are ones that have been published before; there are no doubt many others that are under development and unpublished. The Scandinavian and English examples are included because they represent approaches that are applicable to the U.S.

Ahlstrom Moduli House
(Gullichsen and Pallasmaa, Architects)
Finland

A total component system in limited production; dimensioned for a grid of 2.25 meters.

Of interest for the plan and facade variations that are possible with connector mechanisms that are easy and rapid. (Diagram courtesy of <u>Ahlstrom</u>, Helsinki)

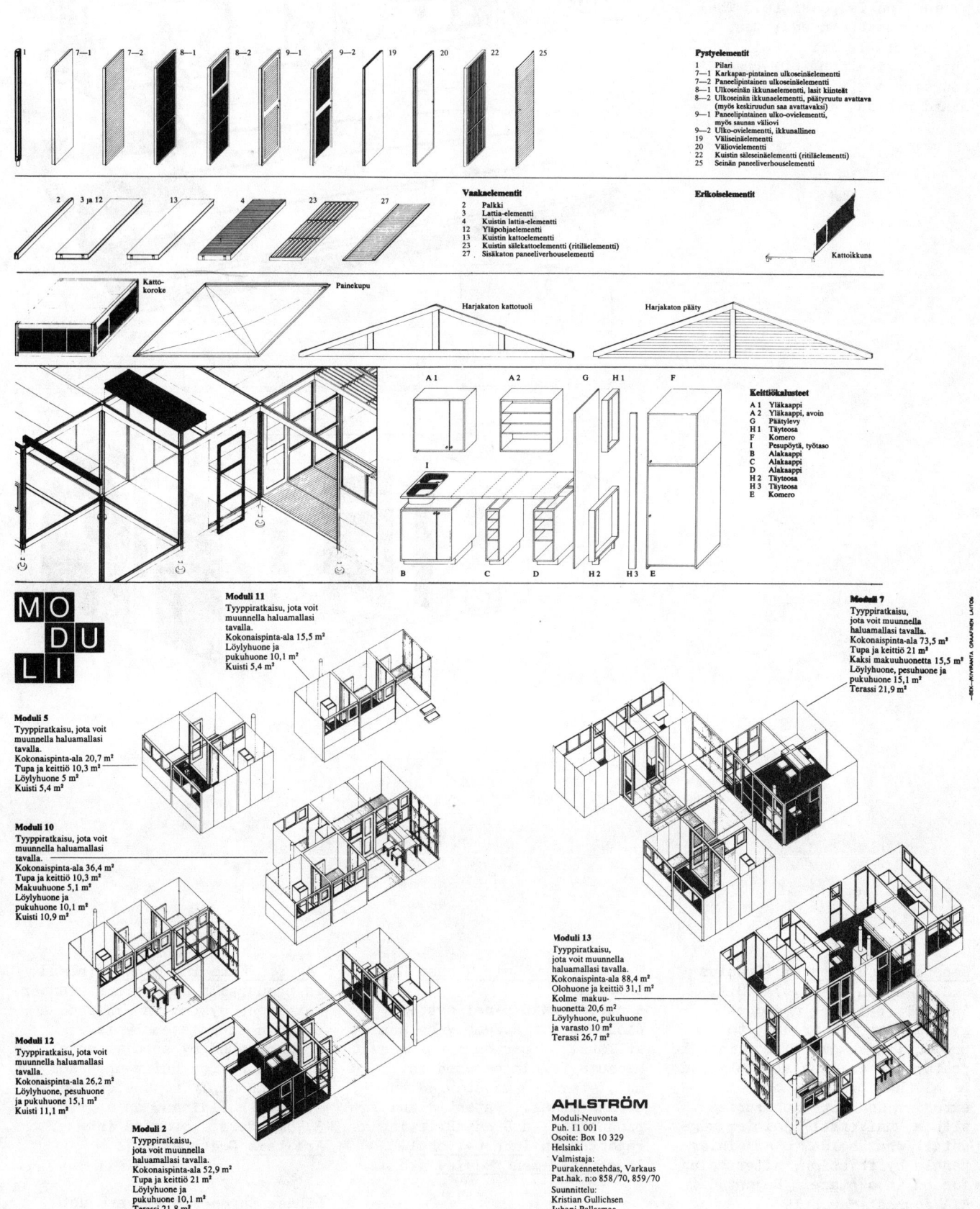

Pystyelementit

1	Pilari
7—1	Karkapan-pintainen ulkoseinäelementti
7—2	Paneelipintainen ulkoseinäelementti
8—1	Ulkoseinän ikkunaelementti, lasit kiinteät
8—2	Ulkoseinän ikkunaelementti, päätyruutu avattava (myös keskiruudun saa avattavaksi)
9—1	Paneelipintainen ulko-ovielementti, myös saunan väliovi
9—2	Ulko-ovielementti, ikkunallinen
19	Väliseinäelementti
20	Väliovielementti
22	Kuistin säleseinäelementti (ritiläelementti)
25	Seinän paneeliverhouselementti

Vaakaelementit

2	Palkki
3	Lattia-elementti
4	Kuistin lattia-elementti
12	Yläpohjaelementti
13	Kuistin kattoelementti
23	Kuistin sälekattoelementti (ritiläelementti)
27	Sisäkaton paneeliverhouselementti

Erikoiselementit

Kattoikkuna

Katto-koroke Painekupu Harjakaton kattotuoli Harjakaton pääty

Keittiökalusteet

A 1	Yläkaappi
A 2	Yläkaappi, avoin
G	Päätylevy
H 1	Täyteosa
F	Komero
I	Pesupöytä, työtaso
B	Alakaappi
C	Alakaappi
D	Alakaappi
H 2	Täyteosa
H 3	Täyteosa
E	Komero

MODULI

Moduli 11
Tyyppiratkaisu, jota voit muunnella haluamallasi tavalla.
Kokonaispinta-ala 15,5 m²
Löylyhuone ja pukuhuone 10,1 m²
Kuisti 5,4 m²

Moduli 7
Tyyppiratkaisu, jota voit muunnella haluamallasi tavalla.
Kokonaispinta-ala 73,5 m²
Tupa ja keittiö 21 m²
Kaksi makuuhuonetta 15,5 m²
Löylyhuone, pesuhuone ja pukuhuone 15,1 m²
Terassi 21,9 m²

Moduli 5
Tyyppiratkaisu, jota voit muunnella haluamallasi tavalla.
Kokonaispinta-ala 20,7 m²
Tupa ja keittiö 10,3 m²
Löylyhuone 5 m²
Kuisti 5,4 m²

Moduli 10
Tyyppiratkaisu, jota voit muunnella haluamallasi tavalla.
Kokonaispinta-ala 36,4 m²
Tupa ja keittiö 10,3 m²
Makuuhuone 5,1 m²
Löylyhuone ja pukuhuone 10,1 m²
Kuisti 10,9 m²

Moduli 12
Tyyppiratkaisu, jota voit muunnella haluamallasi tavalla.
Kokonaispinta-ala 26,2 m²
Löylyhuone, pesuhuone ja pukuhuone 15,1 m²
Kuisti 11,1 m²

Moduli 2
Tyyppiratkaisu, jota voit muunnella haluamallasi tavalla.
Kokonaispinta-ala 52,9 m²
Tupa ja keittiö 21 m²
Löylyhuone ja pukuhuone 10,1 m²
Terassi 21,8 m²

Moduli 13
Tyyppiratkaisu, jota voit muunnella haluamallasi tavalla.
Kokonaispinta-ala 88,4 m²
Olohuone ja keittiö 31,1 m²
Kolme makuuhuonetta 20,6 m²
Löylyhuone, pukuhuone ja varasto 10 m²
Terassi 26,7 m²

AHLSTRÖM
Moduli-Neuvonta
Puh. 11 001
Osoite: Box 10 329
Helsinki
Valmistaja:
Puurakennetehdas, Varkaus
Pat.hak. n:o 858/70, 859/70
Suunnittelu:
Kristian Gullichsen
Juhani Pallasmaa

Versatal Panel System U.S.A.
Based on a machine-inserted
spline which can join plywood
or equivalent material into a
stress-skin panel, the system
is of interest where volume
production is possible. The
spline insertion equipment is
transportable for on-site
shop fabrication. Diagram
courtesy of Versatal Inc.,
Grand Rapids, Michigan.

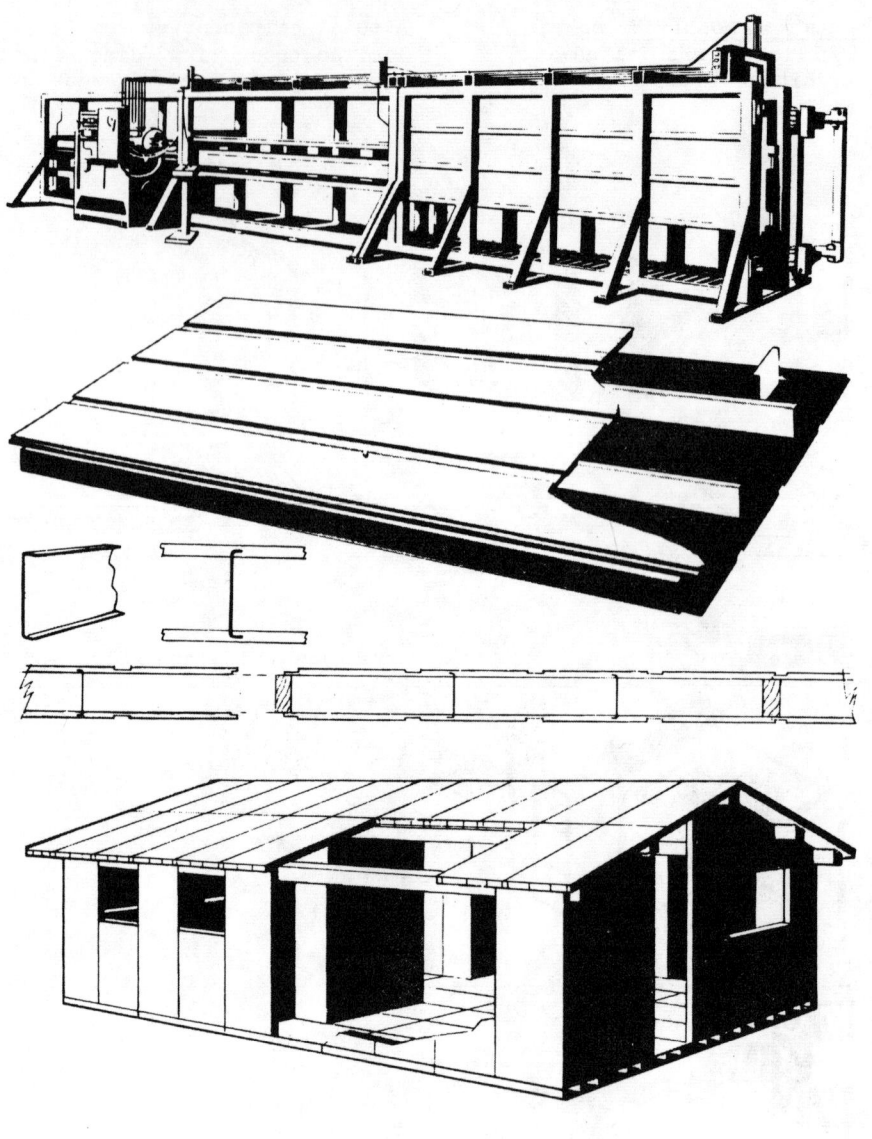

General Panel System (Gropius
and Wachsman 1941-46) U.S.A.
An early building component
system that was brought to the
production stage. The principle
feature is an insulated plywood
panel that provides both
exterior and interior surface,
with a "universal" joining edge
detail that holds to adjoining
panels by friction, after being
forced into place. Reported in
AIA Journal March 1972.

T & N System Built House Great
Britain
A stress-skin panel system
designed for assembly by a crew
of four, of particular interest
because it can be used in
multi-family two-storied row
arrangements. Patented and in
production in Great Britain, as
reported in International
Asbestos-Cement Review #48

"As You Like It" House Finland
In production as a precut lumber
component system in Finland,
easy on-site assembly is
accomplished by special metal
connectors for built-up beams
that can apply to one or two
storied buildings, on a grid of
3.75 meters. Reported in
Arbitare December 1971

(These three systems are not
illustrated here.)

102

"Low Cost" Panel House
(Kallio-Mannilan, Architect)
Finland
Based on a lightweight
insulated panel treated to
provide exterior weathering
and interior surface, and
manufactured as a folded plate,
for use as a wall or as a
roof panel. An interesting
use of new finishing and
joining materials--in this
case fiberglass polyester
fabric on veneered styrox--
as well as the structural
advantages of dome
configurations. Under prototype development, as reported
in Avotakka November 1971.
(Diagram courtesy of Avotakka,
Helsinki.)

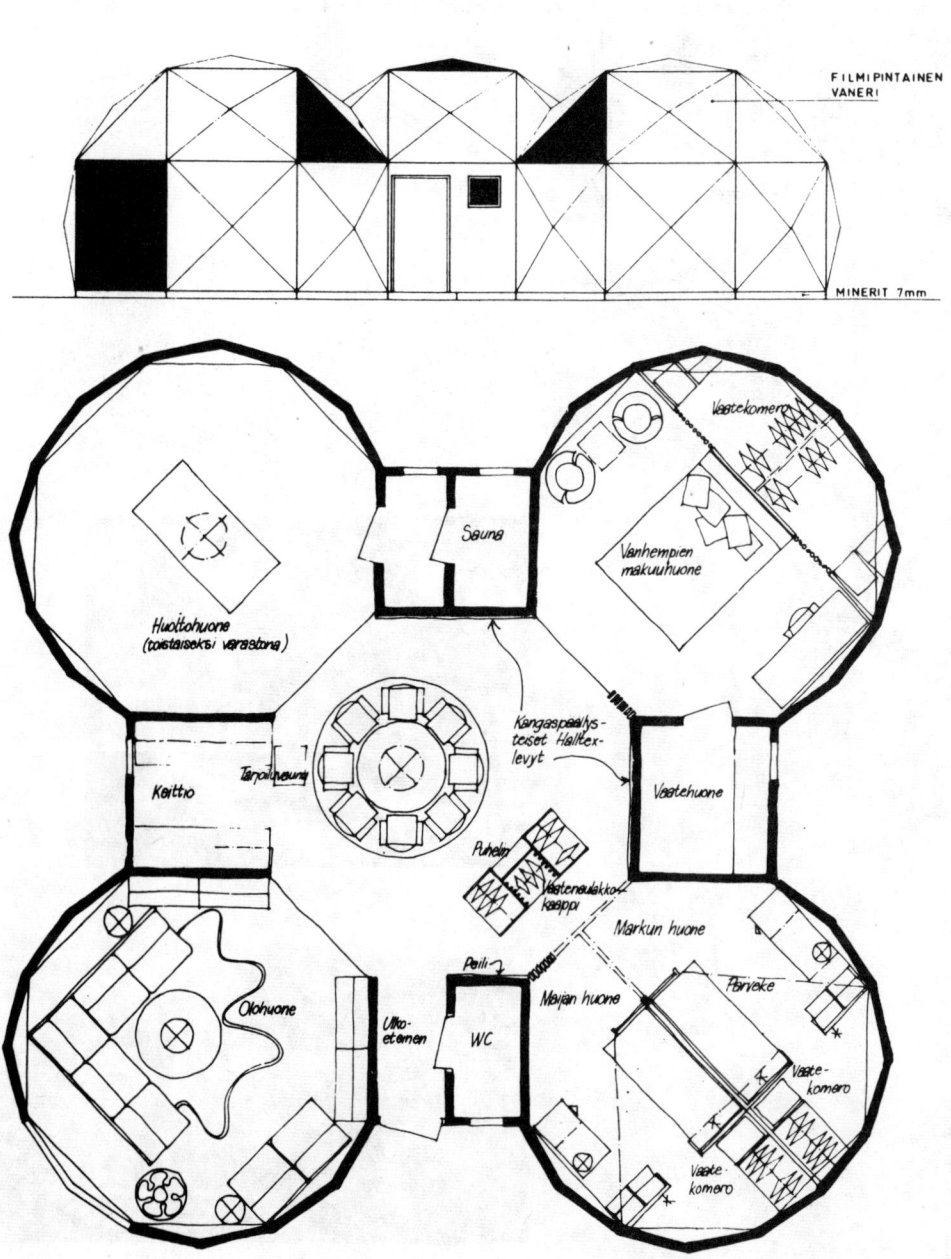

FILMIPINTAINEN VANERI

MINERIT 7mm

Vaatekomero

Vanhempien makuuhuone

Huoltohuone
(toistaiseksi varastona)

Sauna

Kangaspäällys-teiset Halltex-levyt

Keittio

Tamoilhuone

Vaatehuone

Puhelin

Vasteneulakkokaappi

Markun huone

Olohuone

Parveke

Peili

Majan huone

Ulko-etenen

WC

Vaate-komero

Vaate-komero

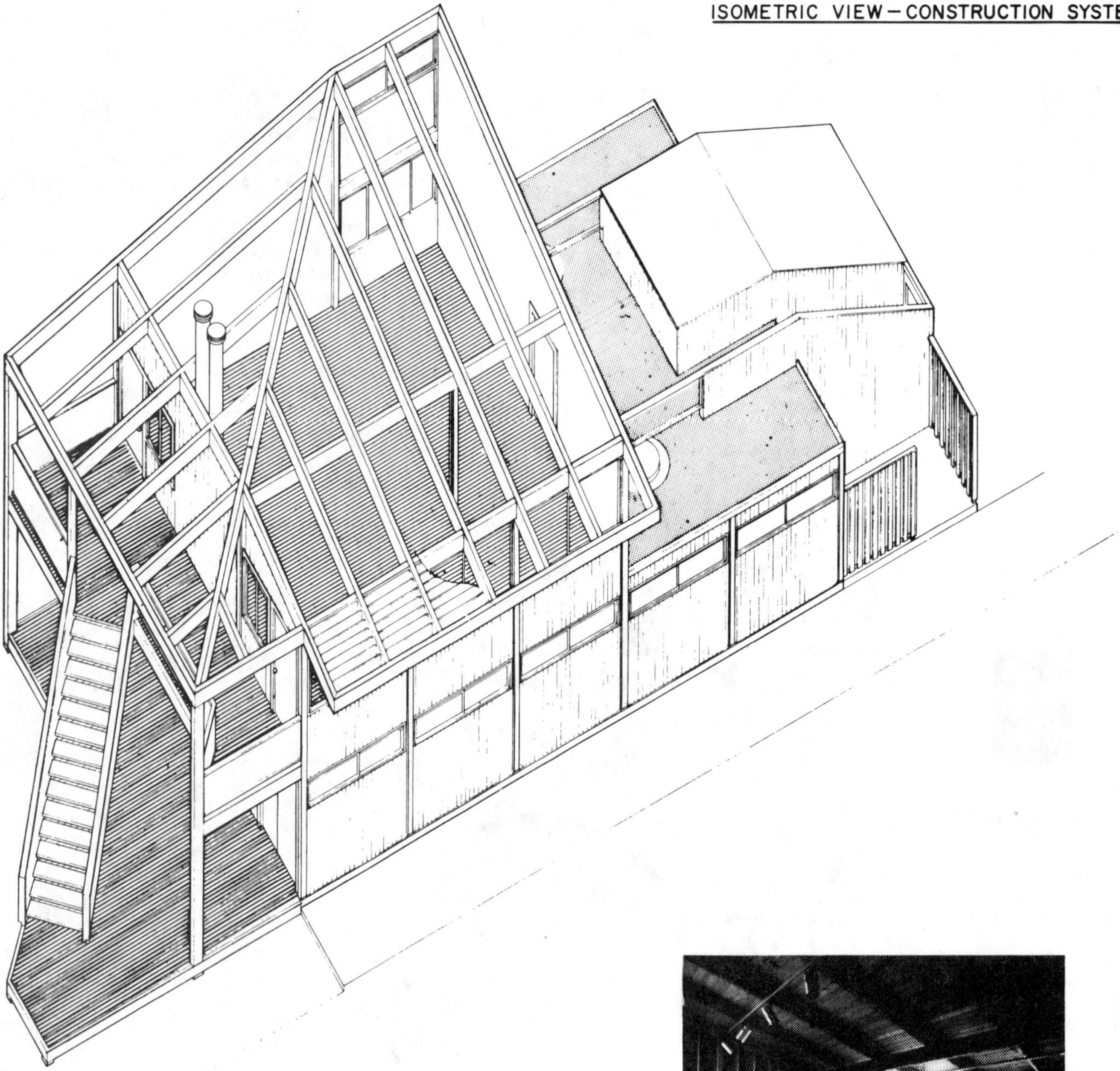

Custom Self-Help Designs
(Donald Watson, Architect)
U.S.A.
A similar approach to Segal's, using stock components as available in the U.S., rationalized for minimum on-site cutting, becomes more feasible if building components are designed to reduce on-site labor and are compatibly dimensioned. In this example, a recreation building built by volunteer labor, the precut structural members are coordinated with stock plywood and window components.

Custom House Designs (Walter Segal, Architect) Great Britain An approach to self-help timber frame construction that utilizes stock building components from available market sources is made possible by a number of design devices, notably, by placing the first floor framing on posts off the ground and providing truss-type framing at the first floor under-level. This enables the columns which are placed at approximately eight foot centers to act as cantilevers, eliminating the need for lateral bracing in the wall panels themselves. In this manner, the wall panel can be selected to provide weathering surface only. The architect provides materials ordering lists, so that the client can process orders for factory delivery and act as general contractor oneself. Reported in The Architects Journal (England) September 1970

Self-Help House (Rudd Falconer, Architect) U.S.A. Of interest as a solution to the need to construct the structural frame quickly, while at the same time relying on labor-intensive methods. In this case structural members are prefabricated by the self-help group using built-up glued lumber, detailed for integral joinery. Room-size structural components are intended to encourage the planning of spaces by the occupant. Under prototype development at the School of Architecture, Washington University, St. Louis, Mo., by the Rural Housing Research Group. Diagram courtesy of Rudd Falconer, Architect.

__Summary.__ The various construction methods can now be ennumerated and evaluated to determine the strategies for component construction which deserve technical research and which hold promise for innovation.

CONSTRUCTION METHOD	Local Materials Low Technology Low Transport Cost	Requires Trade or Construction Skills	Low Technology High Transport Cost	High Technology Low Transport Cost	High Technology High Transport Cost	Variety of Design	Easily changed
1 INDIGENOUS — Local materials, scavanged or re-used.	●	●				●	
2 CONVENTIONAL — Materials as stocked and marketed. Trade skills.		●	●			●	
3 RATIONALIZED — Marketed materials, design coordinated for compatible assembly.		●	●				●
4 ON-SITE PREFAB — Conventional shop assembly of panel components. Trade skills.		●		●		●	
5 PREFAB COMPONENTS — Partial kit, structural, core or skin components					●		
6 PREFAB KIT — Total components delivered for on-site assembly.					●		●

CONSTRUCTION METHOD	ADVANTAGES	DISADVANTAGES	AREAS FOR TECHNICAL INNOVATION
1 INDIGENOUS — Local materials, scavanged or re-used.	ecologically sound local adaption lowest cost	greatest time	recycled material
2 CONVENTIONAL — Materials as stocked and marketed. Trade skills.	self-helper learns trade skills	requires time and skills	easily installed components
3 RATIONALIZED — Marketed materials, design coordinated for compatible assembly.	uses available stock competitive prices	requires technical support	compatibly dimensioned materials connector devices
4 ON-SITE PREFAB — Conventional shop assembly of panel components. Trade skills.	self-helper participates in shop fabrication	mutual self-help only	easily relocated shop equipment
5 PREFAB COMPONENTS — Partial kit, structural, core or skin components	speed in assembly and flexibility	high cost or incompatibility of other subsystems	lower cost, more flexible componen
6 PREFAB KIT — Total components delivered for on-site assembly.	total package delivery factory to consumer	good delivery system and volume required	lower cost, more flexible componen

<u>References</u>

(1) Both organizations at Dupont Circle Building,
     1346 Connecticut Avenue. N.W., Washington, D.C.
     20036 Publications List Available.

(2) Available from ISHA (address as above) $.95, single copy.

(3) "Self-Help Projects: An Evaluation" in <u>South Today</u>
     March 1971

(4) This report available from the National Technical
     Information Service Springfield, Virginia 22151
     $3.00 Order Number PB 185 980. A complete list
     of the ten OSTI/BSD and their order numbers is
     available from Office of Assistant Secretary for
     Research and Technology, U.S. Dept. H.U.D., Washington,
     D.C., 20410

(5) Reports #2, #3, and #4, available $3.00 each from
     National Technical Information Service, order numbers
     respectively PB 196 456, PB 196 375, and PB 196 374.

(6) As reported in <u>Architectural Design</u> November 1971
     (Bloomsbury Way, London) a one page summary of the
     BSD Self-Help Housing Project.

(7) <u>Housing Systems Proposals for Operation Breakthrough</u>
     by U.S. Dept. H.U.D. and available from the Government
     Printing Office $5.25

(8) As reported in <u>RHA Reporter</u> August 1972

(9) Article in <u>House and Home</u> December 1966

Donald Watson is an architect in Guilford, Connecticut,

and a visiting critic at the Yale School of Architecture.

Builders of Bolinas

Lloyd Kahn

(This chapter is a transcription
of Lloyd Kahn's presentation to
the Shirt-Sleeve Session)

In the town of Bolinas there are a lot of builders. I started out by building a shed. The next thing I did was I found
some old redwood logs down at the beach. We went down there,
rolled them into the water, sat on them with kayak paddles, and
paddled them to where we could get them into the truck. Then we
chain-sawed them up and hauled them up from the beach to the site
and split them up into shakes. I got the shakes and made a shake
dome. Right now I am just making myself a place to live.

We went around looking at rather skillful architecture, but
as we went we saw what people were doing with sheds and shacks,
and we got really turned on to the way these people were building,
the way they were thinking, and the beauty of their houses. One
shack we saw was built by a girl for a hundred dollars.

My shake dome is framed with 2 by 3's, about six hundred
board feet. Used wood from a navy base is nailed on top of the frame.
On top of that is tar paper, and on top of that are the shakes.
It's basically just a big open space. The only thing we built in
permanently was the kitchen--the sink and the electric stove.
Otherwise everything is pretty much just grouped around the perime-
ter. It's very difficult to build anything in a curvilinear structure, because you have no straight walls to put anything against,
and almost everything you get, like sinks or stoves, comes with
90° corners. So we just laid the stuff out, put the stove and
refrigerator down on the flooring, looked at the situation for
awhile, and then Sarah and I just figured out where we would put
stuff. Then we laid the flooring, and set up the kitchen so you
move back and forth and don't have to run all over the place to
cook.

The advantage of the dome is that you can totally change the
room around by just moving the stuff around. There are about
thirty plants in the dome. The table is made out of 2 by 12's.
You can usually get these where they are tearing down old buildings;
they were often used as floor joists in old warehouses. The table
is put together with nails--no joinery, no glue. It takes about
three or four hours to make something like this. A lot of the wood
I used is out of chicken coops--a beautifully textured wood.

The thing that is unusual about this kind of space is that you
build the space, move in, and just start sleeping in it. Then you
figure out what you will do inside, rather than the architect's
way of committing yourself on paper and then being tied down to
what you've drawn up. I think this is a lot better way for people
to get personally involved with making their own space the way they

want it. It's probably the easiest way for someone who has little building experience, to just build a big barn, a big space, move in, and start figuring out where you want things.

Lately I've been thinking about small spaces, about sheds. For instance, it's very easy to heat a small space, like the Japanese idea that you heat yourself, instead of the Victorian house where you heat the whole house. So if you carry a little heater around with you, like a little kerosene-burning heater, you can have a couple of heaters and keep yourself warm without having to heat any big volume of space--if you have a small space or shed that you are living in.

The thing that most people in this town are into is building stuff by hand. It's not just a silly romantic notion; it turns out that these people have more time than money. One guy built his house for seventy dollars--a beautiful house with beautiful life in it. It's very small, but shows how you can relate to the outdoors simply through the use of light.

About a week ago we picked up and moved an entire building with about twenty-five people. It was the idea of using human energy rather than machine energy. We were going to jack up the building and put it on trucks and take three days to move it. We would have all been sitting around reading newspapers and screwing around while a few people worked. Instead, we all went over there and it took about an hour. If you can get a bunch of people centering on something, doing the manual labor, it's always a wonderful experience for everyone.

"The first thing I built after we finished Domebook 2. Every piece of lumber was used, much of it picked up off the streets of San Francisco. The door is redwood, but had yellow paint on it when I found it. Total materials cost was about $175. There is fiberglass in the gable ends for light."
(Photo: Arthur Pearson)

A junk sculptor's house along
a road in the South.
(Photo: Bill Beckman)

I started out knowing nothing about building, building post and
beam houses. All I want to do now is get back and finish my own
place. I want to try to build it out of local stuff, out of eucalyptus. I've been trying eucalyptus shakes and they seem to
work. I want to use as much stuff as possible that I can find in
the area I'm in. It's not just a silly notion. I think we will
make better structures out of indigenous materials than any of the
dome stuff I've tried. I don't like the criticism that we are just
romantic and foolish and that we are just diddling around. It's
really the best way to put up a building, to build it out of local
stuff.

A lot of people in town have worked in various ways to break
down the codes. My own approach is to go in and to do whatever
can to get the permit. I'd never play dumb with a building inspector, because my experience is that it tends to set them on edge.
These inspectors are liable for whatever happens, so they're cau-
tious. I get the permit, and then what I build often has little
relation to what I got the permit for. Once you get something built
then it's really hard for the inspector to condemn it.

Other people around the area are going to direct an assault
on the codes, like going to the Board of Supervisors, because it's
good stuff that we want to do, and they should really recognize it
from the top. If we can get the building inspector off the hook,
that's all he wants. In fact the inspector is an experimental
builder himself, but he is locked in to his job. We don't have
trouble from the unions in Bolinas, because we are on the other
side of the mountains from San Francisco.

Lloyd Kahn is the editor and publisher of Domebook 2 and Shelter.

Some weeks after the close of The Shirt-Sleeve Session, Lloyd Kahn
published a retrospective summary of his personal reactions to
what had happened. It is entitled <u>Smart</u> <u>But</u> <u>Not</u> <u>Wise</u>, after the
feeling which Saxton T. Pope attributed to Ishi, the California
Indian, toward the white man: "He looked upon us as sophisticated
children--smart but not wise." Kahn's document is long and complex. It was published originally by his own Shelter Publications,
and was reprinted in several periodicals in this country and in
Europe. Its closing paragraphs are reproduced here:

[but]

"This is an architectural conference,ʌthere are no *people* here,
just *professionals* playing academic futuristic games. No women,
kids, men here to react to your ideas, academic insularity. Moreover, you designers, especially the ones with artistic abilities,
are making plastics and a totally impractical & wierd shelter
outlook appear seductively appealing to those folks who are always
looking for something new and flashy. Spacy air buildings are
deceptive, that's all. No one is ever going to really live that
way, but it's good media. The same thing I learned with domes, they
photograph well."

The planet needs non-polluting energy sources. Solar heat, wind
electricity, methane from compost. Revive waterwheels; sawmills in
New Hampshire were driven by water power. Put 2/3rds of the staff
at MIT on developing clean(er) burning motor vehicles! Create a
mind bank with the Architecture Machine and come up with a solution
to internal combustion before the Chinese have two cars per family!
If successful, you will be national heroes upon graduation, and
receive free non-polluting cars the rest of your natural lives.

¶ Architects, use your skills and desirable positions to assist in
current housing problems. Help people! You don't have to find a
gigantic new solution to housing. The answer may be in our *hands*.
Whisk Whisk Whisk, the sound of 100,000 Chinese brooms sweeping
snow off Peking streets. No snowplows. The shit of Peking
collected and used for fertilizer. No sewage problem.

MIT, architecture schools, have you ever considered that in some
cases, designs get about as good as they're going to get, and then
don't improve for millions of years. Look at your hand! Is there
a need to redesign it? Have architects, builders ever considered
that our grandparents, but more especially the Indians, built far
more sensibly than today's building industry? And that maybe looking for new structures & new materials isn't that important right
now? That you can't think about building, or design unless you
consider the life style? And that the extravagant life style in the
U.S. now can't last, and is in fact maintained at the expense of
subjugated, bombed, harbor-mined people everywhere in the world?

I was particularly disturbed by the vision of the architect sitting
at the cathode tube, drawing his design into the computer, the computer causing the foam truck to build the house. The ultimate in
laziness, machine worship. Machine can do anything better than man
if we develop machine enough, is the premise. Wrong! It's going to
look like shit - guaranteed - it's going to cost too much, it's going
to be ecologically unsound, it will only produce environments that
machines or machine-like people will want to inhabit. . .

¶ So, people out there who are concerned in one way or another with
building shelters, there's a lot of trickery, a lot of hype afoot.
I ran into a good deal of it and wish to pass along my disillusion-
ments for the edification of those who won't therefore have to go
through the same trial & error (much error!) process.

Here is a quick summary of some things I've learned about shelter:

1- Use of human hands is essential, at least in single-house structures. Human energy is produced in a clean manner, compared to oil-burning machines. We are writing for people who want to use hands to build.

2- It took me a long time to realize the formula:

ECONOMY / BEAUTY / DURABILITY: TIME

You've got to take *time* to make a good shelter. Manual human energy. For example, used lumber looks better than new lumber, but you've got to pull the nails, clean it, work with its irregularities. A rock wall takes far more time to build than a sprayed foam wall.

3- The best materials are those that come from close by, with the least processing possible. Wood is good in damp climates, which is where trees grow. In the desert where it is hot and you need good insulation there is no wood, but plenty of dirt, adobe. Thatch can be obtained in many places, and the only processing required is cutting it.

4- Plastics and computers are way overrated in their possible applications to housing.

5- There is a fantastic amount of information on building that has almost been lost. We'll publish what we can, not out of nostalgia but because many of the 100 year old ways of building are more sensible *right now*. There are 80 year olds who remember how to build, and there are little-known books which we'll be consulting in transmission of hand-owner-self-built shelter information.

Before I left Bolinas, Peter Warshall told me to be sure to see the Peabody Museum of the American Indian at Harvard. So the first day of the conference, and twice thereafter that week, we went over to Harvard, and I was truly staggered. Seeing these things in real life rather than pictures - so unbelievably beautiful! Since I like to work with my hands, I usually look at the way objects are made. Chumash baskets! All hunting, religious, cooking implements are incredibly crafted, fashioned and ornamented by men and women in touch with the earth and its streams and breezes. Ingenious shelters! At the museum someone has made fine models of Indian villages with cut-aways showing how their structures were built. There are even miniature baskets in the model settlements.

Walking amidst magnificence of Indian craftsmen with MIT dimly in mind, I realized that there may not be any wondrous new solution to housing at all. That there is far more to learn from wisdom of the past and from materials appearing naturally on the earth, than from any further extension of whiteman technoplastic prowess.

Relics of the past (Indians)
vs
Visions of the future (MIT)

No contest.

We've been losing ground.

Editor's note:

Sim Van der Ryn, Professor of Architecture at the University of California at Berkeley, gave an illustrated talk which, unfortunately, was not recorded, due to the jamming of a tape cartridge. The following pages reproduce a booklet - published by Van der Ryn and his students - which describes his subject.

Outlaw Building News

75¢

making a place in the country

Spring 72

Welcome to this first issue of

Outlaw Building News

an irregular publication sharing news, ideas,
experiments on how to build for yourself
in the city and country. If you want to hear
from us again, or if you have stuff to share,
write us:

Outlaw Building News
Star Route
Point Reyes Station
California, 94956

KITCHEN·COOP

This is a record of a TEN week SECOND year university class in architecture which took place THREE consecutive days a week on a FIVE acre wilderness site

The class was designed to give some actual experience in how to make a place in the country — learning something of the process of building a livable situation in harmony with the setting and ourselves. A learning situation directly connected to life's flow, survival, sharing skills.

There was no place within the university where we could live together, prepare food together, build structures.... so we chose the country. A place where the awareness of change in ourselves, in our place, is too great not to be recognized.

We grew under the sky rather than under a ceiling. We worked to the sounds of nature, rather than the hum of flourescent lights. We came together in a time and place that was right for us as a group.

This publication is about our coming together, what we built, how we lived, and what we thought about the process that determined the form of   "our place in the country."

stew mayer, tom mccoy, david mccracken, tom mcennerney, gail morrison, martha pearson, eric pederson, diane ringeride, rob strauss, steve tisdale, hayden valdez.

begin

A thousand million years ago, there were five forms of photosynthetic life in the ocean: red, brown, green, blue-green, and yellow green. The form colonized the land. We go to the shore and see brown pacific kelp dying in the strange air.

We will never know the experiments which discovered the order principles and integrites which we carry in our genes.

Ever since, lately, the genes have had minds as their co-pilots. The purpose of minds is to help genes make life a success.

Our failures are chiefly failures of the imagination. We know our image of things is wrong. We know this because things go wrong in our own image. If there is a global failure of life, then it will be a failure in our own image, our image of living.

This class is a life architecture class. At its boldest, architecture is a statement of an image of living, for the form living takes is the germ of an architecture. And it has always been so in the way kelp grows its air bladder and floats its long ronds into sunlit water, in the way a fire must be built to cook a morning breakfast. A stovepipe flue that draws well and a floatation chamber inflated under water. Order principles are experimental results. From Wallace Stevens: "The greatest poverty is not to live in a physical world".

We live partly in a physical world and partly in our image of it. Making a place in the country forces a physical world on us and demands some new images of us. Some old images get flushed into the open and we can look at them. This publication has helped us see this process. I hope it shows.

—Rob

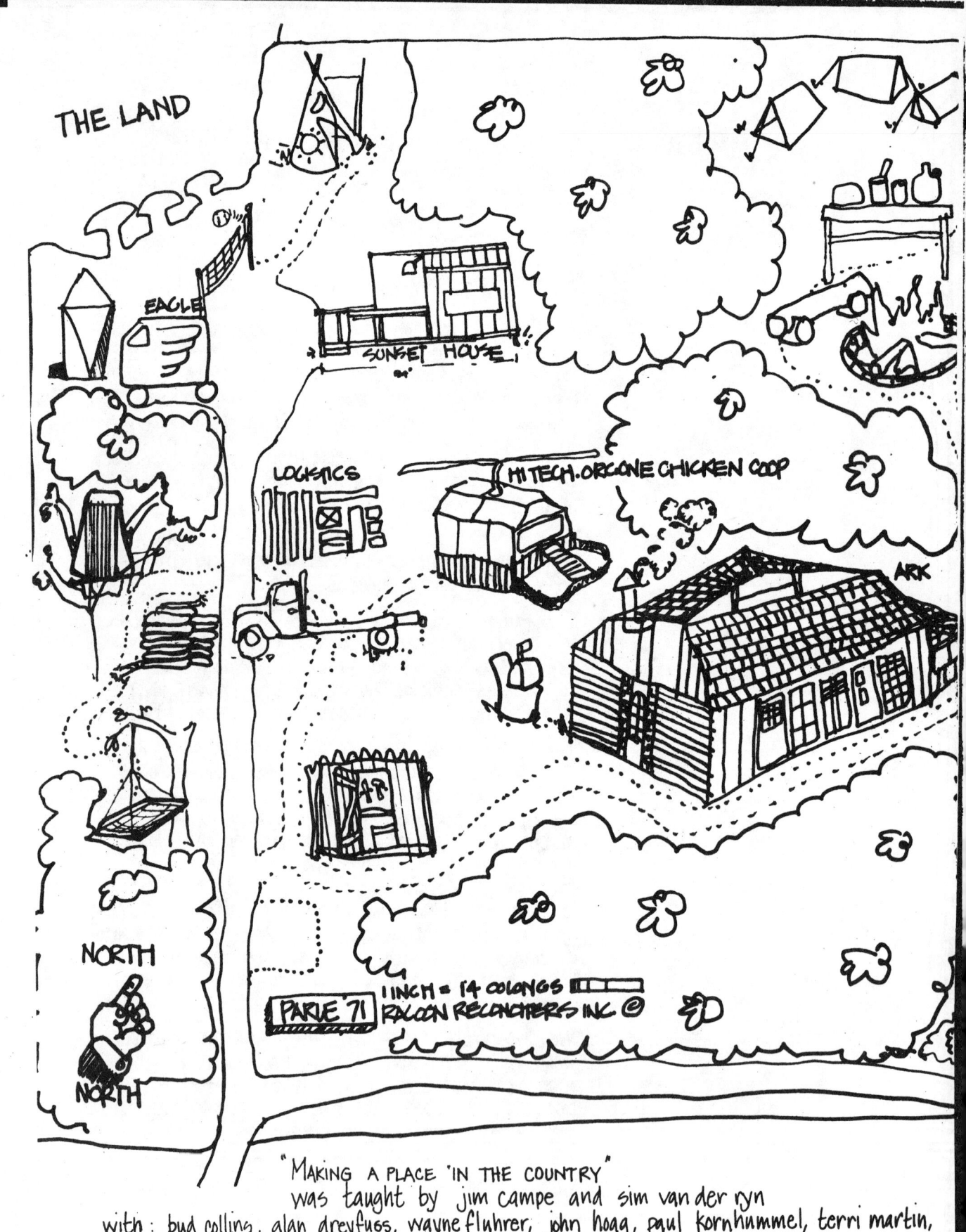

THE LAND

"MAKING A PLACE 'IN THE COUNTRY'
was taught by jim campe and sim van der ryn
with: bud collins, alan dreyfuss, wayne fluhrer, john hoag, paul kornhummel, terri martin,

outhouse

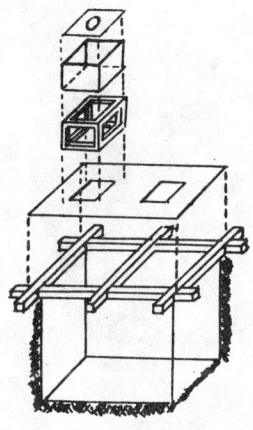

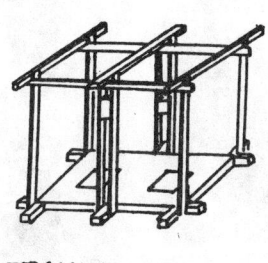

FRAMING

Sure is a lot of eating going on. This seems to be the focus of group activity and organization (all of us), and yet it sure is dis-organized. Hopefully the kitchen will improve that.

kitchen

four weeks later..... the kitchen did improve the cooking process. Now food is more than ever the primary group activity.... the best way to measure any sort of time seems to be breakfast, lunch, and dinner

chicken coops

chicken coop charlie and his gang
lookin so fine doing time
down on the coops
ripping, releasing energy, prying, brushing,
loading 40 years of shit and dust

scorched golden elephant hills
across the fault, back to an "island in
time" of bishop pine by chicken coop
charlie and his gang.

Going to the lumbermarket - supermarket
is one kind of shopping, one kind of learning.
Scrounging is another. Scrounging is
meeting people. Scrounging is no standard
prices, getting to know each other, dealing,
swapping stories and consumating the
sale when everybody's ready

We scrounged the chicken coops from
Tom King. Tom's family used to be in
chicken coop farming and he'd helped
his dad build the coops forty years ago-
before raising chickens was just another
high-tech process. Twenty-five bucks for
eight hundred board feet of clear redwood
You took care taking down what Tom had
built so carefully before, feeling good that
the coops will not die - but only be transformed.

Each man in the market place is alone,
individuals must size up each other and then get along...

Life is basically public relations

During the first two weeks Hayden, John, Wayne and myself took care of buying food and planning meals. We cut up the necessary cash for the purchases which included things such as spices, rice, and onions which were bought in large quantities to last 10 weeks - 50 lbs of brown rice ...$7 - 50 lbs of onions $2.75.

Several weeks later 100 lbs^potatoes were bought. From the end of the first week the food routine was fairly consistent - the cooks for the week rotated - usually two persons per week. They were responsible for buying the food needed for their week of cooking

a process....
A building is not only an idea, a design, and then a physical structure, a building is also a process. A process of laying brick upon brick, beam upon column, roof upon framework. A living process of eating, questioning, resolving, acting. Process, an element even more important than the place for the act, forms, creats, gives birth to the place

building.....
It seems like there must be another, equally cheap, way to build a kitchen than all this shitscraping. I was amazed when I got here to see how much had already been done (arriving a day late), and now I'm amazed at how long this kitchen is taking. Is it easier or more difficult to put something already built back together than to build it from the very beginning?

It seems to me that the group really came together when we raised the chicken coop roof on the kitchen framing. Funny, just a little thing like that. The realization of several days planning and anxieties. But I couldn't have done it by myself.

cooking.....
My week to cook with Tom and Alan. Bad vibes. A lot of more skeptical type feelings manifested themselves this week. Some people are realizing the potential of this class. It is a class in living not just learning. This places a much greater responsibility on the individual.

living....
A warm stove, delightful cooking smells, and cheerful talk feel good after a hard days work. Meals for us are especially warm experiences. A combining of feelings and substance into the best of both.

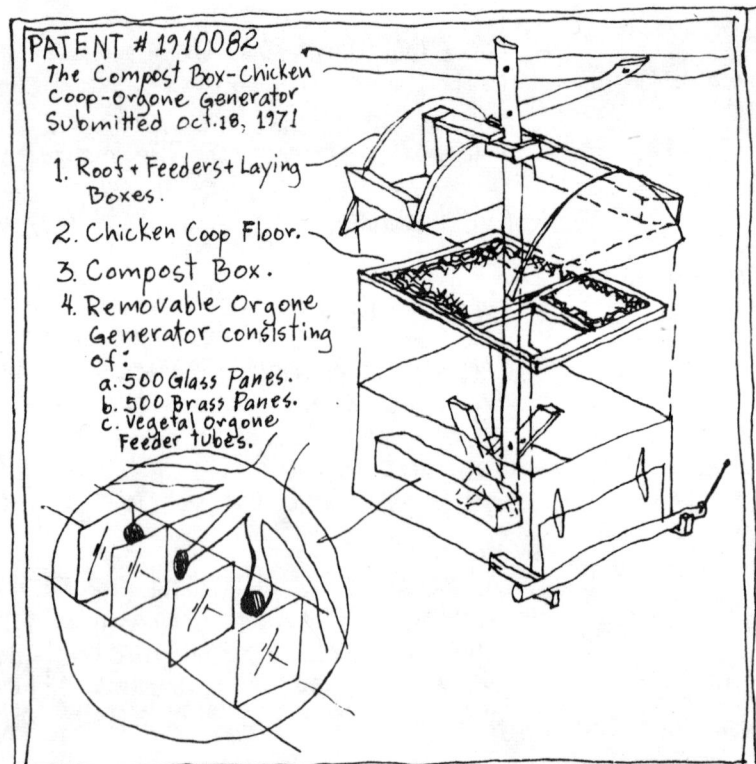

PATENT # 1910082
The Compost Box-Chicken
Coop-Orgone Generator
Submitted Oct. 18, 1971

1. Roof + Feeders + Laying Boxes.

2. Chicken Coop Floor.

3. Compost Box.

4. Removable Orgone Generator consisting of:
   a. 500 Glass Panes.
   b. 500 Brass Panes.
   c. Vegetal Orgone Feeder tubes.

Description.
I. Overall Concept.
   a. Symbiotic multi-function environment.
   b. Spatially stacked components.
   c. Functional orientation in which the orgone generator has easy access to the vegetal orgone components in the composting material; the chicken coop is above the compost to receive the heat rising from the compost and to introduce manure to the compost.
II. Orgone Generator - use of inorganic plates between which are introduced vegetal orgones which are collected and accelerated.
III. Incidentals.
   a. Rotor to turn compost.
   b. Automatic feeder for Chicken Coop.

Design of the Compost Pit - Chicken Coop - Orgone Generator.

We had decided that with the amount of garbage that fifteen people were producing it would be essential to make a compost box. Reading some of the authoritative articles available on composting, I found that no one could agree with anyone else and so I decided to do whatever occured to me that would not oppose the few ideas for which there was a consensus. Averaging the sizes suggested by everyone in the camp, I began building a three by four by five foot box. While building I became aware of the possibility of utilizing the heat of the composting reaction to warm a chicken coop placed above the compost which would symbiotically be enriched by the chickenshit. I had an idea and simple drawings of what was wanted before the building started but the plans for details were developed as I came to them.

This enabled me to receive feedback from the construction as the designing was taking place. The design and construction were simultaneous rather than separate processes. I hope that this sculptural approach to architecture is applicable to more than just building for ourselves.

oven

START - 3:05

1. twigs, split one-by's, split logs, and (white gas) match to all.
2. flame constantly coming out top of mouth.
3. flame out at 5:05
4. temperature at 5:10 — 600° +
5. put in 8 chicks— 30 potatoes at 5:18 and shut door.
6. open at 5:25 — temp 450°
7. open at 5:45 — temp 400° put in brownies
8. open at 6:05 — 300°
9. took out brownies— closed at 6:10
10. chicks out at 6:45
11. potatoes out at 7:15

Volleyball is a linear interaction in which energy oscillates from one side to the other. All concentration is on the side with the ball. The energy must be directed through the net and return on the same path.

Dance is a circular interaction. Our energy flows continuously through a ring of us. We are in constant contact with our personal energy and with the collective energy of the group.

shower

SHOWER
WINDOW
BENCH
WALL
SINK
DOOR
TOWELS
SHELVES

Cheap, lightweight,
quick
    plastic bubble
    with a view of the
                bay
Not a house
    more like a cradle
Doesn't sway much
    even in the wind
    but noisy and damp
    when it rains
It's better-but wetter
    with no plastic,
    as a place for
    an afternoon nap

"You noticed that everything the indian does is in a circle, and that is because the Power of the World always works in circles, and every thing tries to be round."

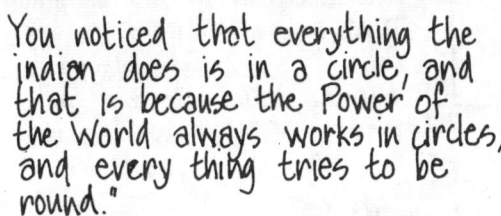

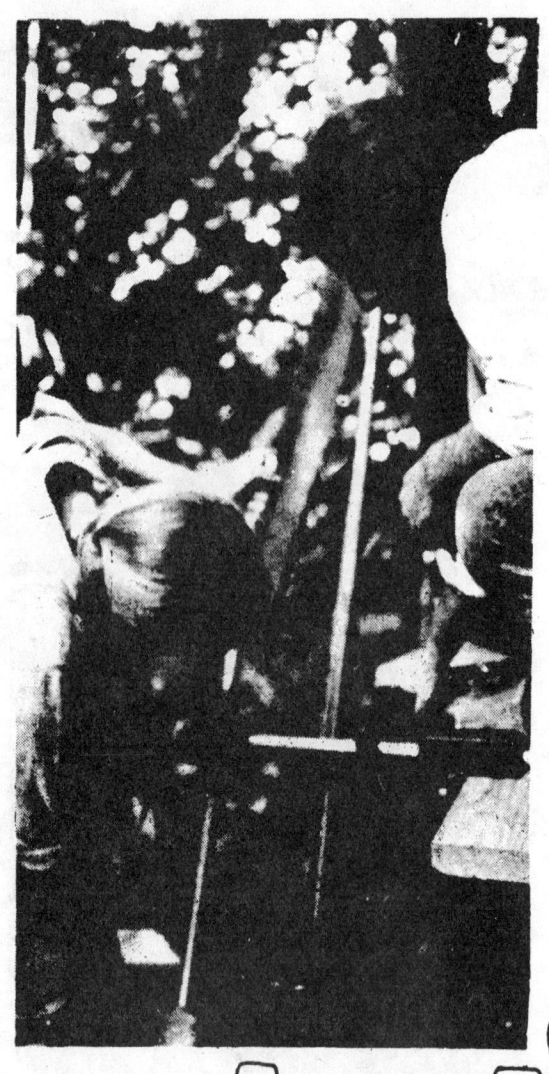

Plumbing too is an art, a sculpture, a wisdom — but only because two days work with metal pipe, exerting sweat and strength to pry loose aged rusted pipe joints, measuring, cutting, threading new pipes, turning them into place finally produces a gentle luxurious stream of water

But it always seems to be "blue thursdays" when we work on the shower. Today — with four people fighting tree branches and wind shingling (or learning to shingle) the roof. Three others working below, we had just about convinced poor Tom that the shower structure was not another hodge-podge mismatched collection of lumber designed and constructed by a clan of dumb derelicts when diane slipped and almost fell off the roof. Wayne missed the nail and hammered his hand, and I dropped my hammer — barely missing tom's head.

I think the most important thing I've learned is that building requires belief in what you are doing, belief that you can do it, belief that it is good.

Building today I turned and saw
what no one had built,
climbing towards the sun
intricate weaving of invisible parts
visible only in enormous mass
yet harmonious, unintruding, peaceful

Do people ever look at trees
and say — I like that, or how does
it use space, or I would have extended
that limb a little farther, or do you
think the space could be used more
effectively
Questions asked only when a tree
is taken apart and put back together
by me.

Juggernaut

(jug'ĕr-nôt') n. San. jagannātha, lord of the world. 1. an incarnation of the Hindu god Vishnu, whose idol so excited his worshipers when it

When I first saw the land I knew that I would build my place in the trees. I noticed that everyone was somewhat afraid of the strength of the trees, so I decided I would show them and myself their power by building a room suspended from a tree.

The design was mostly determined by available materials, which included: hexagonal rings (made a quarter previous), rope, and a cable spool from P.G.&E. It progressed from a flimsy light frame to a massive and over loaded disc.

Here is where it started becoming a problem. Its movement was a major group project causing multiple hernias.

Steve and I worked on it as a part time effort and that is the way it turned out. Today it is the camp swing and plaything for visiting kids

was hauled along on a large car during religious rites that they threw themselves under the wheels and were crushed.

  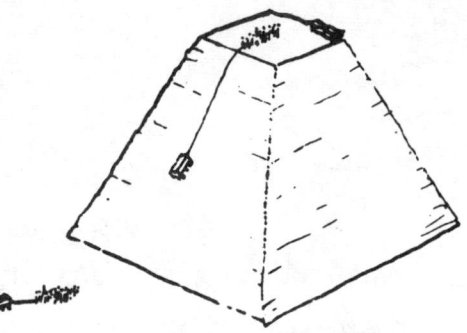

The birds at
Limatour beach
may have been
western grebes
black backs, white underwings
Part of a collective order –
The locking rhythm of wings
part of a larger rhythm.
The center of groupness in each bird
exists only in its simultaneous.
prescence in other birds

Perhaps the greatest pleasure
was seeing others
using and living in places that you
helped build.

Treehouses are simply incredible
when the wind blows -- you rock.
sometimes the squirrels come by,
you're above the trees and under the sky.
the first thing i do when i wake up
is look out at tomales bay to see
if the fog has burned off.
to get down I climb through the hatch
grab the rope -- swing from the oak
to the bay and jump to the ground,
that's all there is to it.

I wanted my own place
sunny, quiet, among trees and ferns,
one tatami mat to sleep on, one to think,
and one more.   Bay poles, stretched plastic,
old coop wood. Uniting rock foundations, a laying frame,
sliding shojii screens defining a space and a porch to extend it.

137

"Today is the day I put up my home
I leave you to the care of the four winds
    Today is the day you see yourself in my lodge
where you can do as you please
We cannot tell you to do this or that;
    We are only men
Help us to think of you everyday we live
in this lodge; guard us in our sleep;
wake us in the morning with clean minds
for the day, and keep harm from us"

"So responsive that it is like an organism – an extension of its inhabitants"

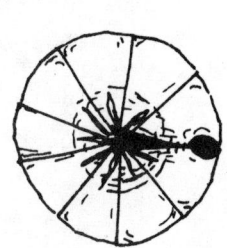

"The tipi is up. The canvas is fitted beautifully. The fittings, smoke-flap, lacing pins, peg loops, so light and effective – a lot like a sailboat."

Living in the city where you pay people
to take care of all of your needs
has instilled a sense of helplessness.

If people are to be confused,
they should be confused by the
myriad of opportunities open to them,
not by their impotence

The class instilled not just a fundamental
knowledge of building, but a confidence
to undertake projects otherwise beyond our
realm. Confidence to follow a vision.

Window Therapy
(critical self analysis)

the need to complete (by myself)
a small detail
that is part of a place
built by everyone

renewing life in old materials

the joy of someone telling me
that they like my window

communicating through a medium
that is a synthesis of my favorite skills

feeling the value of my place
in the group

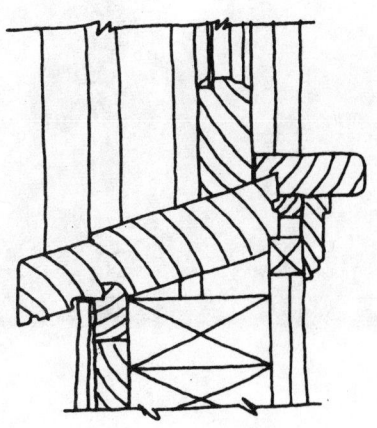

In much the same way
carpentry is architectural dance
One has intentions, and moves
to make them real
One manoeuvres materials and tools
There is the emphatic movement
with a partner or a whole group
as something is moved into place,
steadied or removed. It is a dance
without the exaggerating context
of the stage and audience, it is
movement to balance and transform.
It is a complex of movements
that leaves a work behind
when the dance is over

Yet it is hard when you're involved
in the process of change to think
of it as a series of distinct
modifications of what used to be.
The notion of recording change
implies observing and chronicling
something in discrete bursts of
process – but it was all a continuous process, a dance, a growing craftsmanship.

Lifting a beam or a roof
was always a high
(pure positive energy)

energies touched
through the shared task

somehow afterwards
it seemed easier

to be friends

25

the main building
started as an army surplus tent

IT LEAKED

the 16' x 20' platform
available materials
and Tom's hyperbolic fantasies
determined the form of the building
that replaced the tent

we called it the ARC

details evolved day by day
the roof has changed
lofts have disappeared

the ARC is still growing
                    changing

"The length of the ARK shall be three hundred cubits, its breadth fifty cubits, and its height thirty cubits. You shall make a roof for the ARK, giving it a fall of one cubit when complete; and put a door in the side of the ARK, and build three decks, upper, middle, and lower."

To move ones body in accord with ones thoughts is the highest form of learning I know.

I feel that in most of my education I have not received enough physical contact with the ideas that were being brought forth.

But this class was different.

Our physical actions were first and foremost and ideas followed to explain or label them.

Instead of an inventory of ideas which needed to be filled in separate stacks in my mind I found that I was constantly taking in a stream of concepts and actions which followed in a continuous sequence. My individual actions seemed to be a part of a group action which followed a course of its own, separate from all of the individual ideas.

A building is a whole work,
an integrated sculpture of details,
bits and pieces synthesized by work

Windows are heads, jambs, sills, mullions,
and glass - regular but irregular.

Fitting separate things together
along their separations is boundaries.
Fitting is separate together.
Together is fitting separates.
Separate fits fit together separate.
If it fits, they are separate.

there is not a time for work
there is not a time for play
only one time - no deadlines

the only limit is the time
the work takes - no "wasted time"

Yet I still get frustrated when I learn.
So many repeating actions

INTRICATE SPACE STATION

DESIGN QUESTIONS POSED

Is it acceptable to
provide a station con-
figuration where the
lab space is under
zero gravity condit-
ions, but the living
quarters are under
earth gravity condit-
ions?

Is there a new scale
for social distances
in zero gravity?

What are the psycho-
logical effects of
curtailment of idio-
syncrasies by the de-
mands and limitation
of a space station
environment?

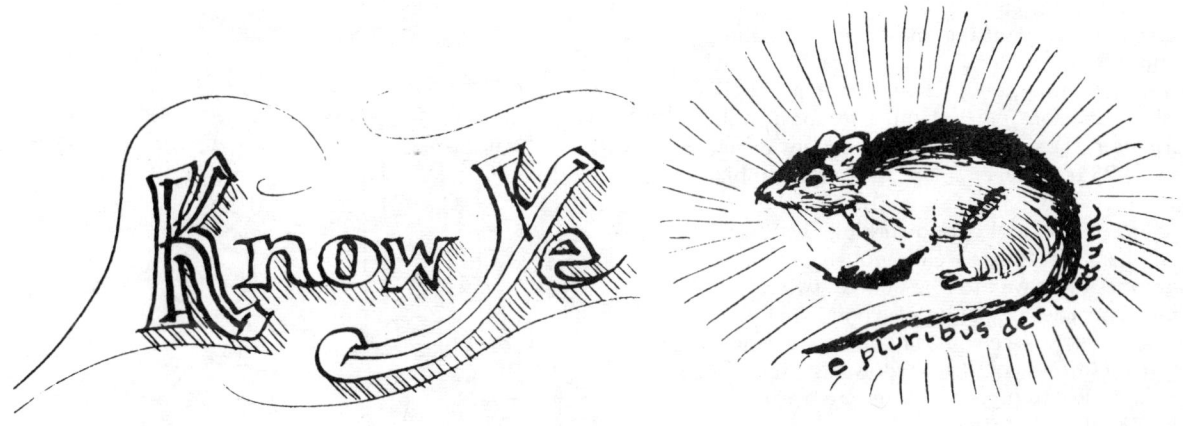

Know Ye

e pluribus derlite

BY THESE PRESENTS THAT

chicken coop charlie

has made a place in the country
and is entitled to be known to all as

OUTLAW BUILDER

with all the rights and priveleges
attached thereto; December 12, 1971

Sim Vanderryn

neotoma fuscipes Rex

James H. Campe

Logistica Maximus

in ten weeks at Inverness I participated in building the outhouse, siding the kitchen, cutting and setting main columns and beams of tent, framing porch and porch rafters, shingling porch, glazing and installing windowpanes, caulking, building an oven, making steel drums, furniture, clearing a snag, tearing down three chicken coops, setting posts for a cabin, building a boardwalk.

tools I used: hammer, saw, chisel, steel tape, chalk line, transit, bushman saw, chain saw, post hole digger, carpenter square, combination square, level, axe, plumb, 3-4-5 triangle, circular saw, brace and bit, sabre saw.

skills: basic hand tool woodworking, can saw, notch, hammer, measure well enough so that everything doesn't have to be done twice—can estimate the properties of a piece of wood, what nails where, mix cement, soil cement, lay bricks, reinforce cement with mesh, cut and fit windowglass, set and plumb a post, wash twenty dishes before breakfast, rig a weighted pulley for a door
and
generally have the background and confidence to be able to figure out how to do anything I want to do.

materials I've learned about using: wood—dimensioning and framing with it joints—nailed, notched, bolted, column seatings, grades, types, rigidity in framing, bracing, bending, grains, redwood.

from clearing and using axe and wedge, I've learned the feel of bay and bishop pine. clearing out the big bishop snag—three or four full days with axe and saw is the most prolonged contact I've ever had with wood and work and myself in work and wood.

Wood, weathered silver redwood, when it rains and the staghorns with their forked tongues go green and lichen stands out in orange and red and black and green, and sitting in the wood on the outhouse seat looking at the silver and the green of leaves.

The chickenhouse cobwebs on the sheathing of the kitchen roof. A pull chain light in the kitchen. Chicken houses that hadn't been used in eleven years. Tomales Bay twelve years ago. Tomales Bay fifty years ago. Driving in the fog. Driving in the fog at night to McClures Beach with coleman and guitars. Deer freeze then run from headlights.

At the beach, dizzy under the stars the sand cave. A coleman and a guitar at one end of a six foot niche in a sand cliff. At the other end the drumming night, the starfield wet sand and somewhere the Ocean.

Wake up early and know that the day will start on its own and fall back to sleep tremendously free.

Rob Strauss

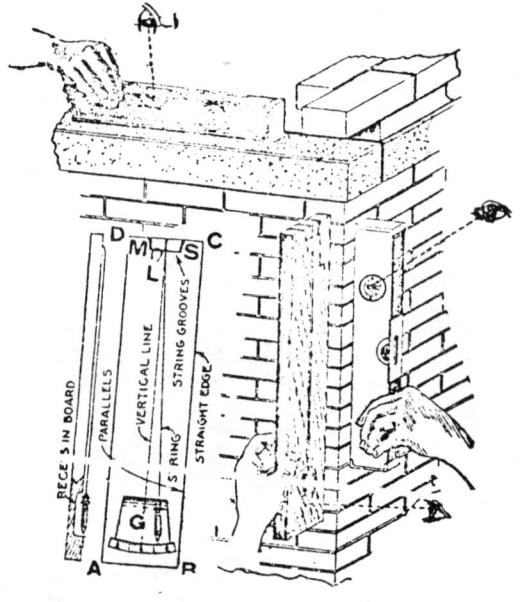

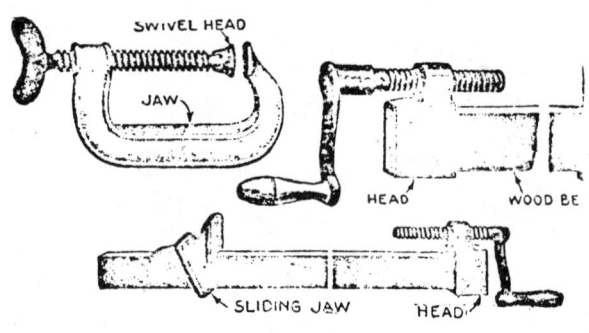

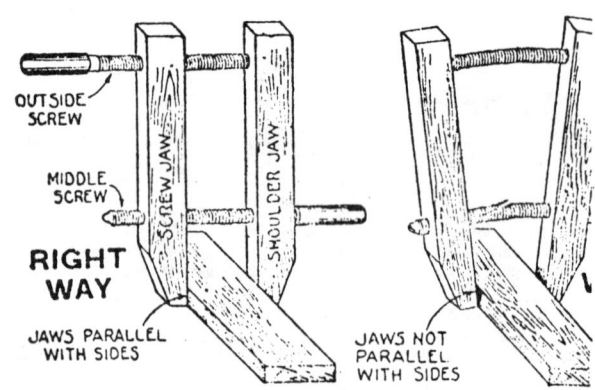

A lot of it has to do with the relationship of design to available space. Out here there's lots of it (relatively) for the few of us, but at best the example is academic. It ignores all the constraints implied in packing people into where they need and want to be, or cannot escape from, while still providing functional and comfortable space which does not stifle. The obvious example is cities. People are still hanging on to rural, frontier individualism, but in the city the frontier is closed. You can't just go out and pick a spot, put what *you* want there, and settle in; it's not that simple. Cities require that a person function as a part of an organism, and yet people still think of themselves as individual people. A few manage to escape to the country, but what do you do with everyone else? Just leave 'em there to choke on each other?

—Gail Morrison

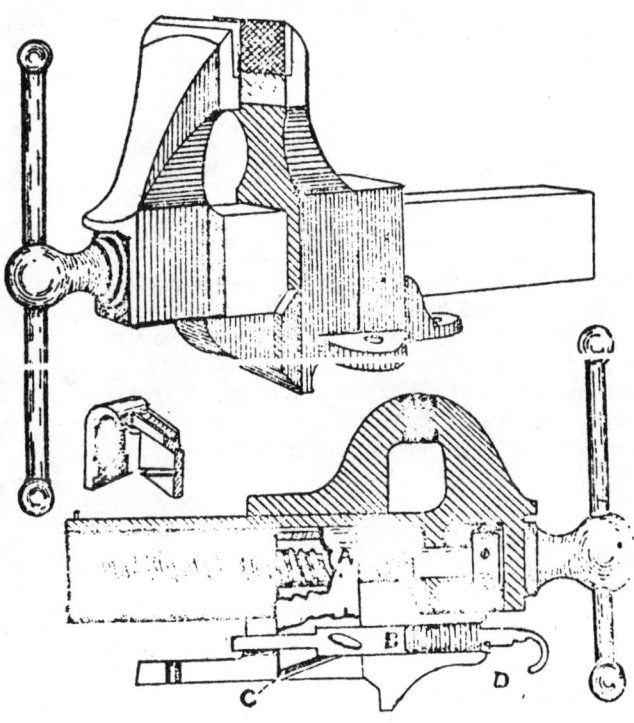

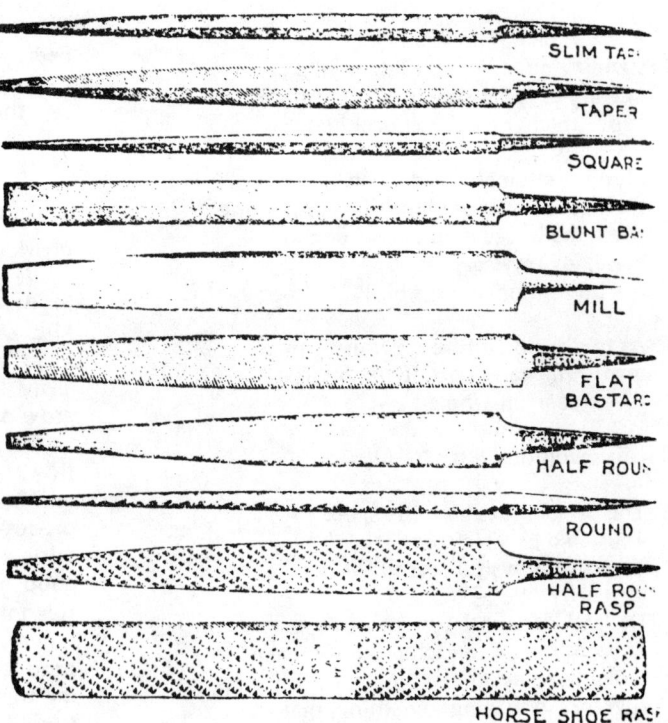

SLIM TAP
TAPER
SQUARE
BLUNT BA.
MILL
FLAT BASTARD
HALF ROU.
ROUND
HALF ROU. RASP
HORSE SHOE RAS.

* Xerox illustrations on these pages from Audel's Carpentry Guide 1929 edition

153

"Since it doesn't seem practical or even desirable to think that direct bloody force will achieve anything, it would be best to consider this a continuing revolution of consciousness which will be won not by guns but by seizing the key images, myths, archetypes, eschatologies and ecstasies, so that life won't seem worth living unless one's on the transforming energy's side."

—*Gary Snyder*

America used the machine as a metaphor
to understand and control nature and man.
America built her institutions
in the image of the machine.
Now the machine has run away with itself,
with America.

Every time has its outlaws
standing outside the ruling myths
moving to tunes others may not hear
above the hymn of the machines.

Myths are transformed when
the tale is no longer beautiful, protecting;
when the image does not sustain life,
hope, joy, mystery,
the fullest human possibility.

As myth and culture are transformed
many will stand outside the old law
to welcome in Whole Law.

Not revolution but evolution.

The primary design is life space
your own psychic space
your own physical space:

Let architecture return to its roots
in each person, in each place we make.
When people can participate, openly and
together in making something, that is
when architecture takes on its ancient meaning.

"It is, unfortunately, only too clear that if the individual is not truly regenerated in spirit, society cannot be either; for society is the sum total of individuals in need of redemption."

—*C. G. Jung*

The course was called "Making a Place in the Country" but some people felt the title a misnomer. In retrospect I agree with them. Making a place implies a commitment to a place, a continuity of process, over many seasons and years—conditions that we couldn't duplicate in a ten-week course. We were more like a half-way house between the abstract halls of the university and total commitment to a way of life. We were strangers starting with some common interests; deepening them through shared experience, but never sharing a common path. We were a class.

What and how we built expresses these realities. Prior to the class, Jim and I had sited the communal buildings, and had obtained many of the materials. The plan for the outhouse and the bath house had been sketched out. We were able to start work on five projects the first day:
—the tent platform (later evolved into the ark)
—outhouse
—bath house
—firie cirrcle
Later that week a chicken house was torn down to provide materials for the cookhouse.

We started with a high burst of energy which grew in intensity for three or four weeks. The plan of the kitchen grew consensually out of the materials on hand and the experience of the cooks. The rotting army tent was abandoned and a building started to grow out of odds and ends.

Working in a consensual, open-ended way, energy flows, things take shape around the form and energy within the group. It is a largely internally-directed process and the products are consistent with the situation. That is not to say the results are necessarily good or satisfying. They are merely consistent with the milieu that creates them. The "ark," still only ninety percent finished, is a confused structure, the products of many hands and minds, often not working together. The lack of consensus over what the ark should be, the hassling over shaping its space, turned many people's interests towards shaping their own private space away from the communal base.

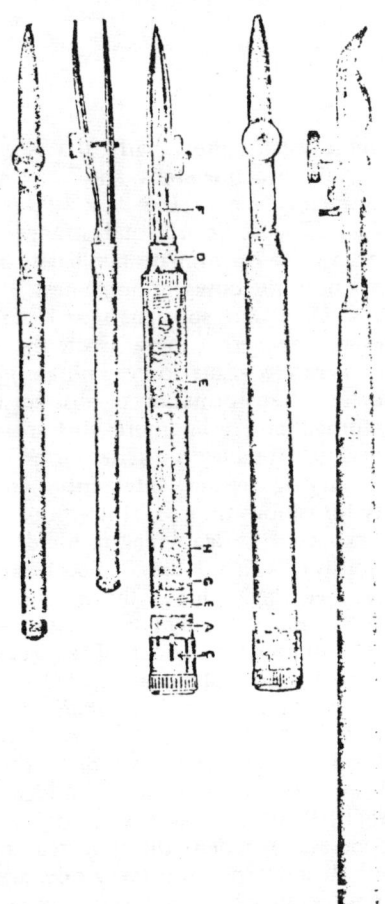

The most important element in building a place, in doing anything is belief in what you are, in believing that it will work, or it is good, or that you are accomplishing something. and that the way not to be lonely is to do what you want, get into your own creative experience, and not worry about what others think, not try to change for them.

looking for nothing in particular I wandered into Wurster Hall and ended up at Inverness. How to make a place in the country. How did you read my mind so perfectly to know what I wanted and needed to know? How to explore the question of how to attain peace of mind and is it attainable and how to be happy seven days a week and not only two and how to really build.

how to, how to, how many how to's did I learn this 10 weeks, embracing a spectrum raning from how to hammer a nail to how to make myself happy, but a spectrum coherent, flowing, unified. I learned how to build a house in which my physical self could live and I learned how to build a consciousness in which my spiritual self could exist.

how to make a place in the country—how to hammer a nail, how to pull and push a saw through wood to make a clean even cut, how to lay a foundation, how to use a level, how to make a window, how to thread a pipe, how to make a sill, how to brace a building, how to shingle a roof, how to build an oven—

and how to live in the country—how to deal with a whole new form of responsibility, how to deal with oneself and freedom, perhaps because I don't know it thoroughly yet—but I know for sure that the most important element in both the building of a physical structure and the building of a peaceful mind and life style is belief. For belief is what binds process to place. A building is not only an idea, a·design—and then a physical structure—a building is also a process—a building is also composed of the invisible process of laying brick upon brick, beam upon stud, roof upon framework. And peace of mind is not only an idea and then a place, but also, and even more so, a process—a living process of eating, sleeping, working, seeing, questioning, resolving, acting. Process—the element which is even more important than the place for the act, the process forms, creates, gives birth to the place. Yet the process is only possible if one believes in the existence of the place.

and realizing now that while belief is directly dependent upon one's own inner strength, self-reliance, peace of mind, these qualities are nourished by a good community of people and a beautiful environment.

This quarter was the first in 13 years of school where community and environment were not contradicted but constructed.

—Terri Martin

and suddenly here in berkely it begins to rain. falling drops of water splashing splattering off concrete sidewalks, asphalt streets, plastered houses, steel structures—inorganic—and with every drop another person crowds into another dry building and another car drowns out the sound of falling rain gunning and choking up and down the street. and more and more—the only way to avoid, escape this crowding is to turn and run until my feet feel the familiar face of bare earth. And returning to Berkeley I know (perhaps) I should try and save try and change try and revolutionize the world, the society, the culture and not try and escape the world of berkeley. But returning and meeting friends on the street, we do not know what to say to each other and the awkward silence needs to be filled with senseless jabber or excused departure.

The best way to measure any sort of time seems to be breakfast, lunch and dinner.

Gordon Ashby spoke of breaking free from linear thought; so far this group has made it impossible to do anything else—too much happening at once. Linear activity implies scheduling; that happens in a large sense (breakfast, etc., last week instead of next week) and in a small one (floors come before windows) but in the middle everything seems to float. Even if it does follow an orderly progression, there is no telling when it will be interrupted by a volleyball game.

The group at first seemed a bit unwieldy—too large for a small group and too small for a large group. The Great Outhouse Dedication Ceremony erased any wonderings, though. A real tribal ritual. Waiting for the Official Sorcerer to bring the moon back was a bit trying, but he was right—he had an unappreciative audience.

Three days doesn't seem enough. Barely enough time to get settled in. But more time here would leave none for supportive forays into civilization, work and rest.

It's hard, when you're involved with the process of change, to think of it as a series of distinct modifications of what used to be. The notion of "recording changes" implies observing and chronicling something in discrete bursts of progress, but it was all a continuous process. Or numerous simultaneous processes, which is even harder to sort out.

You start out to build a kitchen, and in a few weeks it's done. True, there is something on a particular patch of ground thad wasn't there before, but the fact that one day there was a floor, another day a frame, another a roof or siding or windows, is incidental. So try recording process instead of change— that's even more impossible.

There is one change/process of which the intermediate steps remain coherent and clear in my mind (not just the beginning and the "end"—if you can call it that). I arrived in the middle of a night so foggy that I couldn't even see what roads we were taking, crammed in the back seat with a refrigerator, with two total strangers in the front. After a long tramp through the dark woods we came across a campfire, surrounded by a fairy circle of people eating popcorn. That isolated spot of light, so far from everything familiar, seemed completely alone in a universe of darkness. Al offered to show me a place to sleep, and we walked uphill for what seemed like a full five minutes. The next morning I discovered the substance of a giant tent whose shadow I had seen, and the shower frame and the beginnings of an outhouse.

During all these weeks, those paths that seemed so long have shrunk to a few steps, and I can navigate their idiosyncracies even in the dark (if I'm determined enough). The sense of total strangeness has been replaced by one of intimate familiarity and, except for raccoons in the buses, at-home-ness.

Expectatins: Mine were so vague that I can hardly remember what they were. (I never wrote them down because they were too vague to even phrase.) Except for: Architecture: transformation of abstract designs (and, as it turned out, even more abstract mental images) into actual, usable, full-scale forms. Group: intensive (to varying degrees) interaction. How will responsibility for various things sort itself out?

Mostly I expected to learn about building techniques (obviously), which I did. I didn't expect to discover that seemingly little things—DETAILS— would require so much thought, time and decision. As usual, I overestimated the length of ten weeks, and expected to feel a good deal more knowledgeable and (mostly) confident about such things than I do.

I thought very little about "thegroup" beforehand—just another group. But it's a bit unusual—no really dominant leaders (amongst the students) and no dead weight. Perhaps because we're doing such a variety of concrete (wooden) things instead of bullshitting about abstractions. And every now and then it really comes together in some sort of communal or tribal ceremony—the more intense the energy level the more amoeba-like it becomes.

The built environment is a summary of group experience—and is ordinarily at least strange, if not downright hostile. Even skyscrapers are a result of a group of which all of us are a part, but how do you relate to them personally? The whole *is* greater than the sum of its parts. In this class, though, the built environment makes sense, as a result of both group and individual experiences, providing the possibility for the understanding of participation—direct participation—rather than limiting understanding to the usual, after-the-fact observation.

This class, however, is a highly unusual and unrealistic case in terms of the great outside world. It's fine if you can afford it and want to isolate yourself from the beehive that the "average person" lives in, and is fine as a futuristic dream (I suppose). But what about now? My social conscience tells me that I'm playing elitist games, and ought instead to be doing something more useful with anything I get out of this crazy school.

we did not build this place
it is not ours

a frozen vision
a stifling void
awaiting the angels

ceilings without stars weight me down
walls that divide never embrace

together find
a round and fertile place
full of quiet voices
build softly

a house grows from within
tunnels lined with secret thoughts
join chambers filled with light
where we meet
stars in their fullness

no floors cut earth from sky
mind from body.
        —February 1968

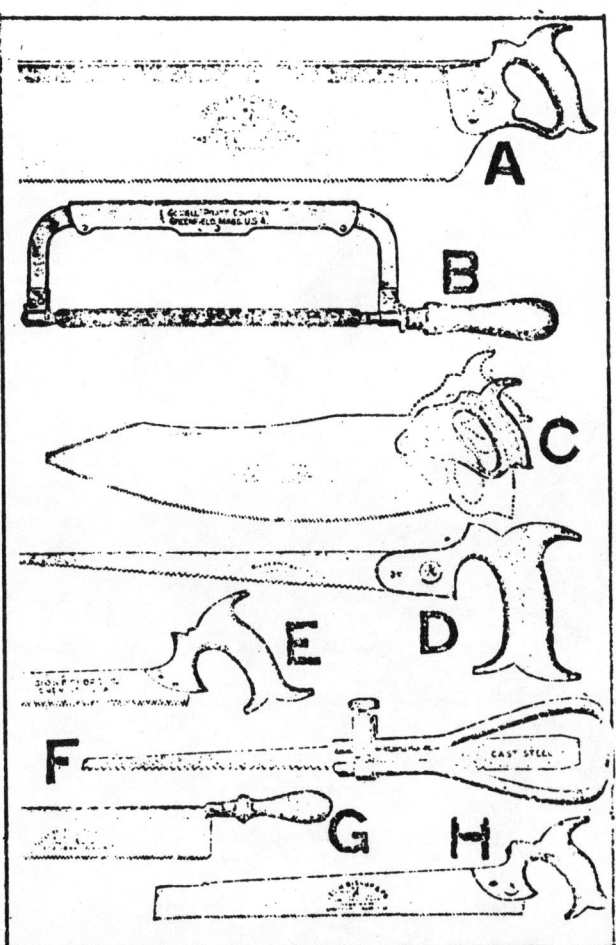

In this transitional culture, people are discovering they must design how to live. In the absence of commonly shared myth and ritual, the act of making one's way is a good part of learning, perhaps more crucial than what is learned in the accepted ways of the university.

People are hungry for the skills, the knowledge that will bring them to another place, but futurist paper utopian visions do not satisfy. People want to directly experience working examples of new environments for living, working, learning. One functioning small-scale example is worth a thousand plans. The tremendous popularity of alternative design information such as the *Whole Earth Catalogue* and *Domebook*, *Mother Earth News*, etc., point up the incredible energy focused on finding alternatives to today's living patterns and design "solutions."

For some years we have heard the extravagant technological promise of good housing at low cost. It has never come to pass. The answer to low cost housing, it seems to me, is to make a break with a "standard of living" that makes us slaves to centralized decision-making and control, to an economy whose values are the magnitude of production and consumption. The dollar is not a reasonable measure of the quality of life or the quality of place.

"I have come to feel that the only learning which significantly influences behavior is self-discovered, self-appropriated learning. Such self-discovered learning—truth that has been personally appropriated and assimilated by experience, cannot be directly communicated to another. As soon as an individual tries to communicate such experience directly, often with quite natural enthusiasm, it becomes teaching and its results are inconsequential."

                                —Carl Rogers

This day I chiseled four mortise joints to receive the tenoned posts that will be the frame of our sauna. In fourteen years of architectural practice I never designed a mortise and tenon joint because it was too much handwork and at carpenter's wages, far too expensive. So now I am learning to make them my self.

It is taking me a long time to get over the guilt of spending days hard at work learning to do things I wasn't trained to do.

It is taking a long time to accept the simple satisfaction of doing what I am doing, living in the present.

                    END
                    --Sim Van der Ryn

We shared little explicit esthetics except perhaps a common regard for the land, and a desire to use as many salvaged and native materials as possible. We share some belief in what we were doing as a way to learn about ourselves and about building.

I didn't keep a journal and I don't remember what happened from deaey to day. I had faith. I felt good, kept an eye on things, and spoke little.

Jim and I had done our homework. When you move from a rigid teacher-centered pattern towards more self-directed learning, far more preparation, planning and orchestration is required, since it is the context that you provide initially that has much to do with what takes place. In our case, it meant having tools and materials on hand, and a clear communication of the rules of the game.

During the summer, we had scrounged materials and trucked around, learning a lot in the process. We had made contact with a number of local people whom we thought could share skills in specific ways, and we had developed a rough "score" to orchestrate events through the quarter. Turns out that our score was still on city time. Once there was a flow going, it was difficult to call a halt to individual activities and have people listen to an outsider discuss ecology, tool care, etc. And in fact, most of the people we had asked to share skills seemed to feel uncomfortable with a format where they came in for a specific performance. In discussing their involvement, many would say, "Well, I'll come by and see what I can do." So Steve dropped by on Thanksgiving and built a table with his chainsaw to put the goodies on. And he stayed. All quarter, people had asked for a demonstration on the care and sharpening of tools. It never came off, because there was no one working with us day to day who had that skill. One morning, weeks after the course was over, while Al and Paul were still spending lots of time on the land, Steve emerged smiling from the house. "Well, we had a tool-sharpening extravaganza yesterday." Learning follows its own path.

Once you have broken away from "learning" supported by the setting and roles and rules of the university, there is another impediment to injecting one-time expertise. How can you know if what someone says makes sense unless you know whether the person who says it makes sense? That takes time, seeing the person in different contexts, seeing him at work.

What I am writing about really centers on overcoming the mechanical fragmentation of time and space that is the hallmark of modern life, of most formal education. Design education is no exception, as it is divided into many small pieces, courses and curricula, all fragmented into discrete pieces of time and space. The rationale and result are measurable, manageable and interchangeable units of production and consumption: so many students, teachers, subjects, classrooms, units, credits, papers, words, grades, projects, degrees, etc. The result is a vast diminuation of coherence, communication, continuity of experience.

There is an old Zen saying that goes "When the student is ready, the master will appear." In the schools and colleges, this wisdom has been turned upside down. Instead of learning by demand according to student needs, we have learning by command according to the needs of the system. Common sense would seem to suggest that most significant things just aren't learned in little pieces at appointed times. Unfortunately, the behaviorists and educationalists have confounded common sense. If you condition kids over years to accpet school learning as the only kind of learning, the consequence is that other, more organic forms of learning will be rejected as non-learning or non-experience. If you can control a person's time and space, you can significantly condition and change his behavior.

My daughter, coming home from public school as I write, comments that maybe they made her school without windows so that teachers and textbooks would make sense.

School—and perhaps urban life in general—rewards people who function well in fragmented time and space. People who are verbal, who can make quick, effective impressions on teachers and others in authority, people who can quickly cue into what responses will be rewarded—these are people likely to win favor with teachers who have little time to get to know students. Such qualities, not bad in themselves, tend to get confused with intelligence or good learning.

The educational establishment has perverted learning processes with the result that many "good students" are neurotic learners. For them, the personal and intrinsic pleasure of learning is suppressed, and the externally useful aspect of learning is all that counts, that is, the approval of teachers and the system. How many "good students" (count me among them) have experience the crippling sensation that when external rewards—and the whip—are removed, they are followed by a sense of shock and loss. In healthy learning, the *process* of learning itself is the reward and its extrinsic value; getting along in society follows without pain.

One of the most damaging effects of neurotic learning is that it tends to condition people to ignore their own expeience, perception and feeling as a source of knowledge in favor of the printed word or other established authority.

It seems as though students become conditioned to expect that a teacher's public face bears no relation to his private life; that there need be no connection between what a man *professes* and what he *is*. In other words, the total bureaucratization and depersonalization of "teacher" usually defended as protecting a teacher's "objectivity"—that is, the teacher as pure role or object.

Recently I gave a talk about this class. Although my discussion of the experiences and beliefs that led me to the place I am now were all prefaced with "I am" or "I believe," it was clear from the comments and discussion that followed that some students heard "I am" as "you should."

Could it be that "I" is heard as "you" because students can't relate what a professor teaches (objective knowledge, fact, truth) to the use of that knowledge in his own life?

The whole man is sought out as a teacher. A major task facing teachers is to repair the rift between the experience of their "objective" and "subjective" selves, to untie the division between the inner and the outer self. It is a division nurtured by the machine metaphor. Its tools are the separation of one's work role from one's identity. This split is aided by the fragmentation of time and space and by the physical settings and enforced rituals that support the role.

Getting myself together started with getting my time and space into one place, to create the possibility and essential condition for wholeness.

Design education provides a fertile field for the seeding of wholeness, for better ways to learn. In design, what one learns becomes apparent in what one creates. The design or material creation becomes a tangible expression of the learning process, a guide and map for future learning. A poorly sited shelter shows inattention to the landscape, a leaky house demonstrates faulty technique. There is no need to design an elaborate evaluative procedure; it is all out front. The learning process is experience itself, as well as preparation for future experience. If your experience of learning is not pleasurable, chances are its future use won't be, either.

The intrinsic purpose of learning is to bring each of us closer to peace, a way of life reflective of peace. Self knowledge is the end of learning. The physicist Erwin Schrodinger has written:

"The reason why our sentient, percipient and thinking ego is met nowhere in our world picture can be easily indicated in seven words: *because it is itself that world picture.* It is identical with the whole and therefore cannot be contained as a part of it."

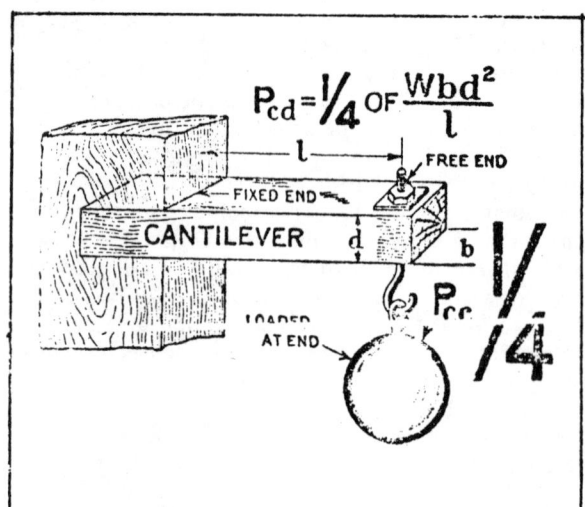

The end of this piece brings me to share some thoughts about alternatives that may make your path easier.

The professor who says, "Do anything you want, I'll be around if you need advice," is not to be trusted. He still holds the power without accepting the responsibility, thus putting students in a double bind. Without context and access to tools and resources, it's hard to learn and the situation usually leads to frustration. Recommended for navel watchers and talkers only.

Instant commitment, instant intimacy is often a path to disaster. People reveal themselves slowly through their actions. Real trust and intimacy in a group grow slowly. It is easier to move forward rather than back. It is easier to start with some explicit structure that can loosen and grow organically, rather than start with no structure and hope something will develop.

The idea of a group, a tribe, sharing work and experience, finding something to believe in, is attractive to people searching for something better than school. But most of us are still children of the institutions we grew up in. We are conditioned to their ways. Most of us have grown up sharing little spiritual experience. We have few rituals that celebrate our unity of body, mind and spirit. We are trying to find our way back into the earth family and there are few guides to show the way. But the magic moments will be there when we need them, and those few moments make it all worthwhile.

A group that comes together around vaguely similar interests (dissatisfaction with where they are now) cannot expect to move. Interest does not equal energy. The physicists define mass as energy at rest. To get energy moving requires at least a common context, an explicit goal. The process of setting an agreed-upon goal starts with a sharing of each person's dreams, expectations, self interest. Groups that form around high-flying goals without sharing nitty-gritty expectations often end in disaster. Ghandi says it for us: the means are the ends in process. Part of where you are going is how you get there.

I tried so many routes
to get to the root
of what it's all about
finding that
what it's all about
is the route.

there is another moon
It orbits in the tunnels
   and caverns of the ocean
It howls for days
   while dark rain scrapes the beach

getting dark, the road shortens
we see downhill! just to a bend
   a spider confused on white gravel
the sound of fish in the voice of a
                companion
   a mark in the sky where the stone
                fell

Outlaw Building News

Star Route
Point Reyes Station, Ca
94956

Jerry Finrow

User - Centered Computer Design Systems: The Evolutionary Designer

This article discusses the Evolutionary Designer, which is a computer system that is intended to enable people to design houses for themselves. This project is partially supported by the University of Oregon Graduate School and the Computing Center. Persons making substantial contribution to the program are Huck Rorick, Donald Baker, Donald Wood, Bill Beal, David Winitsky, Herb Wilkins, Eric James, Bill Merritt and Hussain Mirza. Charles Eastman has supplied us with a version of GSP for use in our system.

I. System Operation

Perhaps the clearest way to understand how the system operates is to give a brief description of a user example. Before I do this I will explain a little about the technical aspects of the system. All of the programs are written in Fortran IV and operate on a PDP-10 time sharing computer. The user communicates to the program via a remote terminal which can be either a teletype or a scope. Most users in our current experiments use a teletype. The user, or client, will sit at the teletype and, following a set of instructions, make connection with the main operating system. All he needs to do is have a user number and type a few simple characters on the teletype, and the computer takes it from there. Once having gotten the attention of the computer, the user will then enter into an interactive dialogue with the program. The computer will begin by asking the user questions. An example of the dialogue is indicated in figure 1 and a sample execution in section III.

On the basis of the dialogue, the computer will begin to assemble user requirements via an "activity-component" unit of analysis. For example, it will discuss sleeping as opposed to the bedroom. The system collects user data and organizes site information before listing spatial components which are gathered into "rooms" along with their dimensions. Suggestions are offered to the user on space size and relationship to other spaces. Finally, the user inputs requirements for space adjacencies, and the program then generates an arrangement of rooms into a floor plan. The user has the opportunity all along the process to alter or change a piece of information and to examine the logic that generated the suggestion that the computer gives. See figures 2, 3, and 4 for illustration of these parts of the program.

At present the system terminates after having drawn a character-generated floor plan. We intend to extend it to accommodate structural and mechanical calculations (we have sketch programs operating on these already, however,

WHO DOES MOST OF THE COOKING IN YOUR FAMILY?
?NANCY

I WOULD SUGGEST THAT THE DINING AREA FOR YOUR EVERYDAY MEALS
BE IN THE SAME ROOM AS THE COOKING AREA SO THAT EVERYONE CAN
SOCIALIZE WHILE MEALS ARE BEING PREPARED AND NANCY   WON'T BE
ISOLATED IN THE KITCHEN. DO YOU AGREE?
?YES

HOW OFTEN DO YOU HAVE GUESTS FOR MEALS? (TYPE 0 IF LESS THAN 1.)
TIMES/MONTH
?7

HOW OFTEN DO YOU WANT THESE OCCASIONS TO BE FORMAL (IE. OTHER THAN
CASUALLY JOINING YOU FOR DINNER)?
                                  TIMES/MONTH
?3

KEEPING THESE ANSWERS IN MIND, DO YOU FEEL YOU NEED A SEPARATE
DINING AREA FOR MORE FORMAL OCCASIONS
?YES

HOW MANY PEOPLE DO YOU WANT THE EVERYDAY DINING AREA TO ACCOMODATE
INCLUDING COMPANY?
?8

HOW MANY PEOPLE DO YOU WANT THE FORMAL DINING AREA TO ACCOMODATE?
?10

NUMBER   COMPONENT SPACE   DIMENSIONS
                                A   B   C   D   E
   1   INFORMAL   DINING   3.50   5.33   5.33   0.00   0.00
   2   FORMAL   DINING   3.50   3.00   10.00   0.00   0.00

        I   A   B   I   A   I

        ************************* -
        ***.CHAIRS & ACCESS....   A
        *****I------I*******..... -
        *******I   I*.....
        *******I TABLE I*......   C
        *******I   I*.....
        *****I------I*******..... -
        ***********************   A
        ****

figure 1

THE FOLLOWING IS A PRELIMINARY ALLOCATION OF COMPONENT SPACES
TO ROOMS:
ROOM 1 AREA= 185.00 SQ.FT.
       FORMAL   LIVING
ROOM 2 AREA= 245.00 SQ.FT.
       FAMILY   LIVING
       RELATIONS
ROOM 3 AREA= 198.96 SQ.FT.
       INFORMAL DINING
                 KITCH SINK
                 RANGE
                 FRIG
                 OVEN
       KITCHEN   CLOSET
                 DISHWASHER
                 COUNTER
                 FREEZER
ROOM 4 AREA= 140.80 SQ.FT.
       MASTER   BED
       JEFF
       NANCY
       JEFF   DRESSING
       NANCY   DRESSING
       JEFF   CLOSET
       NANCY   CLOSET
ROOM 5 AREA= 100.70 SQ.FT.
       BOB   BED
       BOB   DRESSING
       BOB   CLOSET
ROOM 6 AREA= 100.70 SQ.FT.
       SUE   BED
       SUE   DRESSING
       SUE   CLOSET
ROOM 7 AREA= 113.70 SQ.FT.
       GUEST   BED
       GUEST   DRESSING
       GUEST   CLOSET
ROOM 8 AREA=   18.54 SQ.FT.
       JEFF   BATH SINK
       NANCY   BATH SINK
       MASTER   SHOWER
       JEFF
       NANCY
       MASTER   TOILET
       JEFF
       NANCY
ROOM 9 AREA=   23.53 SQ.FT.
       GUEST   BATH SINK
       GUEST   TUB
       GUEST   TOILET
ROOM 10 AREA=   17.50 SQ.FT.
       FAMILY   WASHER
                 DRYER
ROOM 11 AREA= 170.00 SQ.FT.
       FORMAL   DINING
ROOM 12 AREA=   72.00 SQ.FT.
       PUBLIC   ENTRY
       RELATIONS
       VISITORS
       FAMILY
DATE:17-JUN-71   TIME:20:50 00.8

figure 2

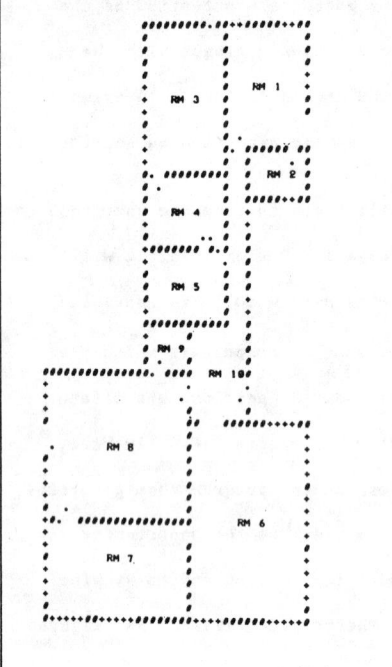

RUN DATE - 17-JUN-71

figure 3

* WOULD YOU LIKE TO CHANGE OR ASK A QUESTION ABOUT A SPACE?
?YES

WHICH SPACE ARE YOU INTERESTED IN FIRST?
(TYPE THE NUMBER)
?1

DO YOU WANT TO KNOW:
1) WHY OR HOW IT WAS CHOSEN
2) HOW DIMENSIONS WERE DETERMINED.
3) BOTH
4) NOTHING
?3

TWO POSSIBILITIES FOR DINING SPACES WERE CONSIDERED: COMBINED WITH
KITCHEN(INFORMAL) AND SEPARATED(FORMAL). EITHER OF BOTH MAY BE CHOSEN
BY CLIENT. IT IS RECOMMENDED THAT AT LEAST ONE BE PART OF THE KITCHEN
(SEE KITCHEN-DINING PATTERN). OUTSIDE AREAS ARE NOT CONSIDERED.
POSSIBILITY OF MORE THAN 2 PLANNED DINING AREAS WAS CONSIDERED LOW.

FOR GROUPS OF 8 OR LESS A CIRCULAR TABLE IS RECOMMENDED, FOR MORE AN
OBLONG. CIRCULAR SEATING CREATES A SINGLE CONVERSATION GROUP. OBLONG
SEATING FOR 6 OR MORE DIVIDES INTO SUBGROUPS. FOR MORE THAN 8 PERSONS
A SINGLE GROUP IS NOT SOCIALLY PRACTICAL. SINCE SUBGROUPS ARE NOT EAS
ILY FORMED ON A CIRCULAR TABLE & THE DIAMETER FOR MORE THAN 8 IS TOO
LARGE FOR CONVERSATION. AN OBLONG IS USED. (NOTE:CIRCULAR SEATING DOE
NOT HAVE A SPOT AT "HEAD OF THE TABLE".) 2'/PERSON IS NEEDED TO SIT
TABLE (GRAPHIC STANDARDS). DIAM OF CIRC TABLE=(# OF PEOPLE)(2)/(PI).
RECT. TABLE=3'X(# OF PEOPLE). WIDTH=3'. CHAIR CLEARENCE=3.5'(IBID) A

DO YOU WANT TO:
1) KEEP SPACE AS IT IS
2) ELIMINATE IT
3) CHANGE ITS DIMENSIONS
?1

DO YOU HAVE ANOTHER QUESTION ABOUT A SPACE?
?NO

DO YOU WANT TO CONTINUE?
?YES

figures 4

figure 5

they are not integrated into the operating system), perspective and working
drawings, and finally, materials listing. The intention, in the long run, is
to have a designer-independent system that will give the client all that is
required to then approach a contractor and initiate the construction phase
of his house. The system is very easy to operate, It converses in plain
English, although the conversation takes the form of an interview schedule.
We have tested the system on a number of users and have found them fascinated
and very much involved in the process of designing their own houses. They
seem to be thoughtful and careful in their decision making (see figure 5),
treating the program as if it were a designer.

II. Theoretical Issues

There are a number of theoretical issues that we have explored in our work
that may be worth some elaboration. They are a) the development of a communal
and evolutionary body of knowledge in design, b) the technical questions at
stake in large scale computing design systems, c) user involvement in design
of their environment. While the last issue is the most germane to the intention of this publication, the other two set the framework for the latter,
therefore, I will discuss each in some detail.

a. Communal and Evolutionary Design Knowledge

To begin I will outline a very simplistic history of design. In indigenous
(as opposed to primative) cultures, the question of design of the environment drew from cultural heritage and ritualized activity patterns that en-
sured a rather tight fit between the needs of the culture and the form of
the environment. This is not to idealize these cultures, rather it is just
a fact (Rappoport, House Form and Culture, 1969). Users were designers and
builders, to a very real extent.

As societies became larger and more complex, we begin to discover the
emergence of the village carpenter, or master builder, who was also a sort
of architect. His value lay in knowing how to make something that fitted
into the tradition of the community. He was not far, physically or psychologically, from the people he worked with, which insured a somewhat closer
relationship between the environment and user than was known to be is the case
today. The industrial revolution brought much more complex systems of social
organization into being, and hence the development of highly specialized roles
in society. It is at this time that we begin to see the emergence of the
architect as a really special person with a special role. While this
specialization has had certain advantages, it has had the effect of removing the experience and values of the designer from the experiences and

values of his clients, hence hastening a breakdown in the fitness of the
environment to the behavior and needs of people. While this is a
very simplistic historical picture of our current dilemma, it does illustrate a problem that concerns a number of people; that is, how to restore the
proper fit between users in an environment and the environment itself (Walkey,
the Fit Environment, 1969). I am not interested in developing this trend
of thought in more detail, but rather in suggesting a setting for the development of one model for "de-specializing" of the role of the designer.

I will now propose some criteria for the establishment of user-centered
design systems. When I look at the earliest physical environments, what
strikes me is the relative anonymity of the "common" environment (Rudolfsky,
Architecture Without Architects). I am stuck in current practice by the
glaring search for identity of both buildings in the environment, and the people
who designed them. I would conclude from this that at least one basic requirement for any design process that would seem to be more appropriate to
user involvement would be design characterized by anonymity.

A second problem in developing ways of getting users more involved in the
design of their own environments has to do with the "sociology" of design
professional roles. Perhaps the most significant barrier between professional
roles and "common" people has to do with language and concepts. In seeking to
define a special role, we designers (and other professionals) have developed
languages that bar "common" people from understanding us. For example, it is
not uncommon to have a designer tell a client that the "flow of space from
the center nodes of activity in a house give rise to a sense of self
worth; an uplifting feeling...." What does all this mean? How can a
client possibly challenge such a statement; who understands it anyhow? A
second requirement for user-based design systems would be that any person
could make use of them without developing special knowledge or language.

A third issue which a user-centered design system would need to be built
around is change. Traditional indigenous design practice, while achieving
good fit, often cannot survive change. The result has been the destruction
of such design systems and a reliance on practice that is far removed from
understandings of indigenous problems. A good example could be the current
influence of Mies in Egyptian architecture. The traditional Arab city
structure and dwelling unit responded well to the intense heat and glaring
sun. Current design practice tends to see the indigenous solution as being
'outdated' when, in fact, it has evolved on much more "rational" basis

(from talks with Hussein Sharkawy). The result is destructive of the culture in which traditional practice has been operating. While the indigenous practice has been successful, new information and changing social values have been occuring. These problems, along with dwellings becoming unsafe for habitation, have caused new structures to need to be built. The dilemma is apparent, how to capture traditional knowledge and extend and apply it through more current technologies. It would then seem important that the system should be capable of change and development over time. These three criteria, anonymity, common language, and capacity for change, form the foundation for user-centered design theory. I am sure that there are additional criteria which might be outlined; however, I feel these to be of primary importance, and they are the ones that the Evolutionary Designer is built around.

In terms of the Evolutionary Designer concept, let us look at the first of the requirements; that of anonymity. Innumerable persons have made contributions to the concept through development of specific parts of the program. Most of the work has been done by students and faculty at various locations. Each person working on a part of the system has been less interested in his own achievements than in building something that seems of greater worth. The programs that make up the Evolutionary Designer are, by and large, quite anonymous. The system is not closed to change and development by others, rather it is expressedly designed to be changed in ways that support a communal effort. Specific parts of the program are actually the work of perhaps 3-4 persons separated in space and in time. For example, one person may outline a program on bathroom design, a second will extend it, and a third may put very much more detail into it and perhaps rewrite it to work more efficiently. This has had the effect of insuring anonymity, which is a functional premise of the program.

Concerning the second point, that of language, while the programs themselves suffer very much from specialization (by being written in a programming language) the actual operation of the program is quite simple. The dialogue of the program is in common English, and is quite easy to understand on the part of the user. Eventually we hope to be able to have persons not knowledgeable in Fortran or any of the file and editing routines (Line-Ed and Teco) to be able to add to and modify the program. That is, we would put together a system whereby users can get instructions about how to write simple programs. At the moment the system can accept comments from a user about some aspect of the program he may not agree with or on which he wishes to comment. The system manager can review these comments and take appropriate corrective actions if they seem reasonable.

This leads us into the third area, that of the evolving nature of the program. We have developed a basic structure to the system which allows any part of it to be pulled out and examined in detail and for comments and additional considerations to be brought into the program. Basic operating versions of the program can be reviewed by persons with various interests. A behavioral scientist may have new information on possible relationships between the social and private areas of the house, he could add these new awarenesses to the program and hence make the program more knowledgable. The capability for change and updating is central to the system's existence and evolving nature. We have recently added a number of routines that were not previously in the program and have modified others. For example, we now have a section on financial counseling ~~counciling~~ which previously did not exist. It is illustrated in figure 6.

An example of a routine that has been remodelled and expanded deals with the design of the conversation area in the general social space (figure 7).

We feel that the system we have been developing responds to some degree to all three of these criteria. We realize that the system is incomplete, however, it does indicate at least a first generation computer model for integrative evolutionary design knowledge that is user based. For further reading see ~~an~~ the article by Huck Rorick published in the summer 1971 issue of the Journal of Architectural Education, and an article by myself in the Proceedings of the Kentucky Workshop on Computer Applications to Environmental Design, Michael Kennedy, editor.

b. Technical Questions of Large Scale Computing Systems

The program itself raises an interesting issue for computing theory, that is, how do you design and construct large scale complex operating systems which perform a number of diverse tasks? To begin to discuss this issue, I should first outline just how the system currently works, and point out some of the problems that we have been having with it. The program is illustrated in figure 8. The various sub-programs that are meant to solve specific problems are designed to be removable. Modules that are dotted in are not yet operational, but have been sketched and are running independently of the main operating system.

The basic goal is to develop a system that is like a large tinkertoy, so that it can be taken apart and remodeled and built up in a number of different ways. This is to ensure that the program can be changed in order to make the system flexible enough to accept new knowledge. This is the core of the evolutionary idea.

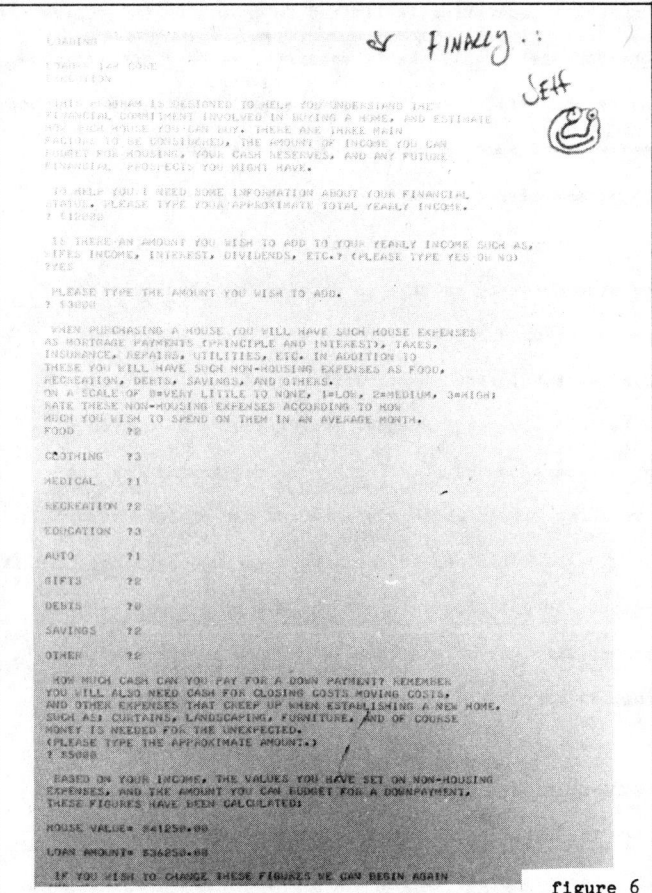

figure 6

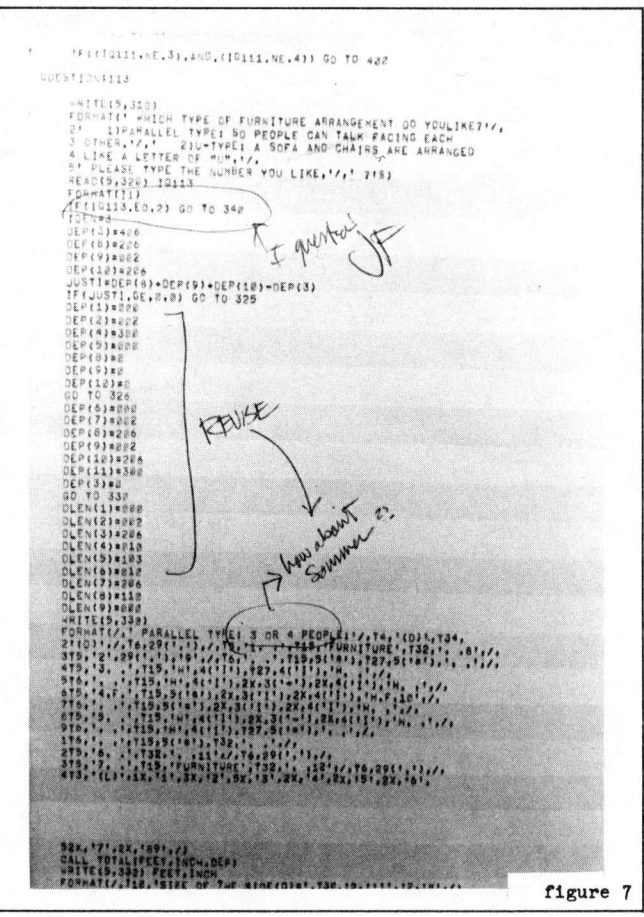

figure 7

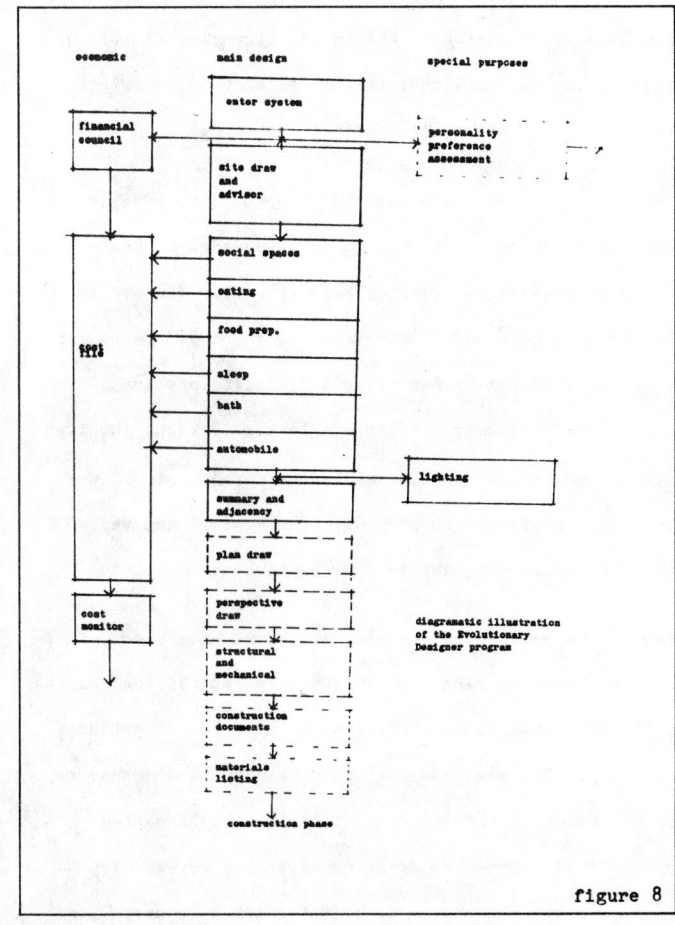

diagramatic illustration
of the Evolutionary
Designer program

figure 8

The program as it is currently operating is limited in growth by the active storage capacity of our computer (24K). As is apparent, we have not yet completed all of the modules that will make the system a really adequate designer. This is our major problem, that such computer programs are very large and that conventional storage and computing systems are not of sufficient size, at least at our facility, to accommodate this kind of an experimentation. In conventional operating procedures to solve this problem, we have been conducting some studies aimed at converting a version of a space planning language (GSP by Charles Eastman) into a programming technique known as "chaining." Using this technique, we put together the various programs in a sequence such that their access in and out of the computer storage is controlled by a program known as "Driver," which swaps in and out of core memory and disc storage the various subroutines in the program. The magnitude of this problem is best illustrated by pointing out that the GSP program is being modified to work in tandem with our main system as the spatial allocation routine, and it is larger in size than our current system. In addition, the perspective and working drawings modules will also be larger than our present main program. This means that we will be working with an enormous program, passing data from one to the other in extremely complex ways. While this problem is very great, it is essential to solve in order to actually build a comprehensive user-based design system. It is interesting to note the inefficiency of computing systems by coming to the realization that all of the processes we are modelling (design decisions, spatial allocation, drawing, etc.) are achieved by the brain in considerably less effort and space!

There are additional problems that we are beginning to encounter. One is the lack of a full-time systems manager who is able to fully understand the main system and help to relate individual programming effort to it. The way in which the program is written is intended to enable a number of people to work at creating or modifying parts of the system to solve specific problems. This means that the programming is relatively inefficient because persons involved and interested in design issues are often less skilled at programming. We feel that it is important for designers to work with the system; however, persons with more extensive programming skills must be involved.

In addition to the size of the system and its complexity, we have been finding that there is a real need to make its operation as interesting and engaging as possible. We find that users sitting at a teletype for extended periods of time become bored. The question that this raises has somewhat to do with the ergonomics of man-machine relations. This is an interesting point, because in some ways what we need to do is to develop a personality for the program, a way of getting the system to be interesting to users, just as a human designer is interesting to a client. Perhaps the terminal should have clothes or there should be flashing lights or some other form of "decoration". In the site program we find that the visual generation of a site plan holds user attention much more firmly than does terminal-based operation. (figure 2 has some scope hard copy examples). The visual "happenings" would seem to be important to the success of the system to a user.

A final problem we have been having with the system has to do with the reliance on a machine to design in general. We find ourselves at the mercy of a mechanical device. To some extent you could look at it like power steering. If it works everything is fine. If it does not, you cannot do anything at all. We have had several system crashes in mid-programming and have often lost substantial amounts of work. We have learned to make copies of all programs on card decks so that if anything happens to the system, we still have the previous version. We also have two backup copy tapes containing the whole program, which we can use if the main system is destroyed.

These issues are significant to the success of any computer system that is user-oriented and hopes to be inclusive and complete. We feel that we are confronting many of these problems with computer-aided design for the first time, and we know that they need to be solved in order for any meaningful development to proceed. Continued research on the programming problems would seem to be important to many people interested in comprehensive design systems, as well as of direct interest to the user.

c. User Involvement in Design

In our work, we feel that user involvement in design decision-making is the main goal. All of the programming effort and the theoretical notions revolve around the issue of getting into the hands of the user the decision making tools for determining his own environment. We feel that the direction we have taken will contribute to the generally evolving body of research aimed at this same problem. In order to place our concern, it would be useful to explain the more specific intention of our work.

Let me preface this discussion by stating that we do not see our work as "the answer" to the problem, but rather as one model for dealing with one aspect of the general concern. We wish to extend the use of sound design decision making to groups of persons who have little access to such knowledge.

We have chosen, as an initial vehicle, a relatively simplistic problem area - the individual detached single family dwelling unit. We are fully aware that this problem is not necessarily the most pressing problem in user-based design.

We have decided to look into it partly because we think we can handle it, and partly because the housing market is little impacted by designer involvement. As you might have guessed by now, we feel that designers potentially have something to offer the user, and we feel that well-considered design decisions are important to building better and more well fitting environment. We also feel that the relationship between a client and a designer is one in which the logic of the problem solution is the primary tool of the designer, and that the final decision about the nature of the environment rests with the user. The designer's role is to point out opportunities that can be taken advantage of, and problems that might arise from certain decisions; an altogether modest view of design.

In looking into the problem of single family housing, it is our opinion that users in the conventional tract-house-buying situation are given much less chance to be involved in the design of their dwelling than in the traditional designer-client model. We have conducted a number of rather casual studies into single family dwelling market in our own area. We have found that the kinds of involvement that users can have in design of their own home revolves around two types of activity. Firstly, they can select from a range of house plans (usually 3-5 in a 50-100 home subdivision), and secondly, they can add certain "options" such as brass hardware, interior wall panelling, etc.. While they have some level of choice, little is done to educate or point out the implications of these choices on their particular family situation. We did discover, however, that the state landgrant college (O.S.U.), through its extension service, conducts short-courses on house buying. We visited these courses and found them to be excellent. Persons attending the course are given a whole set of evaluative guidelines and charts which cover such things as site, construction, possible problems, choice, etc.. We are interested in studying these people to see if the course has had an impact on their home-buying behavior. But Even if these courses do have an impact, few people attend them. Generally speaking, the user is very much at the mercy of the marketing professional, and of the hard slick-sell techniques developed to maximize sales rather than satisfaction of user needs. At the same time, we realize that the industry is fulfilling the need for housing which has traditionally not interested the designer.

It is interesting, in reviewing the history of architecture, to note the lack of historical documentation of "common" habitation. History has enshrined those monuments to power than seem least relevant to the needs of man. In current practice somehow we must become more concerned about "common" problems.

In thinking and being concerned about this problem, we have felt that a computer design system which can be used by contractors, architects, or developers, in working with potential housing users, seems to make some sense. That is, the users design their house via the system, and the builder uses the output as his plan of action. This enables the user to be more involved in the process of determining his own environment, and the builder still has in-house designer capability. This model is illustrated in figure 9:

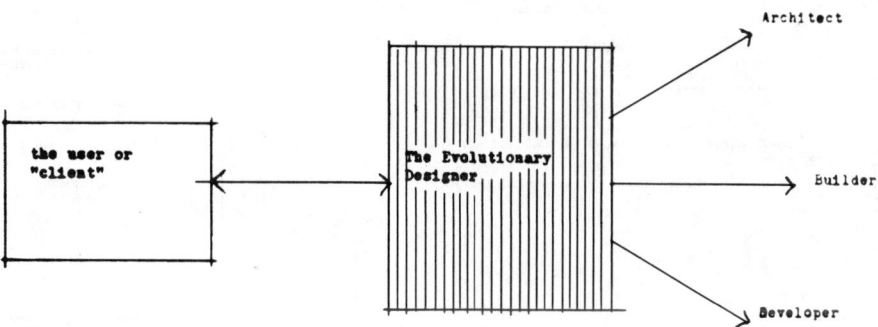

figure 9

There are two ways in which the user can be directly involved in using the program. The first is by designing a house. The user, after accessing the program, then answers the questions put to him and quizzes the program about its suggestions, making whatever changes seem appropriate. The result of this process is, presumably, much more related to user needs. This would seem to be the most personally important level of program involvement. The second level is at the level of system modification. This presumes an enlightened user. For example, perhaps a user detects a problem in the way in which the program determines the size of the sleeping space for children. He can change the dimensions so that it makes more sense to him, but beyond this, he can also suggest changes to the whole system as well as specific portion of the program. The system manager can catch the suggestions, and perhaps discover a better method to determine the answer to this problem. This is an attempt to support the evolving nature of the program so it is not reserved for just the designer or programmer eletists. Anyone can make any comments and have them considered by the system. We have already had cases where users have made specific suggestions about the operation of some portion of the program. We have reworked those portions of the program, and have greatly benefited from these comments. We have assumed an assimilative rather than defensive posture where program change is involved; we seek to maximize the learning capability of the system.

173

```
        TIME
        01:46
        02:10
        KILO-CORE-SEC=1628

.EX #ARCHIT
FORTRAN: QUIZ
LOADING

LOADER 18K CORE
EXECUTION

HAVE WE MET BEFORE? (TYPE YES OR NO.)
?NO

    THIS IS PART OF AN AUTOMATED ARCHITECT. I MAKE RECOMMENDATIONS ON
WHAT SPACES YOU NEED AND THEIR DIMENSIONS. WOULD YOU LIKE INSTRUCTIONS
FOR TALKING WITH ME? (TYPE YES OR NO.)
?NO

WHAT IS YOUR NAME?
?ALFRED

ARE YOU THE HEAD OF THE HOUSEHOLD ALFRED      ?
?YES

WHAT IS YOUR MARITAL STATUS:
1) SINGLE
2) MARRIED
3) DIVORCED OR SEPARATED
4) WIDOWED
?2

HOW MANY PEOPLE WILL BE LIVING IN YOUR HOUSE OTHER THAN YOURSELF?
?5

WHAT ARE THEIR NAMES? (TYPE ONE NAME PER LINE.)

?MARY

?MARK

?MIKE

?SUSAN

?BETH
```

```
I NEED A LIST WITH THE AGE, SEX, & RELATION TO YOU OF EVERYONE
IN THE HOUSEHOLD. WOULD YOU FILL IN THIS INFORMATION AFTER EACH
PERSONS NAME? UNDER SEX TYPE 1 FOR MALE, 2 FOR FEMALE. UNDER
RELATION: 1=HEAD, 2=SPOUSE, 3=CHILD, 4=OTHER.

NAME     AGE SEX  RELATION

ALFRED    27 1 1

MARY      24 2 2

MARK       5 1 3

MIKE       4 1 3

SUSAN      3 2 3

BETH       2 2 3

I NEED AN ESTIMATE OF THE BUDGET FOR BUILDING YOUR HOUSE. WOULD YOU
LIKE SOME ADVICE ON FINANCING, ETC. BEFORE GIVING ME A FIGURE?
?NO

WHAT IS THE TOTAL BUDGET FOR YOUR HOUSE NOT INCLUDING LAND?
?24000.

IF YOU'RE TIRED NOW, WE CAN TAKE THIS UP AGAIN LATER.
SHALL WE CONTINUE?
?YES

    SHALL WE DISCUSS THE DINING AREA?
?YES

    WHO DOES MOST OF THE COOKING IN YOUR FAMILY?
?MARY

I WOULD SUGGEST THAT THE DINING AREA FOR YOUR EVERYDAY MEALS
BE IN THE SAME ROOM AS THE COOKING AREA SO THAT EVERYONE CAN
SOCIALIZE WHILE MEALS ARE BEING PREPARED AND MARY      WONT BE
ISOLATED IN THE KITCHEN. DO YOU AGREE?
?YES

    HOW OFTEN DO YOU HAVE GUESTS FOR MEALS? (TYPE 0 IF LESS THAN 1.)
TIMES/MONTH
?2

    HOW OFTEN DO YOU WANT THESE OCCASIONS TO BE FORMAL (IE. OTHER THAN
CASUALLY JOINING YOU FOR DINNER)?
                                TIMES/MONTH
?1
```

```
    KEEPING THESE ANSWERS IN MIND, DO YOU FEEL YOU NEED A SEPARATE
DINING AREA FOR MORE FORMAL OCCASIONS
?YES

    HOW MANY PEOPLE DO YOU WANT THE EVERYDAY DINING AREA TO ACCOMODATE
INCLUDING COMPANY?
?8

    HOW MANY PEOPLE DO YOU WANT THE FORMAL DINING AREA TO ACCOMODATE?
?8

NUMBER    COMPONENT SPACE                 DIMENSIONS
                              A       B       C       D       E
  1   INFORMAL  DINING      3.50    5.33    5.33    0.00    0.00
  2   FORMAL    DINING      3.50    5.33    5.33    0.00    0.00



   I   A I B I   A I

..................... -
.....................
...CHAIRS & ACCESS.... A
.....................
........1------1.......
........1      1.......
........1 TABLE1....... C
........1      1.......
........1------1.......
.....................
..................... A
.....................
..................... -


    WOULD YOU LIKE TO CHANGE OR ASK A QUESTION ABOUT A SPACE?
?YES

WHICH SPACE ARE YOU INTERESTED IN FIRST?
  (TYPE THE NUMBER)
?1

DO YOU WANT TO KNOW:
1) WHY OR HOW IT WAS CHOSEN
2) HOW DIMENSIONS WERE DETERMINED
3) BOTH
4) NOTHING
?3


TWO POSSIBILITIES FOR DINING SPACES WERE CONSIDERED: COMBINED WITH
KITCHEN(INFORMAL) AND SEPARATED(FORMAL). EITHER OF BOTH MAY BE CHOSEN
BY CLIENT. IT IS RECOMMENDED THAT AT LEAST ONE BE PART OF THE KITCHEN
(SEE KITCHEN-DINING PATTERN). OUTSIDE AREAS ARE NOT CONSIDERED.
POSSIBILITY OF MORE THAN 2 PLANNED DINING AREAS WAS CONSIDERED LOW.
```

```
FOR GROUPS OF 8 OR LESS A CIRCULAR TABLE IS RECOMMENDED. FOR MORE AN
OBLONG. CIRCULAR SEATING CREATES A SINGLE CONVERSATION GROUP. OBLONG
SEATING FOR 6 OR MORE DIVIDES INTO SUBGROUPS. FOR MORE THAN 8 PERSONS
A SINGLE GROUP IS NOT SOCIALLY PRACTICAL. SINCE SUBGROUPS ARE NOT EAS
ILY FORMED ON A CIRCULAR TABLE & THE DIAMETER FOR MORE THAN 8 IS TOO
LARGE FOR CONVERSATION, AN OBLONG IS USED. (NOTE:CIRCULAR SEATING DOE
NOT HAVE A SPOT AT "HEAD OF THE TABLE".) 2'/PERSON IS NEEDED TO SIT
TABLE (GRAPHIC STANDARDS). DIAM OF CIRC TABLE=(# OF PEOPLE)(2)/(PI).
RECT. TABLE=3'X(# OF PEOPLE). WIDTH=3'. CHAIR CLEARENCE=3.5'(IBITDATA

DO YOU WANT TO:
1) KEEP SPACE AS IT IS
2) ELIMINATE IT
3) CHANGE ITS DIMENSIONS
?1

    DO YOU HAVE ANOTHER QUESTION ABOUT A SPACE?
?NO

IF YOU'RE TIRED NOW, WE CAN TAKE THIS UP AGAIN LATER.
SHALL WE CONTINUE?
?YES

SHALL WE DISCUSS THE KITCHEN?
?YES

I WOULD LIKE TO ADDRESS MY QUESTIONS ABOUT THE
COOKING AREA TO THE PERSON WHO WILL BE DOING THE COOKING. IS MARY
HERE TO ANSWER THESE QUESTIONS?
?NO

DO YOU FEEL YOU CAN ADEQUATELY ANSWER QUESTIONS ABOUT THE COOKING AREA?
?YES

GOOD THEN WE CAN CONTINUE

DURING THE WORK WEEK HOW MANY COMPLETE MEALS PER DAY WILL YOU GENERALLY
BE PREPARING?
?3

HOW MANY PEOPLE IN YOUR HOUSEHOLD ARE USUALLY PRESENT FOR THESE MEALS?
?
5

DO YOU:
1. ENJOY COOKING?
2. NOT MIND COOKING
3. DISLIKE COOKING
?1

WOULD YOU MIND SPENDING MOST OF YOUR DAY IN THE KITCHEN AREA?
?YES

DO YOU PLAN ON SPENDING ANY MORE THAN THE MINIMUM AMOUNT OF TIME IN THE
KITCHEN?
?NO

WHEN YOU ARE WORKING IN THE KITCHEN DO YOU LIKE TO TAKE TIME OUT
FOR SMALL TALK AND FAMILY AFFAIRS?
?YES

WILL YOU BE ATTENDING ANY SMALL CHILDREN WHILE YOU ARE PREPARING
MEALS FOR YOUR FAMILY?
?YES
```

WOULD YOU LIKE TO INCLUDE AN AREA ADJACENT TO THE COOKING AREA WHERE
CHILDREN CAN PLAY AND YOU CAN WATCH, YET STILL COOK?
?YES

A FREE STANDING SINK COUNTER PENNISULA SEPARATING THE COOKING AREA
FROM THE MULTI-PURPOSE AREA IS A GOOD WAY OF ACCOMPLISHING THIS. DO
YOU WANT THIS TYPE OF PLAN?
?YES

DO YOU WANT TO BE IN CLOSE CONTACT WITH YOUR FAMILY WHEN YOU ARE
IN THE KITCHEN?
?NO

DO YOU FEEL THE COOKING AREA SHOULD BE SOMEWHAT ISOLATED FROM THE
REST OF THE HOUSE?
?YES

DO YOU WANT THE KITCHEN TO BE MORE THAN JUST A PLACE TO COOK? (EX.
A PLACE TO DO IRONING, MENDING, TAKE CARE OF HOUSEHOLD BUSINESS, ETC.)
?YES

BESIDES COOKING WHICH OF THE FOLLOWING ACTIVITIES SHOULD OCCUR
IN YOUR KITCHEN IF ADEQUATE SPACE WERE PROVIDED.)
1. LAUNDRY
2. SEWING
3. IRONING
4. HOBBIES
5. DESKWORK
6. MIXING DRINKS
7. VISITING
(TYPE THE NUMBERS
?1237

WOULD YOU LIKE TO INCLUDE A FREEZER
1. IN THE COOKING AREA (IF SPACE WERE PROVIDED)?
2. IN ANOTHER SPACE?
3. COMBINED WITH THE REFRIGERATOR?
4. OR IT IS NOT FINANCIALLY WORTHWHILE?
?1

DO YOU WANT THIS FREEZER TO BE:
1. AN UPRIGHT VERTICAL MODEL
2. A HORIZONTAL CHEST TYPE
?2

DO YOU FEEL A SECOND OVEN IS NEEDED AND WOULD BE WORTH ITS COST TO YOU?
?YES

WOULD A DISHWASHER BE WORTH ITS COST TO YOU?
?YES

NUMBER	COMPONENT SPACE		A	B	DIMENSIONS C	D	E
3	KITCH SINK		1.50	2.00	1.50	2.10	3.00

```
I A  I B I C I

I-----I----I-----I   -
ICNTR ISINKI    I   D
I-----I----I-----I   -
.................
.....ACCESS......   E
.................
.................   -
```

| 4 | | RANGE | 2.00 | 2.50 | 2.00 | 2.10 | 2.50 |

```
I A  I BI CI

I-----I----I-----I   -
ICNTR IRNGEI    I   D
I-----I----I-----I   -
.................
.....ACCESS......   E
.................
.................   -
```

| 5 | | FRIG | 2.50 | 0.00 | 2.00 | 2.50 | 0.00 |

```
I A I BI

I----I----I   -
IFRIGICNTRI   C
I----I----I   -
..........
...ACCESS..   D
..........
..........   -
```

| 6 | | OVEN | 2.00 | 2.00 | 2.10 | 3.00 | 0.00 |
| 7 | SECOND | OVEN | 2.00 | 2.00 | 2.10 | 3.00 | 0.00 |

```
I A  I B I

I----*----I   -
IOVENICNTRI   C
I----I----I   -
..........
...ACCESS..   D
..........
..........   -
```

| 8 | KITCHEN | CLOSET | 2.00 | 2.10 | 2.50 | 0.00 | 0.00 |

```
I      A      I

I--------------I   -
I    CLOSET    I   B
I--------------I   -
................
......ACCESS......   C
................
................   -
```

| 9 | | DISHWASHER | 2.00 | 2.10 | 2.50 | 0.00 | 0.00 |

```
I A I
   I----I   -  '
I D.W.I   B
I-----I   -
.......
.ACCESS   C
.......
...... -
```

```
0
A5)
```

| 10 | | FREEZER | 5.00 | 2.20 | 3.00 | 0.00 | 0.00 |

WOULD YOU LIKE TO CHANGE OR ASK A QUESTION ABOUT A SPACE?
?NO

IF YOU'RE TIRED NOW, WE CAN TAKE THIS UP AGAIN LATER.
SHALL WE CONTINUE?
?YES

MAY I ASK SOME QUESTIONS ABOUT THE SLEEPING AREAS?
?YES

DO YOU AND YOUR WIFE PREFER: 1) A DOUBLE BED OR 2) TWIN BEDS?
?1

HOW MANY GUEST BEDS DO YOU WANT?
?3

NUMBER	COMPONENT SPACE		A	B	DIMENSIONS C	D	E
11	MASTER	BED	2.00	5.50	2.00	6.80	2.00
12	MARK	BED	2.00	3.00	2.00	6.50	2.00
13	MIKE	BED	2.00	3.00	2.00	6.50	2.00
14	SUSAN	BED	2.00	3.00	2.00	6.50	2.00
15	BETH	BED	2.00	3.00	2.00	6.50	2.00
16	GUEST	BED	2.00	5.00	2.00	6.50	2.00
17	GUEST	BED	2.00	5.00	2.00	6.50	2.00
18	GUEST	BED	2.00	5.00	2.00	6.50	2.00

```
I A I  B  I C I

....I---------I....   -
....I         I....
....I         I....
....I  BED    I....
....I         I....   D
....I         I....
....I         I....
....I---------I....   -
......ACCESS......   E
.................   -
```

19	ALFRED	DRESSING	6.00	6.00	0.00	0.00	0.00
20	MARY	DRESSING	6.00	6.00	0.00	0.00	0.00
21	MARK	DRESSING	6.00	6.00	0.00	0.00	0.00
22	MIKE	DRESSING	6.00	6.00	0.00	0.00	0.00
23	SUSAN	DRESSING	6.00	6.00	0.00	0.00	0.00
24	BETH	DRESSING	6.00	6.00	0.00	0.00	0.00
25	GUEST	DRESSING	6.00	6.00	0.00	0.00	0.00
26	GUEST	DRESSING	6.00	6.00	0.00	0.00	0.00
27	GUEST	DRESSING	6.00	6.00	0.00	0.00	0.00

```
   I     A     I

       ...    -
   ............
   ............
   ....DRESS....   B
   ............
   ............
       ...   -
```

28	ALFRED	CLOSET	8.00	2.00	3.00	0.00	0.00
29	MARY	CLOSET	8.00	2.00	3.00	0.00	0.00
30	MARK	CLOSET	8.00	2.00	3.00	0.00	0.00
31	MIKE	CLOSET	8.00	2.00	3.00	0.00	0.00
32	SUSAN	CLOSET	8.00	2.00	3.00	0.00	0.00
33	BETH	CLOSET	8.00	2.00	3.00	0.00	0.00
34	GUEST	CLOSET	6.00	2.00	3.00	0.00	0.00
35	GUEST	CLOSET	6.00	2.00	3.00	0.00	0.00
36	GUEST	CLOSET	6.00	2.00	3.00	0.00	0.00

```
   I      A      I

   1---------------1   -
   1    CLOSET     1   B
   1---------------1   -
   ..................
   ......ACCESS......   C
   ..................
   ..................   -
```

WOULD YOU LIKE TO CHANGE OR ASK A QUESTION ABOUT A SPACE?
?NO

DO YOU NEED A WASHER & DRYER IN THE HOUSE?
?YES

IF YOU'RE TIRED NOW, WE CAN TAKE THIS UP AGAIN LATER.
SHALL WE CONTINUE?
?YES

SHALL WE DISCUSS THE BATH AREAS?
?YES

DO YOU AND YOUR SPOUSE WANT TO HAVE SEPARATE WASH BASINS?
?YES

WHAT TYPE OF SINKS WOULD YOU PREFER:
1) SINKS WITHOUT COUNTERS?
2) SINKS WITH COUNTERS ON ONE SIDE ONLY?
3) SINKS WITH COUNTERS ON BOTH SIDES?
?3

DO YOU WANT EACH OF YOUR CHILDREN TO HAVE SEPARATE LAVATORY FACILITIES?
?YES

WHAT KIND OF SINKS DO YOU WANT FOR YOUR CHILDREN:
1) SINKS WITHOUT COUNTER SPACE?
2) SINKS WITH COUNTER SPACE ON ONE SIDE ONLY?
3) SINKS WITH COUNTER SPACE ON BOTH SIDES?
?2

HOW MANY GUEST LAVATORIES DO YOU WANT?
?2

IN A GUEST LAVATORY, WHAT KIND OF SINK DO YOU WANT?
1) THE TYPE WITHOUT COUNTER SPACE?
2) THE TYPE WITH COUNTER SPACE ON ONE SIDE ONLY?
3) THE TYPE WITH COUNTER SPACE ON BOTH SIDES?
?1

DO YOU PREFER:
1) SHOWERS
2) BATHS
3) EITHER OR NO PREFERENCE
?1

NUMBER	COMPONENT	SPACE	DIMENSIONS				
			A	B	C	D	E
40	ALFRED	BATH SINK	0.00	1.60	0.00	1.25	2.33
41	MARY	BATH SINK	0.00	1.60	0.00	1.25	2.33
42	CHILD	BATH SINK	0.00	1.60	0.00	1.25	2.33
43	CHILD	BATH SINK	0.00	1.60	0.00	1.25	2.33
44	GUEST	BATH SINK	0.00	1.60	0.00	1.25	2.33
45	GUEST	BATH SINK	0.00	1.60	0.00	1.25	2.33

```
   I  A  I  B  I  C  I

   1-----1----1-----1    -
   1CNTR 1SINK1     1    D
   1-----1----1-----1    -
   ..................
   ......ACCESS......    E
   ..................
   ..................    -
```

46	MASTER	SHOWER	2.60	2.60	2.30	0.00	0.00

```
   I     A     I

   1-------1   -
   1 SHOWER1   B
   1       1
   1-------1   -
   ..........
   ..ACCESS.   C
   ..........   "
```

47	CHILD	TUB	5.50	2.50	2.30	2.50	0.00
48	CHILD	TUB	5.50	2.50	2.30	2.50	0.00
49	GUEST	TUB	5.50	2.50	2.30	2.50	0.00
50	GUEST	TUB	5.50	2.50	2.30	2.50	0.00

```
   I    A   I

   1-------1   -
   1  TUB  1   B
   1-------1   -
   .........
   ..ACCESS.   C
   .........
   .........   "
```

51	MASTER	TOILET	1.80	2.25	1.50	0.00	0.00
52	CHILD	TOILET	1.80	2.25	1.50	0.00	0.00
53	CHILD	TOILET	1.80	2.25	1.50	0.00	0.00
54	GUEST	TOILET	1.80	2.25	1.50	0.00	0.00
55	GUEST	TOILET	1.80	2.25	1.50	0.00	0.00

```
   I  A  I

   1----1   -
   1TOI 1   B
   1----1   -
   ......
   ACCESS   C
   ......   "
```

WOULD YOU LIKE TO CHANGE OR ASK A QUESTION ABOUT A SPACE?
?NO

IF YOU'RE TIRED NOW, WE CAN TAKE THIS UP AGAIN LATER.
SHALL WE CONTINUE?
?YES

SHALL WE DISCUSS THE LIVING AREAS?
?YES

IN ADDITION TO THE COMPONENTS ALREADY LISTED, THE
HOUSE REQUIRES SOME UNSPECIALIZED MULTIPURPOSE SPACE,
GENERALLY DESCRIBED AS "LIVING". THIS SPACE MAY OCCUR IN
MANY PARTS OF THE HOUSE (EG. KITCHEN-LIVING,FAMILY RM,
REC. RM, LIVING RM, ETC.). SOME OF THE ACTIVITIES IN THIS
CATEGORY ARE:
1) CONVERSATION
2) TV WATCHING, HIFI
3) PLAYING PIANO OR OTHER LARGE INSTRUMENTS
4) KIDS PLAY
5) DANCING
6) PARTIES
DO YOU FEEL YOU NEED ANY OF THESE ACTIVITIES TO OCCUR IN A MORE FORMAL
SPACE WITH A SEPARATE SPACE FOR MORE INFORMAL USE?
?YES

WHICH ACTIVITIES WILL OCCUR IN THE MORE FORMAL SPACE. (TYPE THE NUMBERS.
THE SAME ACTIVITY MAY OCCUR IN OTHER SPACES.)
?1236\632\,2,3,5

HOW MANY PEOPLE DO YOU USUALLY HAVE FOR PARTIES OR DANCING?
?20

WHICH ACTIVITIES OCCUR IN THE INFORMAL AREA?
?2,4

NUMBER	COMPONENT	SPACE	DIMENSIONS				
			A	B	C	D	E
56	FORMAL	LIVING	12.00	21.25	0.00	0.00	0.00

NUMBER	COMPONENT	SPACE	DIMENSIONS				
			A	B	C	D	E
57	FAMILY RELATIONS	LIVING	12.00	25.42	0.00	0.00	0.00

WOULD YOU LIKE A SUMMARY OF THE SPACES CHOSEN?
?YES
```
```

THE FOLLOWING IS A PRELIMINARY ALLOCATION OF COMPONENT SPACES
TO ROOMS:

ROOM 1 AREA= 255.00 SQ.FT.
        FORMAL   LIVING
ROOM 2 AREA= 305.00 SQ.FT.
        FAMILY   LIVING
        RELATIONS
ROOM 3 AREA= 198.96 SQ.FT.
        INFORMAL   DINING
                     KITCH SINK
                     RANGE
                     FRIG
                     OVEN
        SECOND   OVEN
        KITCHEN   CLOSET
                     DISHWASHER
                     FREEZER
ROOM 4 AREA= 140.80 SQ.FT.
        MASTER   BED
        ALFRED   DRESSING
        MARY   DRESSING
        ALFRED   CLOSET
        MARY   CLOSET
ROOM 5 AREA= 100.70 SQ.FT.
        MARK   BED
        MARK   DRESSING
        MARK   CLOSET
ROOM 6 AREA= 100.70 SQ.FT.
        MIKE   BED
        MIKE   DRESSING
        MIKE   CLOSET
ROOM 7 AREA= 100.70 SQ.FT.
        SUSAN   BED
        SUSAN   DRESSING
        SUSAN   CLOSET
ROOM 8 AREA= 100.70 SQ.FT.
        BETH   BED
        BETH   DRESSING
        BETH   CLOSET
ROOM 9 AREA= 173.80 SQ.FT.
        GUEST   BED
        GUEST   BED
        GUEST   DRESSING
        GUEST   DRESSING
        GUEST   CLOSET
        GUEST   CLOSET
ROOM 10 AREA= 113.70 SQ.FT.
        GUEST   BED
        GUEST   DRESSING
        GUEST   CLOSET
ROOM 11 AREA= 18.54 SQ.FT.
        ALFRED   BATH SINK
        MARY   BATH SINK
        MASTER   SHOWER
        MASTER   TOILET
ROOM 12 AREA= 23.53 SQ.FT.
        GUEST   BATH SINK
        GUEST   TUB
        GUEST   TOILET
ROOM 13 AREA= 23.53 SQ.FT.
        GUEST   BATH SINK
        GUEST   TUB
        GUEST   TOILET

LIST OF COMPONENT SPACES

MMMMMMMMMMMMMMMMMMMMMMMMMMMMMMMMMMMMMMMMMMMMMMMMMMMMMMMMMMMM

NUMBER	TYPE		DIMENSIONS			
		A	B	C	D	E
1	INFORMAL DINING	3.50	5.33	5.33	0.00	0.00
2	FORMAL DINING	3.50	5.33	5.33	0.00	0.00
3	KITCH SINK	1.50	2.00	1.50	2.10	3.00
4	RANGE	2.00	2.50	2.00	2.10	2.50
5	FRIG	2.50	0.00	2.00	2.50	0.00
6	OVEN	2.00	2.00	2.10	3.00	0.00
7	SECOND OVEN	2.00	2.00	2.10	3.00	0.00
8	KITCHEN CLOSET	2.00	2.10	2.50	0.00	0.00
9	DISHWASHER	2.00	2.10	2.50	0.00	0.00
10	FREEZER	5.00	2.20	3.00	0.00	0.00
11	MASTER BED	2.00	5.50	2.00	6.80	2.00
12	MARK BED	2.00	3.00	2.00	6.50	2.00
13	MIKE BED	2.00	3.00	2.00	6.50	2.00
14	SUSAN BED	2.00	3.00	2.00	6.50	2.00
15	BETH BED	2.00	3.00	2.00	6.50	2.00
16	GUEST BED	2.00	5.00	2.00	6.50	2.00
17	GUEST BED	2.00	5.00	2.00	6.50	2.00
18	GUEST BED	2.00	5.00	2.00	6.50	2.00
19	ALFRED DRESSING	6.00	6.00	0.00	0.00	0.00
20	MARY DRESSING	6.00	6.00	0.00	0.00	0.00
21	MARK DRESSING	6.00	6.00	0.00	0.00	0.00
22	MIKE DRESSING	6.00	6.00	0.00	0.00	0.00
23	SUSAN DRESSING	6.00	6.00	0.00	0.00	0.00
24	BETH DRESSING	6.00	6.00	0.00	0.00	0.00
25	GUEST DRESSING	6.00	6.00	0.00	0.00	0.00
26	GUEST DRESSING	6.00	6.00	0.00	0.00	0.00
27	GUEST DRESSING	6.00	6.00	0.00	0.00	0.00
28	ALFRED CLOSET	8.00	2.00	3.00	0.00	0.00
29	MARY CLOSET	8.00	2.00	3.00	0.00	0.00
30	MARK CLOSET	8.00	2.00	3.00	0.00	0.00
31	MIKE CLOSET	8.00	2.00	3.00	0.00	0.00
32	SUSAN CLOSET	8.00	2.00	3.00	0.00	0.00
33	BETH CLOSET	8.00	2.00	3.00	0.00	0.00
34	GUEST CLOSET	6.00	2.00	3.00	0.00	0.00
35	GUEST CLOSET	6.00	2.00	3.00	0.00	0.00
36	GUEST CLOSET	6.00	2.00	3.00	0.00	0.00
37	FAMILY WASHER	2.50	2.00	3.00	0.00	0.00
38	DRYER	2.50	2.00	3.00	0.00	0.00
39	PUBLIC ENTRY	6.00	6.00	6.00	0.00	0.00
	RELATIONS					
	VISITORS					
	FAMILY					

ROOM 14 AREA= 17.50 SQ.FT.
        FAMILY   WASHER
                     DRYER
ROOM 15 AREA= 152.11 SQ.FT.
        FORMAL   DINING
ROOM 16 AREA= 72.00 SQ.FT.
        PUBLIC   ENTRY
        RELATIONS
        VISITORS
        FAMILY
LIVING ROOM AREA = 255.00 SQ.FT.
FAMILY ROOM AREA = 305.00 SQ.FT.
DATE: 9-JUN-71   TIME:21:30 28.5

ADJ= 12   3
ADJ= 12   9
ADJ= 16   4
ADJ= 16   4
ADJ= 16   5
ADJ= 16   6
ADJ= 16   7
ADJ= 16   4
ADJ= 16   5
ADJ= 16   6
ADJ= 16   7
ADJ= 16   8
ADJ= 16   8
ADJ= 16   9
ADJ= 16 10

     THE MINIMUM POSSIBLE AREA FOR YOUR HOUSE WITH
THE CURRENT SPACES & ALLOCATION TO ROOMS IS 2457.27

THAT'S ALL I CAN DO FOR YOU NOW. WHEN I SAY GOODBYE, THE COMPUTER
WILL TYPE SOME EXECUTION INFORMATION, THEN TYPE "EXIT", SKIP DOWN 2
LINES, AND TYPE A PERIOD. IF YOU WOULD LIKE TO SEE A PRELIMINARY
FLOORPLAN OF THE ROOMS, PLEASE TYPE THE FOLLOWING COMMAND TO THE
COMPUTER WHEN THAT HAPPENS:
     RUN DSK PENDR
FOLLOWED BY A CARRIAGE RETURN.
GOODBYE, ALFRED

EXECUTION TIME: 26.18 SEC.
TOTAL ELAPSED TIME: 57 MIN. 19.80 SEC.
NO EXECUTION ERRORS DETECTED

EXIT

.TIME
48.80
02:59
KILO-CORE-SEC=2401

40	ALFRED	BATH SINK	0.00	1.60	0.00	1.25	2.33
41	MARY	BATH SINK	0.00	1.60	0.00	1.25	2.33
42	CHILD	BATH SINK	0.00	1.60	0.00	1.25	2.33
43	CHILD	BATH SINK	0.00	1.60	0.00	1.25	2.33
44	GUEST	BATH SINK	0.00	1.60	0.00	1.25	2.33
45	GUEST	BATH SINK	0.00	1.60	0.00	1.25	2.33
46	MASTER	SHOWER	2.60	2.60	2.30	0.00	0.00
47	CHILD	TUB	5.50	2.50	2.30	2.50	0.00
48	CHILD	TUB	5.50	2.50	2.30	2.50	0.00
49	GUEST	TUB	5.50	2.50	2.30	2.50	0.00
50	GUEST	TUB	5.50	2.50	2.30	2.50	0.00
51	MASTER	TOILET	1.80	2.25	1.50	0.00	0.00
52	CHILD	TOILET	1.80	2.25	1.50	0.00	0.00
53	CHILD	TOILET	1.80	2.25	1.50	0.00	0.00
54	GUEST	TOILET	1.80	2.25	1.50	0.00	0.00
55	GUEST	TOILET	1.80	2.25	1.50	0.00	0.00
56	FORMAL	LIVING	12.00	21.25	0.00	0.00	0.00
57	FAMILY	LIVING	12.00	25.42	0.00	0.00	0.00
	RELATIONS						

WHAT IS THE INPUT DATA SET NAME?
?SET11

WOULD YOU LIKE A SUMMARY OF THE PRESENT ROOMS, DIMENSIONS, AND ADJACENCIES?
?NO

WOULD YOU LIKE YOUR FLOORPLAN TO BE DRAWN 1) ON A GRAPHICS TERMINAL, OR 2) ON YOUR TELETYPE
?2

VOICI -

RUN DATE - 10-JUN-71

WOULD YOU LIKE A SUMMARY OF THE ADJUSTED ROOM DIMENSIONS?
?YES

ROOM	TYPE	DIMENSIONS
1	BEDROOM	14.33
**		
2	BATHROOM	10.00
**		
3	BEDROOM	11.00
**		
4	ENTRY	7.00
**		
5	KITCHEN	11.00
**		
6	LIVING RM	20.00
**		
7	DINING RM	15.00
**		

TOTAL AREA= 941.33 SQUARE FEET

EXECUTION TIME: 1.42 SEC.
TOTAL ELAPSED TIME: 4 MIN. 21.85 SEC.
NO EXECUTION ERRORS DETECTED

EXIT

.

IV. FUTURE

The program is unfinished. Perhaps, as would seem appropriate to an evolving
effort, it will never be completed. We feel that we have begun to develop a
system that (can potentially) bring design expertise and user influence to a
level of utilization in the mass housing market unknown in the past. We have
had a number of inquiries from industry concerning our design machine. They
seem quite interested in the possibility of such a system's being available for
general use. With the support of industry we hope to set up further experiments which will see the system tested on real user situations, as a way of
probing its effectiveness in actual marketing conditions. In addition to these
interests, we will be working at expanding the program to enable it to become
more inclusive, capable of rendering complete design services. This will require continued effort at program development. We have also found that, by
having to work within the discipline of computing, we are asking ourselves
fundamental questions concerning the nature of design. What is a criterion
statement, how do you know when you have one, what does it look like, what is
an appropriate process for determining the minimum exterior openings in a
room, etc.. To a very real extent we have discovered new questions in the
process of building our design system, and it has been a good critique of
usual design practice. Perhaps a brief discussion of some of the more interesting future issues we will be dealing with is in order.

First, we are interested in just how a potential buyer makes decisions about
what "style" house to buy. Our experience has led us to conclude that all
buildings have this elusive (?) quality and that, in housing, it is of particular
significance to the user and his image of himself. We have begun to experiment
with the implications of style, and with treating this in our computer. One interesting possibility lies in understanding all of the implications of style.
For example, when one thinks of "colonial," one likely has in mind some vague
notion of a colonial exterior and also some sense of spatial organization.
In our colonial example, a central hall with two rooms off of either side at
both the first and second level could be the "mental map" of "colonial". We
are very much interested in the adjacency models and room sizes that these
images suggest, and would like to be able to generate design based on such
stylistic rule systems. We feel that this might assist our effort in dealing
with users' personal "taste" (needs?). We also have some ideas about building
a computer modelling technique to handle these style questions.

A second issue that will concern us in our continued development relates to
effective cost control. Designers are rather well known for their lack of
ability to deal effectively with the economics of building. What does something

179

actually cost? What economic tradeoffs make the most sense for the user? How can a budget be continuously monitored to avoid costly re-design at later stages? ~~etem.~~ We have spent a great deal of time in trying to develop models for controlling cost. We feel that we have come close to a solution. Currently we employ three economic systems. One advises the user just how much he can spend on a house. Another monitors cost and checks decisions made all along the way against a standardized model for certain cost ranges, as a way of alerting the users to decisions inconsistant with his projected budget. Another program adds up the costs of items in very specific ways to generate an accurate cost estimate. All of these programs are in various states of development and are in need of considerable revision. We hope to pull all of these sketch programs together into a coordinated cost and economic assessment routine.

Thirdly, we hope to get both an improved spatial allocation program operating, and also to link up the output of this to a perspective routine. This would give the user a "picture" of what the implications are of his decisions, enabling him to make modifications if they seem necessary. In addition to these two programs (both of which are operating sketch programs), we need to develop both a construction documents package and a materials listing segment. Both of these programs will likely be the last to get attention; however, they are quite important to the comprehensiveness of the Evolutionary Designer scheme. We feel that it is possible for us to get to this stage of development within two years.

Finally, and most importantly, we need to reorganize the entire program for operation on limited-size machines. The chaining technique discussed earlier seems promising, yet it has not demonstrated its ability to pass data in the right sequence. Tests we ran on the GSP revision indicate that we can expect some major problems in trying to implement this concept. The idea of having such a large operating system is intriguing, and will likely require a number of alternative approaches. For example, we may have to write the whole system in a more simple computer language, and experiment with other kinds of swapping techniques for running programs together. It is likely that little progress can occur in this area without the assistance of more sophisticated computer specialists. In any event, the development of this part of the system is absolutely central to the capability of the concept to work, and we will be giving it top priority in the future.

V. CONCLUSION

It has been asked of us if we feel that a computer based system is consistent with humanizing design; is it really possible to see the use of computing as a liberating mechanism? This is a tough question to answer. In many ways computers are not liberating. The reliance on a special program language and on mechanical processes of problem solving gives one a sense of uneasiness about one's dependencies. At the same time, the demonstrable power and potential capabilities one senses from the work we have engaged so far gives us a kind of faith that such a system will ultimately prove useful. Perhaps one of the most significant insights on computer-aided design comes from the Architecture Machine work at MIT, and which has been coroborrated in our studies. Persons seem to be able to talk more readily to a friendly machine than to another person. We feel that by providing a vehicle for communication between the user and himself, we have made a contribution to the theory of design, and ultimately to constructing a more well-fitting environment.

Jerry Finrow is Associate Professor of Architecture and Director of the Center for Environmental Research at the University of Oregon, Eugene, Oregon.

Section III

<u>RESPONSIVE DWELLINGS AT</u>
<u>HIGHER DENSITIES</u>

"In the savage state every family owns a shelter as good as the best . . . . in modern civilized society not more than one half the families own a shelter . . . . when the farmer has got his house, he may not be the richer but the poorer for it, and it be the house that has got him. . . . our houses are such unwieldy property that we are often imprisoned rather than housed in them . . . "
--Henry David Thoreau

Daria Bolton Fisk

HOMEGROWN
An Evolutionary Housing Game

HOMEGROWN is a game about flexible, user-controlled housing, in which players construct housing for their growing and changing families. Earlier theoretical work on dynamic, participatory housing had been undertaken by a group of us, and had met with only limited success. Our work had been primarily verbal and hypothetical, and we found that while it frequently ellicited enthusiastic response, it was just as often greeted with skepticism, as the real world operation of our ideas lacked definition and credibility, and we were sorely in need of concrete physical examples of our so-called "evolutionary housing."

HOMEGROWN was born out of that dilemma, with the hope of translating the idea of "evolutionary housing" into more readily understandable and realistic terms, beginning to test its feasibility and acceptability, and perhaps elliciting some support for its practical application. The attempt was to put the argument into directly interactive, physical-psycho-social terms, so its implications, problems, and potentials could be exposed, explored, and ellucidated. The hope was especially to engage people with no special architectural expertise or previous exposure to the idea of self-directed and manipulable housing.

The players are addressed, then, in their capacities as home owners, users, and dwellers, and each player assumes the role of the head of a family. The game then provides a system of building parts set up to encourage variety and change, while the stage is set in the context of changing times and a myriad of personal circumstances.

With the hope of engaging players at a level and scale with which they are most likely to be familiar, HOMEGROWN centers with some detail on individual houses and the relations between immediate neighbors. To this end, players build with a scaled-down model system where $\frac{1}{2}$" represents 1 foot, so that details like furniture can be readily included, making the model environments relatively believable and easy to visualize.

Although the building system and components have no exact counterparts in our present-day full scale world, the attempt has been to abstract operational characteristics of actual and possible building systems and housing components, with special emphasis on changeability. The use of considerable abstraction has the further advantage of freeing the players from previous and particular housing associations, which might otherwise tend to cloud the issues to which the game speaks. One danger with a system which is both abstract and highly flexible, is that players may easily be overcome if they are given neither reference points nor limits. Our

attempt, therefore, was to provide relatively high degrees
of choice and variability, but within a system of definate
limits, and at the same time to provide enough detail so that
the game operationally reflects some essential characteristics
of real world environments.

For ease of future transformability, for example, houses
are generally supported by columns and beams, rather than with
load bearing walls. One can remove or replace wall panels, then,
with relative ease, as higher panels are not dependent on lower
ones. At the same time, however, it is possible to forego the
use of columns and beams, building exclusively with panels, and
discovering through experience the advantages and limitations of
either structural system. In both cases members are held together
by a fabric zipper, which wraps the columns and beams and lines
the edges of the panels. This allows members to be easily attached
or detached, and might be seen as similar to a tongue-in-groove
edging system, but without the problems of slippage or the conventional limitations of right angle connections.

Some initial order is introduced by reference to the grid
system for the columns. A nine foot grid was chosen, suggested
by dimensions commonly in use in housing today, and by the convenience of using identical members for columns and beams. This
system provides us with the additional and useful dimension of
12.7 feet, whose grid is found by consistently following the
diagonals of the 9' grid. Given columns and beams of these two
basic dimensions, and a flexible joint to connect them, there is
both a clear frame of reference, and the possibility of a wide
variety of three dimensional configurations.

Complementing the flexibility of the structural system, the
infill panels offer a range of sizes and consequent configurations,
dimensionally coordinated through the Fibonacci series to fit our
square/diagonal system. With the same panels serving as walls,
floors, or rooves, again direct reference to reality is lost,
while interchangeability and economy in the number of parts produced, is gained. The materials used are wood, foggy glass, and
clear glass, exhibiting a series of reactions to light and visual
privacy, from the opaque to the translucent, to the transparent,
or the range of distinctions between a wall and a window.

The effects of some of these components and configurations
can be experienced more particularly during the evaluation period
which follows each building round. Environmental forces then
come into play, as the sun and the wind machines are activated,
and reactions to them are observed, or as the earthquake test
for structural stability is rendered. Players readily discover
for themselves the structural advantages of triangulation, and
the importance of stabilizing infill panels in the absence of
diagonal bracing. And just as readily they discover the interactive relationships between their houses and the sun and the
wind, or with other devices checking for privacy and view by
measuring sight lines, testing for circulation through the use
of flexible model figures, and so forth. In this way some of
the characteristics of the buildings take on added meaning as
their particular consequences are examined and illustrated.

Tests of the housing environments in operation become especially important, as they are used to determine whether any stress exists in the environment. Any player responsible for either internal (within his own household) or external (outside his own household) stress will be charged with either altering the physical environment to elliminate the stress, or with paying a tax. This is even more important when one realizes that the winner of the game is the lowest tax payer, and the game is considered over only when no one is being taxed, or, when all stresses have been elliminated. Thus the incentive to cooperate in HOMEGROWN is relatively high, with the hope that players will learn to create mutually satisfying environments.

As it is, however, not all players are cooperative in HOMEGROWN. In a number of games played to date, in fact, there has been a distinct relation between the amount of cooperation vs. competition and the density of the game. With the board providing a constant site, the density is varied by changing the number of players in the game. So far we have played with as few as two players, and as many as twelve, and our experience to date indicates that at very low densities cooperation is high, largely because there is enough space to avoid conflicts. At slightly increased densities, the game often takes on a competitive complexion, with land scarcer, interdependencies and confrontations greater, yet players still often striving for single family, detached homes, with open space, etc. At very large densities, the game may often become cooperative again, as players quickly recognize that single family, detached dwellings are out of the question, so that collaborative efforts may be established at the outset. Another observable, if unsettling, effect of density appears in players' attitudes to family and household changes. At low densities, with relatively low constraints on expansion, family or household changes are often greeted as opportunities to explore new possibilities, improve existing environments, and so forth. As the density increases, however, and interdependencies mount, changes are often feared as threats to the existing, often hard-won and tenuous, equilibrium, so that new arrivals may be treated as problems to be avoided, not as enriching new inputs. In one such circumstance a player chose to collect insurance money as a consequence of his wife and prospective baby dying during childbirth, in preference to accomodating a live wife and baby, and no financial bonus.

Similar examples of the sort of perversions higher densities and interconnections may cause can be found by reactions to stress. Under low densities, players are often immediately prepared to alter their environments to elliminate stresses and avoid taxes. Under higher densities, however, a more frequent reaction is to try to tax your neighbor if your neighbor decides to tax you. At this middle density, again, one of the most frequent conflicts arises over different attitudes toward public and private. While hinted at under lower or extremely high densities, it is more often in the middle ground that distinctions between the public and the private become an issue. Interestingly

enough, this particular issue is often prompted by the
proximity of families with either large differences in
income or substantial differences in family size and composition. A community of couples and single individuals
has been known to balk at sharing common open space with
a family with three young children, for example. In other
cases players with very low incomes have applied social and
psychological pressure to rich players, in the hope of converting private land to common public "park." In some cases
such suggestions are seen as personal affronts, if not outright invasions, by the landholders. In one such case no
formal tax was levied, but the actual stress was so great
that the parties involved chose to abandon the game rather
than to openly confront the conflict.

Such extremes are hardly the rule, however, and one
happy observation about HOMEGROWN is that it seems to create
friendships more often than not. Particularly heartening
is the fact that despite the diversity of players, the game
has so far illustrated a consistent and growing enthusiasm
in people for making and remaking their own housing environments. Frequent initial hesitancy is replaced time and again
with gradual confidence, experimentation, and then sudden
surprise and delight in the discovery that one can, in fact,
control and even build one's own house successfully.

Even more encouraging is the observation that with no
previous exposure to the game, or particular architectural
expertise, players seem surprisingly adept at making their
own environments. In fact, one game of architectural experts
has indicated that professional knowledge and concern may
actually be a handicap if one's hope is to create environments
to live and discover delight in. It was at least made clear
in that game that the concerns of the professional architect
are in some cases very far from those of the home dweller.
In that particular game, it turned out that the one player
who was unanimously agreed to have created the most enjoyable
environment, and was the only one to avoid being taxed, was
also the one player without a professional affiliation with
architecture. His house seemed so elegantly simple and free
of conflict, in fact, that the others suggested taxing him as
an outsider.

Another interesting observation is that players seem
to fare quite well without being told what or how to build,
and often use the game parts in completely unanticipated ways.
Players quite often prefer large wall panels to combinations
of small ones, for example, despite the fact that the smaller
panels more closely approximate the sizes of housing components presently in use (such as windows, doors, etc.). Similarly, I was surprised to find people more interested in
varying ceiling heights than in departing from the regular
grid for wall placements. Triangles were also in much higher
demand than originally anticipated, as players did not hesitate
to use the diagonals of the grid in combination with the regular
squares. Thus we found frequent departures from the boxes that

we are so used to in our usual housing experience. At the same
time, the variations on our usual right angle schemes seemed to
arise quite naturally, rather than in conscious opposition to
the real world norm.   A number of triangular houses have been
built, for example, and in no case were these considered even
radical enough to warrent comment.   It is also encouraging to
see the ease with which players articulate personal concerns and
biases in their houses, giving them ready physical expression.
One player delighted in the pattern made by her roof panels,
while her neighbor took great pains to make sure that she would
always be able to see the stars from her bed. This same player
remarked that while their families were small, the women had
been mainly concerned with large kitchens with plenty of light
and work space; whereas once they had children, the interest
shifted to the securing of ample and private bedrooms, now seen
as the woman's only refuge from her children. And the list
could go on and on, for one man's meat is another man's . . . ,
and one man's meat at another time is the same man's . . .

        Our hope is to allow and encourage just such things to
be expressed, using tools such as games and building systems
as vehicles through which people may begin to enter into processes going on around them, without fear of their lack of expertise, and with a growing awareness of their own creative
potentials.

Daria Bolton Fisk teaches architecture at the University of

Texas in Austin.

Excerpts from the Oral Presentation

"This was the game played by the professionals, and they

were <u>incompetent</u> at this game. Their concerns had very little

to do with what the places were like to live in. This guy,

for example, just refused on principle to use columns and

beams, so he started to build only with panels... He couldn't

make any opening for his stairway. He had planned an entire

second story which he never built. He couldn't get the

whole thing to stand up properly... Meanwhile, he tried to

convince this guy over here that <u>he</u> shouldn't use columns,

either. This other guy, who was a structural engineer, had

an environment which was falling down, literally. He couldn't

pass the stability test... In this game, the one player who

was unanimously agreed to have been the most successful,

was the only nonprofessional in the game."

The hardware store, with its
supply of building components.

A view of the environment created by the group of building professionals.

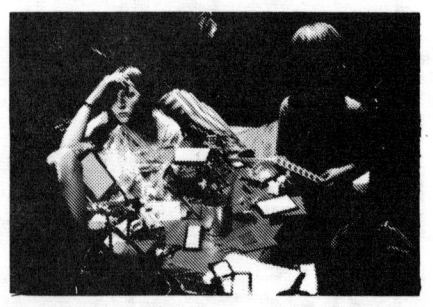

A group of players from a farm commune in Missouri.

"This guy was a mechanical engineer. He built this tri-angular house, which had terrific orientation to the sun and wind, giving him real advantages, but consequently screwing the rest of the community."

"This guy was a structural engineer. After the first round he was completely unconcerned with building a place to live in. He was more interested in exploring the structural potentials of the system. This is a shot of the tower that he started to build -- it was cantilevered and post-tensioned. It was a complete failure."

"This guy was very concerned that any open space should be common space. Meanwhile, this girl had a plan that certain areas would be private courtyards, and her whole house was oriented toward these areas. As it turned out, a lot of tension developed over this one area. Here's where his house opens onto the land where she wanted to have a private courtyard... The girl felt personally affronted because he was posing this use of open space as a kind of duty... As it turned out, both left the game. They were supposed to come back later and finish the game. She gave about sixteen excuses why she couldn't, and finally she said, "I'm just having such a hard time with my neighbor." When I talked to him about it, he said, "It wasn't that I wasn't getting along with someone in the neighborhood.""

"This girl was sharing two walls with two neighbors, which in a normal situation is quite difficult to handle. As it was, every time she planned a change, she'd say to her neighbors, "Well, I'm planning to do this, what do you think about it?" Usually they had no objections, and she went ahead. She generated very few conflicts, where in similar situations, as in the professional game, there were endless arguments."

"It often happens that the people who enter the game with the least money are the most cooperative."

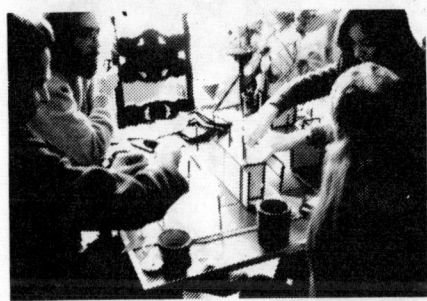

This group finally arrived at a solution of placing their entire community under an inflated roof, simulated by a dry cleaner's bag inflated with a vacuum cleaner.

A player adds a law office to her house.

"This group finally decided that they should put a common roof over the whole thing, and save everybody some money. So somebody went out and found a cleaning bag, and they found a hair dryer to blow it up, and they put a bubble over everything. But then nobody could get in and move around in his individual house. So suddenly all these people who had been so self-concerned were having to look at the whole environment. Some people had never seen what was going on next door to them.... the whole complexion of the game changed... One guy's hypothetical family was expecting twins. Everyone said, 'Gee with this common roof and no walls, there'll be an incredible noise,' so immediately one person turned on a record player, and another a radio, trying to simulate what would happen to their environment. This was immediately annoying to everyone, so they began to try to figure out how to cope with the noise... they planted some trees... They began to be incredibly responsive to community problems. One guy's two kids got very sick, so that he had to hire a live-in nurse, which was an incredible financial burden. The rest of the community agreed to take over his share of community expenses... Another time a card turned up saying that the schools were overcrowded, so that a child could attend for only three hours a day. Then this fellow whose wife had left him offered a spare room of his house as a classroom... I've never seen another game as cooperative."

"There's no question that people react in the construction of their houses to what they see their neighbors doing."

"One of the points of the game was to see if people who had no previous exposure to architecture would be interested in making their own environments."

"This girl, in the beginning, built a huge house, while this other person was building a tiny little efficiency apartment. After she built the whole thing, she began dismantling it, saying 'Well, now that I know what it's like, I'll have no trouble building what I need for the moment.'"

"It was interesting that people were very much exhibiting concerns for their own private houses early in the game, and only after their houses were well-established did they begin to be concerned with public places."

"This woman was always concerned that she should be able to see the stars from her bed."

User-Responsive, Adaptable, Industrialized Housing

Neil Pinney

The study which this paper summarizes* was an exploration towards a possible new base and method for design and a direction whereby industrialized shelter could be made more responsive to the constantly changing immediacy of users' needs. This paper reflects the study, which is in three parts:

I.   Rationale for participation and adaptable housing.
II. Evaluation of existing industrialized housing systems and participatory design methods.
III. Suggestion for a new method and system.

The contents do not focus on issues of technology and do not directly examine innovations in the technology of housing but on use of technology and on design process.

I. RATIONALE

The fulfillment of this country's enormous housing need has been anticipated by recent studies and indicators. It is foreseen that at least fifty percent of this housing will be supplied by industrialized housing corporations (Enzer 1971). It is also projected that a growing portion of the new housing will be provided in the form of multiunits. The reasons given in the projections for this emphasis away from single detached units to multiunit housing is:

1. Limited supply of money. Insurance companies, long a major source of mortgage money, are now seeking equity investments which are by law obtainable in rental housing (usually built as multiunits); or in the land on which condominiums or cooperatives are built. The private money sources see this as an opportunity to share in the rising real estate values.
2. The higher cost of land, which in many instances, especially in urban areas, will cause more intensive use of it.
3. Large scale producers long range plans seem to emphasize the production of this type of housing because they see a rising market for it.
4. Environmental and ecological considerations. Higher dwelling unit densities (usually resulting in multiunits) have the potential for better conservation of natural systems (such as drainage patterns, vegetative cover, etc.) on the unbuilt portion of the land

The Problem

A common public image in this country of multiunit housing is one of repetitive, institutional and unresponsive characteristics. Multiunit housing can be different. If this kind of housing is produced on a large scale and if many families are expected to accept it as a way of living then the image of multiunit housing has to be changed. Most people have a strong need for individuality in their housing environment.

---

* Neil Pinney (project director), Participatory Dwelling Design - Process and Product, M.I.T. School of Architecture and Planning, Community Projects Laboratory, 1972.

Human Needs

It has been suggested, following Carl Jung (1964), that the dwelling
as symbol of the "self" represents the association between deeper levels
of psychological needs and the built environment. A conventional house
and a rigidly static concept of self are mutually supporting. Psychological needs then, emerge into a self concept and are expressed in the
dwelling (Cooper 1971). With tight limitations on housing choice, this
personalization process is now most easily expressed in interior design
and choice of bric-a-brac (Laumann and House 1971). While this is an
important mode of expression, if given wider housing options and the
capability to express oneself in the total dwelling design itself, there
would be a stronger link between the expression of one's psychic needs
and the built environment.

Maslow (1960) conceives of levels of human needs and arranges them in
a hierarchy from lower needs to higher needs:

. Physiological needs,
. Need for belongingness and love,
. Esteem needs,
. Self-actualization needs.

When human needs are more than summarily considered, then the definition
of the problem and the resulting options must be different from traditional
ones. This interpretation of housing stresses the qualitative aspects in
terms of both physical and psychological needs. Individual dwellings with
identifiable, specialized characteristics need not be precluded from a
dense multiunit complex. The potential capability of industrialized housing
production lines to economically produce one of a kind variations of
standard components and the growing interest in users to participate in
the process of planning and design are both resources. If properly coordinated they could bring richness and house identity capability to mass
housing.

But there are major limitations in existing housing systems for participation
and flexibility (see evaluation summaries). Their application to date has
also been, for the most part, unimaginative. These limitations may eventually
prove to be a serious environmental problem. With the cited housing forecasts for the 1970's, we can soon expect to be confronted with many square
miles of industrialized housing. If it continues to be as dull and unresponsive as it now is, people may come to regard it as more environmental
junk, and the next generation of slums may be the result.

With both capability for house identity and easy component replacement,
some of the forces of physical degeneration instrumental in the creation
of slums would be relieved. With the availability of used shelter units
or components, the present filtering process in housing could be maintained with the important difference that the filtering of housing components would not be location-specific, but mobile. This would vary the
age of what some call the housing stock in individual areas which again
would tend to keep those areas from becoming obsolescent.

Housing Space Needs

Housing space needs may be considered in several scales:

. Personal space,
. Social space,
. Dwelling space,
. Neighborhood space or "interstitial space".

At the scale of dwelling space, the space contained by the home is one
of the most important qualities in deciding to move (Rossi 1955).
(Moves to other areas being motivated by economic and other factors.)
Heretofore, moving has been practically the only way to bring space demands resulting from changing family size or changing life styles up to
the reality of available space. Moving destroys friendly neighboring
patterns that usually take a considerable time to develop, and uproots
children from highly valued peer group attachments. This suggests that,
all else being equal, a family's continuity within a neighborhood would
be a desirable circumstance; and that dwelling adaptability is a largely
untried but potentially valuable way for a family to preserve its neighborhood continuity.

At the scale of neighborhood space, different social groups conceptualize and respond to space differently. Where one group will find channels, pathways and open spaces between the dwellings important, and personally identify with this space more than with the privacy of the dwelling, another group will do just the opposite. This attitude toward space is dependent on a number of variables such as personality type, physiological disposition, social roles and class, cultural experience and environmental realities. (Fried and Gleicher 1961, Hartman 1963, Hall 1967, Wilner and Baer 1970).

These attitudes are not static. Lifestyles are more in a state of flux now than they have been since the Second World War due to rising expectations, some upward mobility from lower socio-economic stratas, the woman's liberation movement and alternate lifestyle experiments. Together, these factors indicate the all too temporal nature of descriptive data taken from traditional life-style, stratification, or cultural studies. This indicates the need for very different kinds of neighborhood space for different kinds of people, but yet space that can change as the people themselves, their attitudes and behavior, change. This, in turn, again calls for a high level of personal participation in the determination of these spaces and a high level of component adaptability.

Some participatory design methods that have been tried (see evaluation, Table 3) have clearly shown that users and potential users

1. have definite opinions and thoughts about their dwellings and environment;
2. are capable of suggesting changes and improvements;
3. are willing to communicate these ideas in various ways; and
4. can manipulate, in varying degrees, models, building kits and other materials to design dwellings and neighborhoods. (Mitchel 1968 and Nigro 1971).

The first part of this paper can be roughly summarized by two postulates yet to be tested.

1. Initial participation: If given the opportunity, prospective users will become engaged and participate in the design of their dwellings and shared spaces and facilities in an effective way (as measured by the degree of satisfaction of fit of their designs as compared to others produced by remote professional designers). Or, stated differently, higher satisfaction will occur from the dwellings designed by users with guidance than ones produced by remote professional designers.
2. Capacity to be altered: Once the dwelling has been designed and installed, additional needs will emerge as the original or new user's requirements change through time, and his continued participation would assist in reshaping his dwelling and the shared neighborhood facilities.

II.
EVALUATING
THE HOUSING SYSTEMS

The following guidelines were established for selecting the systems for evaluation:

1. Company produces, is planning to produce, or has produced multiunit housing. (Because of the forecasts of heavy use of multiunits.)
2. Quality of technical information received was adequate to allow evaluation. (This elimated many systems.)
3. Product of a large company of a company with a viable future in housing production. (Because these companies will make a large impact on quality of design.)
4. Or a product that has been designed to facilitate participatory design.
5. System may be representative of other similar systems not included in the evaluation. (This eliminated many systems.)

It was observed that the systems fall conveniently into two major categories: Macrosystems and Microsystems. The classification and terms developed for this evaluation are as follows:

Macrosystem is the system by which the sub-units agglomerage. They are:
I.   Self Supported Boxes (offset bearing)
II.   Self Supported Boxes (aligned bearing)
III. Wall Supported Floor
IV.   Column Supported Floor (single level)
V.   Column Supported Floor (multi level)
VI.   Open Frame Support

Left table — Evaluation Summary Box Systems

* SCORE
5 = very high
4 = high
3 = medium
2 = low
1 = very low
For explanation of scoring see Footnote 3.

SYSTEMS	MACRO-SYSTEM TYPE						MICRO-SYSTEM TYPE				PRIMARY STRUCTURAL MATERIALS				STATE OF DEVELOPMENT			COMP. SCORE *	
	I SELF SUPPORTED BOXES (OFFSET BEARING)	II SELF SUPPORTED BOXES (ALIGNED BEARING)	III WALL SUPPORTED FLOOR	IV COLUMN SUPPORTED FLOOR (SINGLE LEVEL)	V COLUMN SUPPORTED FLOOR (MULTI LEVEL)	VI OPEN FRAME SUPPORT	A FRAMED BOX	B OPEN BOX	C MONOCOQUE BOX	D PANEL	WOOD	METAL	CONCRETE	PLASTIC	CONCEPTUAL	PROTOTYPE	PRODUCTION	EXISTING PARTICIPATION	POTENTIAL PARTICIPATION
1. Behring																		1	4
2. Boise Cascade																		2	4
3. Development International Corp.																		2	4
4. Freuhauf																		1	4
5. Grumman																		2	4
6. Levitt.																		2	4
7. National Homes																		2	5
8. Pemton.																		2	4
9. Rationelt.																		2	4
10. Scholz.																		2	4
11. Stirling Homex																		1	4
12. Townland																		2	5
13. TRW.																		-	5
14. Variel.																		2	3

*Average of total scores from evaluation, Appendix

Evaluation Summary Box Systems

Right table — Evaluation Summary, Panel or Slab Systems

* SCORE:
5 = very high
4 = high
3 = medium
2 = low
1 = very low
For explanation of scoring see Footnote 3.

SYSTEMS:	MACRO-SYSTEM TYPE						MICRO-SYSTEM TYPE				PRIMARY STRUCTURAL MATERIALS				STATE OF DEVELOPMENT			COMP. SCORE *	
	I SELF SUPPORTED BOXES (OFFSET BEARING)	II SELF SUPPORTED BOXES (ALIGNED BEARING)	III WALL SUPPORTED FLOOR	IV COLUMN SUPPORTED FLOOR (SINGLE LEVEL)	V COLUMN SUPPORTED FLOOR (MULTI LEVEL)	VI OPEN FRAME SUPPORT	A FRAMED BOX	B OPEN BOX	C MONOCOQUE BOX	D PANEL	WOOD	METAL	CONCRETE	PLASTIC	CONCEPTUAL	PROTOTYPE	PRODUCTION	EXISTING PARTICIPATION	POTENTIAL PARTICIPATION
1. Balency																		2	4
2. Bison																		2	4
3. Coignet																		2	4
4. Componoform																		2	4
5. Echo																		2	3
6. Igeco																		2	4
7. Jesperson-Kay																		2	4
8. Lift Slab**																		2	4
9. Mitchel																		3	5
10. Nenk**																		-	3
11. Rouse-Wates																		2	4
12. Sepp Firnkas																		2	4
13. Siporex																		2	4
14. Tracoba 1 & 4																		1	4

*Average of total scores from evaluation, Appendix C
**Microsystem is unspecified

Evaluation Summary, Panel or Slab Systems

Microsystems are those unit types that agglomerate into the macrosystems. They are:

A. Framed Box
B. Open Box
C. Monocoque Box (integrally rigid shell)
D. Panel

Other viable systems such as domes, inflatables, foam, and suspension were not evaluated because of the limiting five criteria above.

From the considerations described in the first part of this paper (and other considerations), a set of evaluation categories was developed. They are:

1. Privacy
2. Communality
3. Continuity
4. Accessibility
5. Identity
6. Variety
7. Adaptability
8. Comfort
9. Economy

From this, a set of 22 evaluation variables was developed that was used to rate the systems. The evaluation summaries follow.

Conclusions of the Evaluation

The analysis in this part shows that the housing systems which were rated are generally "low" to "very low" in their present capability to contribute to user satisfaction in housing by participatory means. The potential of the systems to improve is seen as "medium" to "high" with only four out of twenty-eight receiving "very high" potential totals. One explanation of why so few rate "very high" is that none of the systems (except one) were originally designed to facilitate user participation, and that a housing system would almost have had to be designed specifically for user participation in order to receive very high ratings here. The systems, of course, did not have to be evaluated at the level of detail that they were to understand the almost non-existent allowance for user participation.

The existing design and procurement processes are not conducive to the realization of responsive systems. Part of the difficulty is that managers of the present processes perceive their systems' potential for flexibility and adaptability differently than designers or users.

		METHODS					
		1. CHARRETTE	2. FLATWRITER (FRIEDMAN)	3. GAME (MITCHEL)	4. GAME (PANTEK)	5. HOME-A-MINUTE KIT	6. PLANNING & DESIGN WORKBOOK (PRINCETON)
COMPONENTS OF METHODS	INPUT — USER WORKING ALONE		●	●	●		
	TRAINING KIT OR GUIDEBOOK				●		●
	PERSONAL PRE-INTERVIEW & ORIENTATION . . .	●					●
	DESIGN CONSULTANT WORKING WITH USER . . .	●		●			●
	PRINTED PLANS	●		●		●	●
	OUTPUT — MODEL OF DWELLING (physical model) . . .	●				●	●
	MODEL OF NEIGHBORHOOD (physical model) . . .	●					●
	COMPUTER ACCOUNTING & SCHEDULING . . .		●	●			
	COMPUTER OPTIMIZATION		●				
	COMPUTER ON LINE INTERACTION . . .		●	●			
	COMPUTER GRAPHIC DISPLAY		●	●			
	FULL-SCALE ENVIRONMENTAL SIMULATION . . .	●					
HOUSING TYPES	SINGLE FAMILY	●	●	●	●	●	●
	MULTI-FAMILY	●	●	●			
STATE OF DEVELOPMENT	CONCEPTUAL		●		●		
	PROTOTYPE				●		
	IN USE	●		●		●	●
	REALIZATION OF OUTPUT INTO ACTUAL HOUSING .	NO	NO	NO	NO	?	?
COMPARATIVE EFFECTIVENESS SCORE*. (average of total scores from evaluation Appendix E)		3	2	3	2	2	3

* SCORE:
5 = very high
4 = high
3 = medium
2 = low
1 = very low
For explanation of scoring see Footnote 3.

Evaluation Summary Participatory Design Methods

The analysis of the participatory design methods showed generally scores of "medium" to "high". Most of these scores have to be seen as potential, since very few of the methods have been used in the actual realization of housing. The highest scores are reached by those who allow personal pre-orientation and a direct dialogue between design consultant and user during the process of design.

III.
PARTICIPATORY DESIGN METHOD

The remainder of the paper does not suggest another housing system, of which there are already many, but describes a systematic approach to carry out a process and a possible system for describing, visualizing and manipulating dwelling space which could be used with many building systems, both existing and future.

The participatory design method suggested in this section has management as well as design functions and is divided into two levels of operation: Level 1 Metadesign and Level II Contact Design.

Metadesign is the level of general decision making that requires a high degree of competence and knowledge in all dwelling design areas. Contact Design is the level of specific decision making that requires a high level of sensitivity to the needs, requirements, and preferences of the users. Metadesign, with its focus on macrosystems, and Contact Design, with its focus on microsystems are convenient in recognizing and dealing with the difference in the time cycles of change.

Metadesign

Metadesign is basically a planning operation which determines the demographic characteristics and anticipates the needs and preferences of the potential users. It also recommends fundamental control policies affecting decisions on project economics and performance criteria. Since the specific users are most often unknown at this stage, the decisions on criteria can be made using information such as surveys of sample groups with similar attributes to the target user group. Sometimes, however--as can be the case in a local development corporation where the potential users are both shareholders and managers--the users have an opportunity to engage the process at both levels.

197

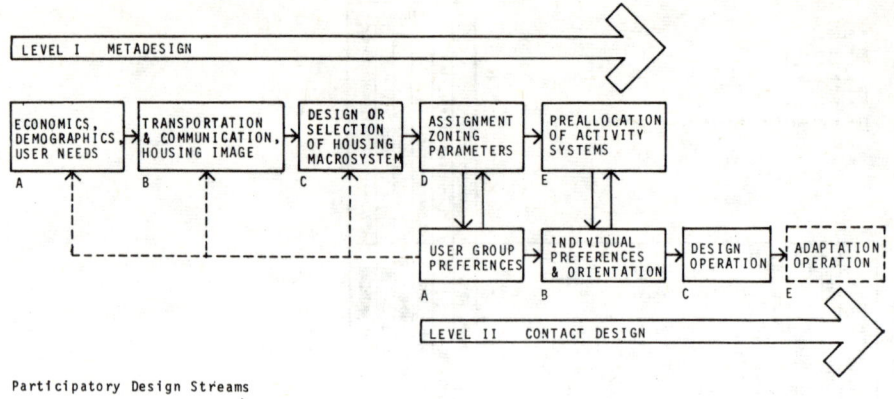

LEVEL I METADESIGN

ECONOMICS, DEMOGRAPHICS USER NEEDS	TRANSPORTATION & COMMUNICATION, HOUSING IMAGE	DESIGN OR SELECTION OF HOUSING MACROSYSTEM	ASSIGNMENT ZONING PARAMETERS	PREALLOCATION OF ACTIVITY SYSTEMS
A	B	C	D	E

USER GROUP PREFERENCES	INDIVIDUAL PREFERENCES & ORIENTATION	DESIGN OPERATION	ADAPTATION OPERATION
A	B	C	E

LEVEL II CONTACT DESIGN

Participatory Design Streams

Contact Design

Contact design is the stage where the actual users naturally appear in the design process and become engaged in the physical design of those spaces and facilities that affect them on an individual basis. Contact design consists of the following steps:

A. Sensitizing Level I to the real user group values and preferences (which may influence decisions on zoning and the design or selection of the housing macrosystem). This information may be gathered through a survey.

B. Mutual orientation between individual users and operators of the method (which may influence decisions on activity system allocation).

C. Design operation with users and operators.

To enrich the user's cognitive and expressive potential, he is introduced to scale models of the housing components which he is encouraged to manipulate, to copy or alter standard plans, or to create his own. Basic stimulus material of this kind, together with assistance from the consulting designer, forms the basis of design actions, the costs and consequences of which are relayed to the user by computer operator and model builder through the consulting designer. As new developments in computer design and visualization methods become available these can be introduced in this process.

D. Adaption operation. This can occur at any future time with individuals or groups of users requiring change. They would again interact with trained representatives of the building management who might be recalled from the previous metadesign group if the adaption operation is complex. This allows for a built-in mechanism which can facilitate change within the housing complex.

Model of building complex or neighbourhood

USER MODELS HOUSE TO SCALE W/ STOCK PARTS

DESIGN CONSULTANT

COMPUTER GRAPHICS

COMPUTER OPERATOR

Contact Design Operation, Methods of Assistance

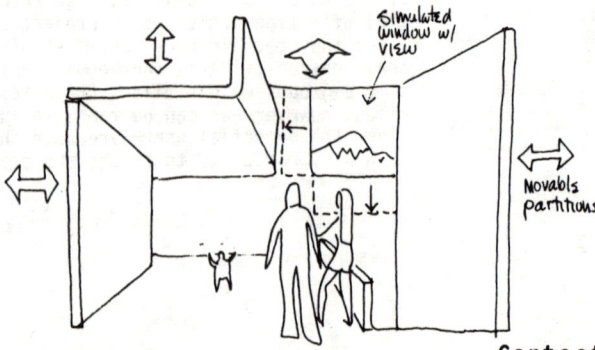

Simulated window w/ VIEW

Movable partitions

Contact Design Operation, Enclosure Simulation

Building Envelope and Zoning System

A building envelope is defined which (i) recognizes differential attributes of the site such as segments with better views or more attractive end areas, and (ii) can allow maximum flexibility of dwelling layouts within it. Zones are determined for each level to accomodate accessibility, dwelling adaptability and community facilities. A typical level has primary zones, for dwelling space use exclusively, and secondary zones for dwelling expansion, access and other uses. A user (with assistance) would decide what area relative to views, access, relationship to basic ground, and distribution of service and activities in which he desires to be located, and how many modules of space he needs. This is his primary zone which he buys or leases and which he can, if he wishes, fill 100% with his dwelling. Surrounding this primary zone are "expansion zones", on no less than one but possibly on four sides of his primary zone, which can be utilized in various ways immediately or in the future. Portions of the expansion and access zones are relegated for both private housing expansion, and semi-private and semi-public interstitial systems and spaces.

An example of zoning rules that might be applied would (working from a basic primary zone) have the frontal expansion zone allow for a maximum of 50% use by each dwelling, 25% of which would be free, and 25% of which must be traded for space in the user's primary zone. (The application of a rule such as this would be intended to retain a three dimensional web of open spaces throughout the complex to insure adequate light and ventilation in the deeper areas of the dwelling as well as usable outside areas.) The lateral expansion zones (at least one for every two dwellings) would allow for 50% of space utilization for private use by each of the neighboring dwellings).

An important function of the expansion zones would be to allow for <u>future</u> adaptability of dwellings. To save part of these zones for this function and to discourage a user from buying or leasing the bare minimum of primary space, if he thinks that he can utilize 100% of the "free" expansion space allowed him; a rule can be applied which would require the user to accomodate a large portion (70% to 90%) of his <u>present</u> space needs exclusively within his primary space. This complex space and cost accounting operation can be facilitated by an interactive computer program. Under the system of zoning described here, these or other dwelling designs would have the capability of expansion (or reduction), and transformation (e.g., change from single family dwelling to cooperative or commune).

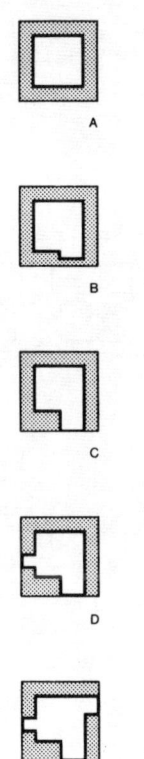

Expansion by Zoning Rules

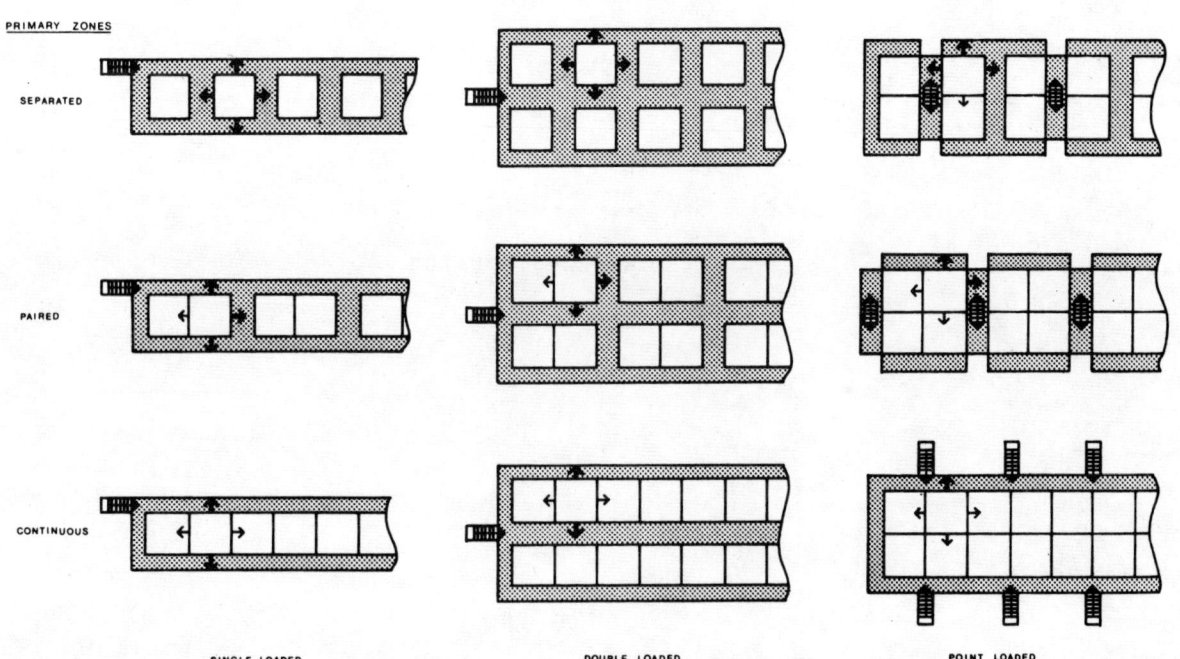

Alternate Zoning Possibilities

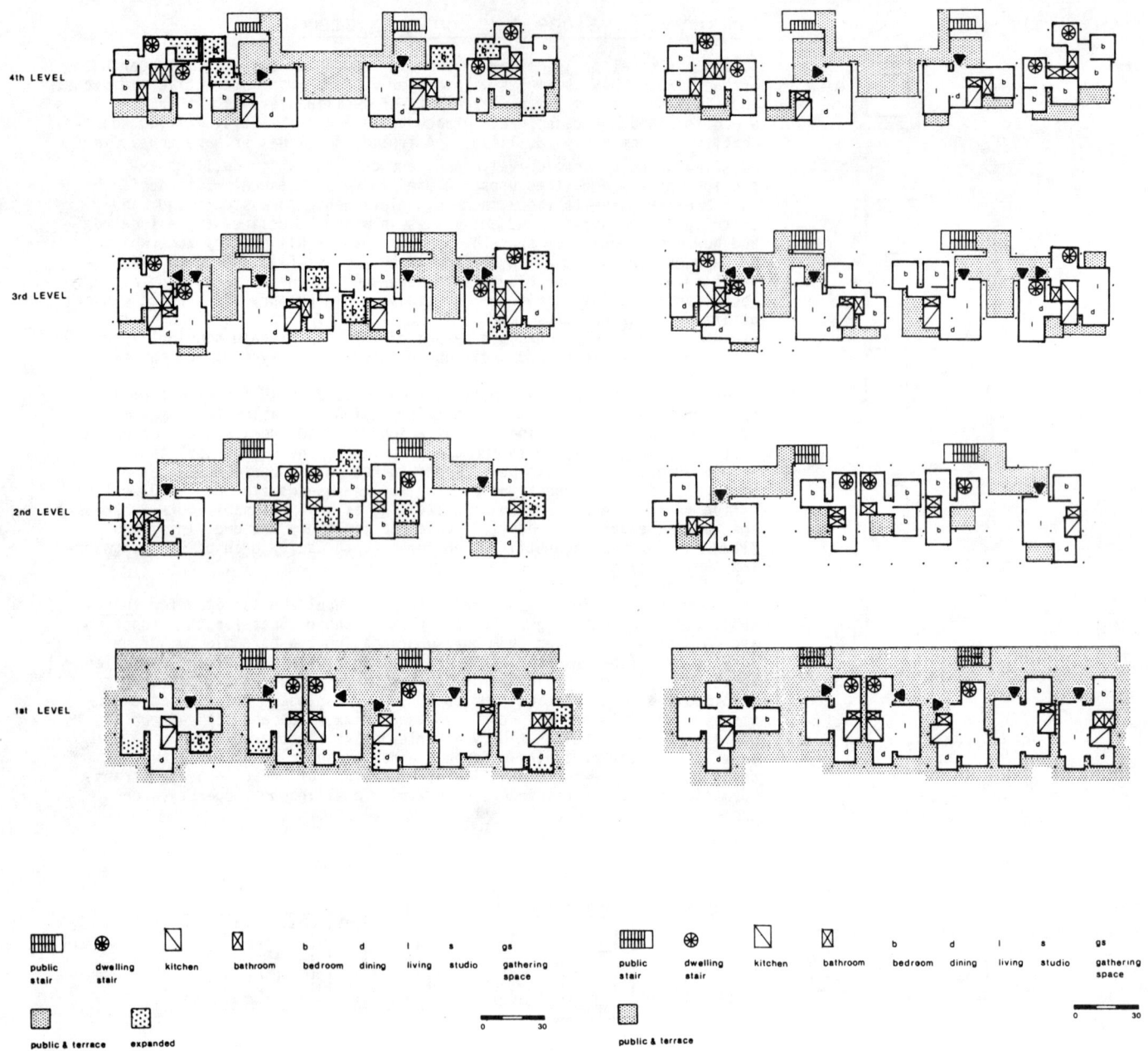

4th LEVEL

3rd LEVEL

2nd LEVEL

1st LEVEL

public stair	dwelling stair	kitchen	bathroom	b	d	l	s	gs
				bedroom	dining	living	studio	gathering space

public & terrace space expanded space

0   30

Expanded Housing

public stair	dwelling stair	kitchen	bathroom	b	d	l	s	gs
				bedroom	dining	living	studio	gathering space

public & terrace space

0   30

Reduced Housing

Since the users are generally unknown to each other previously and arrive on the scene at different times, how can the interests of later users be secured? Noting several potentially difficult problems of group decision making on shared facilities (or conditions which would affect more than one user) might help to answer this question. Some of the problems could be:

1. Visual and acoustic privacy from dwelling to dwelling, especially outside spaces.
2. Determination of shared dwelling entrance courts or play areas.
3. Placement of tot lots, meeting courts, convenience store, etc. Handling of solid waste collection, maintenance, etc.

Each of these examples represents a different class of design problems deriving from problems and opportunities of interaction.

1. Privacy of a new dwelling design from the conspicuous surveillance (or noise) from previously executed designs.
2. Communality desired for a dwelling in which it is anticipated that other subsequent dwellings will join. Or a dwelling which joins in communality with existing ones.
3. Allocation of services and activity areas.

These problem categories may be handled in the following way:

1. For Privacy:

   CONDITIONAL SPECIFICATIONS
   a. Openings on lateral facing sides of dwellings to be conditionally specified as slit type and/or with obscure glass so as to avoid transmission of direct views from dwelling to dwelling.
   b. Openings on "front" to be conditionally side baffled so as not to impose direct views on neighbors. Baffling and fenestration to be adjustable for future varying conditions upon mutual agreement with those neighbors who are affected.

2. For Communality:

   RIGHT OF INITIATION
   Shared neighboring courts to be initiated at the request of one user. Subsequent users may relegate part of their expansion zone space for this purpose or not as they desire.

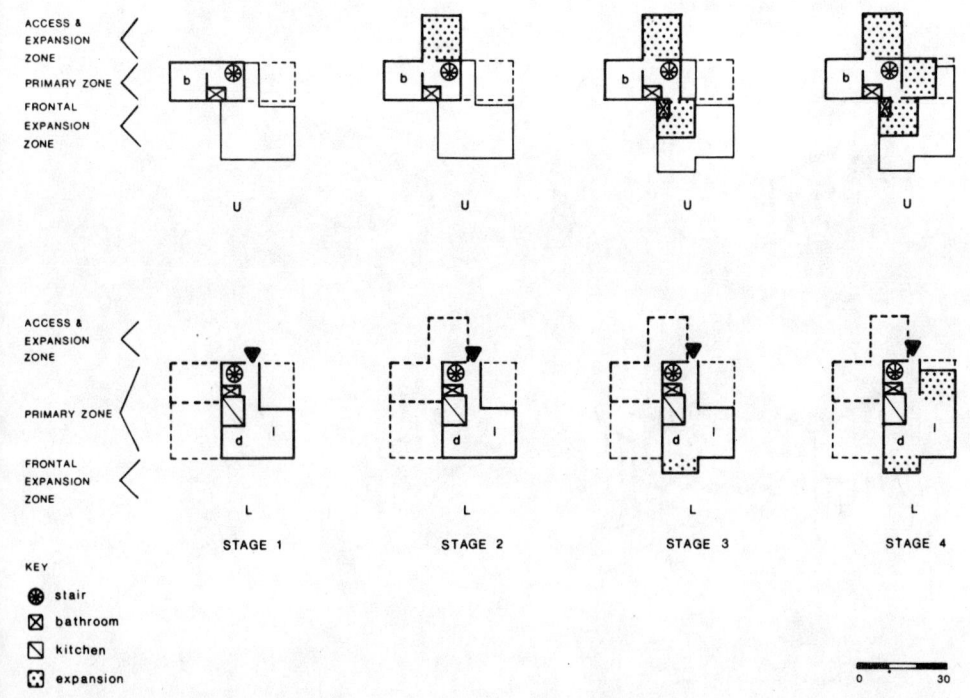

KEY

🞘 stair
⊠ bathroom
◺ kitchen
⊡ expansion

0 — 30

Dwelling Expansion

3. For Allocation:

ALLOCATION MODIFICATION
Initial allocation of service and activity function areas to be determined at metadesign level. A future function, when assigned a
predetermined location, acts as a magnet to attract users with needs
that would be satisfied by the function to choose a space close by.
The building management should have procedures available for allowing
users to influence redistribution of functions at future times after
moving in, so as to adjust to changing needs.

Types of Management Actors

There are five basic types of actors, any of which could assume responsibility of management of a participatory design and delivery operation.
They are:

1. Combination producer-developer
2. Producer only
3. Developer only
4. Independent agency
5. An organized user group like a Community Development Corporation

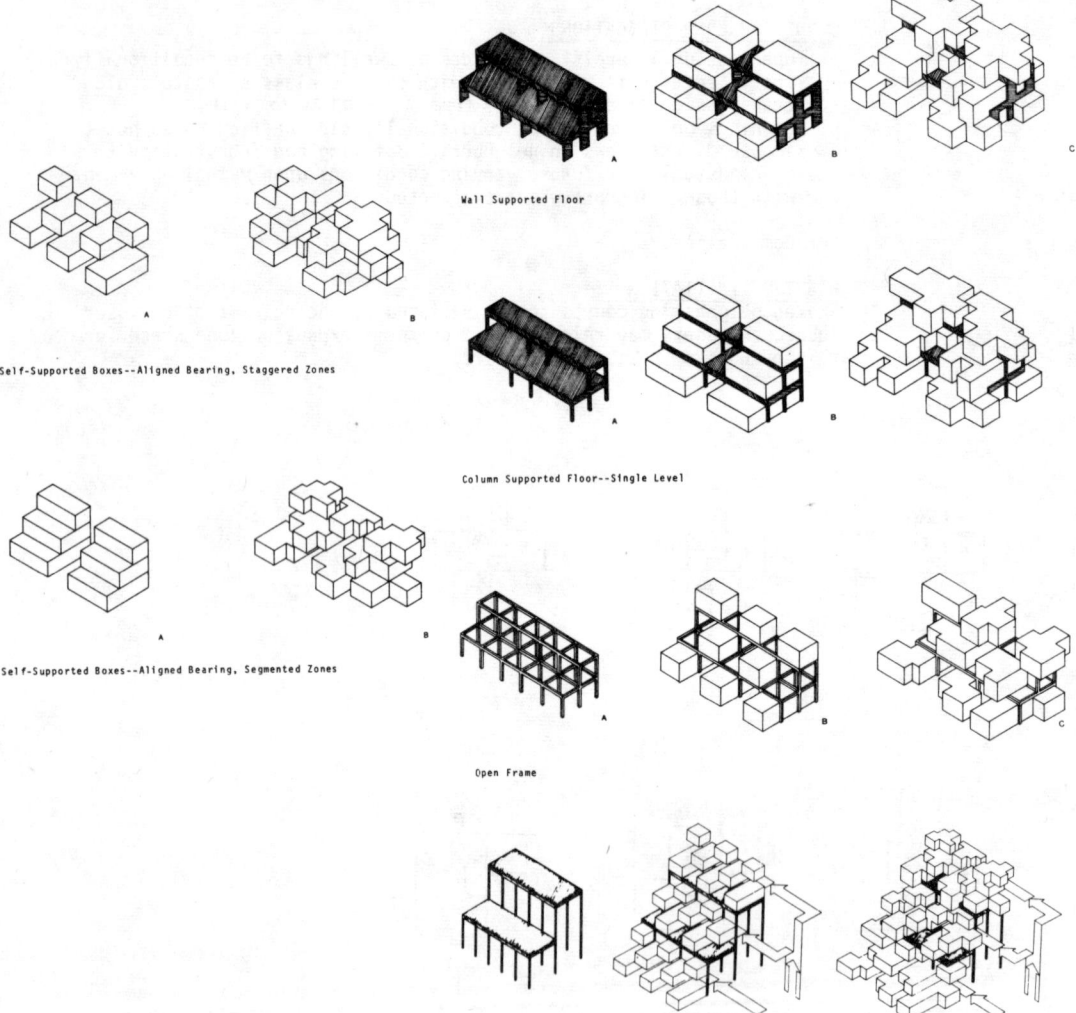

Self-Supported Boxes--Aligned Bearing, Staggered Zones

Self-Supported Boxes--Aligned Bearing, Segmented Zones

Wall Supported Floor

Column Supported Floor--Single Level

Open Frame

Column Supported Floor--Multi Level

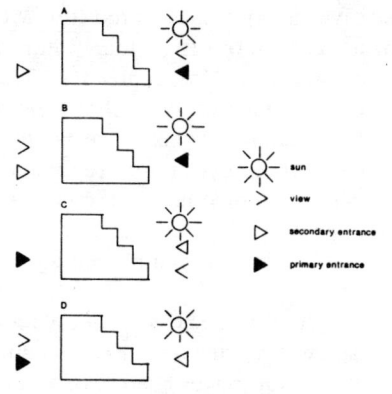

Hypothetical Combinations of Site Conditions

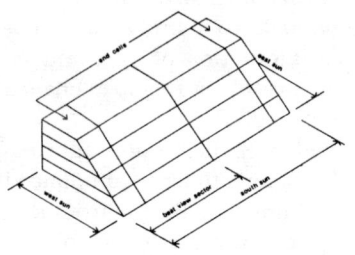

Building Envelope

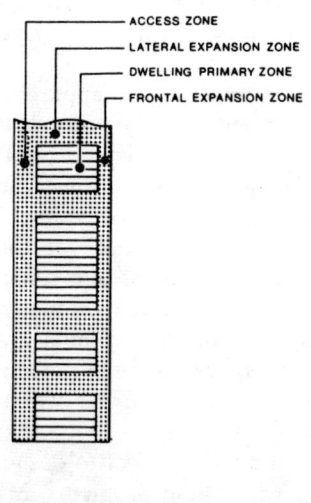

ACCESS ZONE
LATERAL EXPANSION ZONE
DWELLING PRIMARY ZONE
FRONTAL EXPANSION ZONE

Zoning

Conclusion

Regarding adaptability, it is not necessarily intended that in advocating an increased capacity of dwellings to adapt to change should discourage individual areal mobility. Rather, it would offer families and individuals the option to remain in a single location if they so wish. Dwelling adaptability can therefore by thought of as responding not only to changing individual housing needs within a generation, but also to the requirements of shifting housing market conditions over periods of considerable length and in unknown future states of society.

As this is a report on the first or conceptual stage of the study, areas of practical concern that have thus far not been studied and will require future attention should be mentioned. These include (i) legal and economic analysis; (ii) laboratory testing (with real or surrogate users) of the procedures outlined in Part 3; (iii) further design development with attention to structural and construction details, and alternate agglomeration geometries; and (iv) study of procurement management and scheduling.

Since this paper is primarily concerned with how the emerging industrialized housing systems may be made more responsive through the development and application of user participatory methods, illustrative design simulations were simple and conventional so as not to detract from the primary objectives. Hence, orthogonal geometry and standard floor plans were used for purposes of illustration. It cannot be emphasized strongly enough, however, that the participatory method proposed would leave ample potential for including practically any possible element of architectural and technological spectra. Thus, an environment of enormous user-determined richness and individuality could be achieved. The steps involved in that further progress lie ahead of the immediate stage of this study.

REFERENCES

Cooper, Clare. The House as Symbol of Self. Working paper No. 120. Institute of Urban and Regional Development, University of California, Berkeley.

Enzer, Selwyn. Some Prospects for Residential Housing by 1985. The Institute for the Future, 1971.

Fried, Marc, and Gleicher, Peggy. "Some Sources of Residential Satisfaction in an Urban Slum". Journal of the American Institute of Planners, Vol. XXVII, No. 4 (November, 1961).

Hall, Edward T. The Hidden Dimension. Garden City, N.Y.: Doubleday, 1966.

Hartman, Chester W. "Social Values and Housing Orientation". The Journal of Social Issues, April, 1963.

Jung, Carl G. Man and His Symbols. Garden City, N.Y.: Doubleday & Co., 1964.

Laumann, Edward O.; House, James S. Living Room Styles and Social Attributes: The Patterning of Material Artifacts in a Modern Urban Community. University of Michigan, 1971.

Maslow, Abraham. Toward a Psychology of Being. New York: Van Nostrand, 1968.

Mitchel, Neal B. "Design Firm Uses Gaming Simulation for Urban Problem Solving". Professional Engineer (May, 1969).

Nigro, Viviana. "Residential Evaluation of Apartments: An Explorations Study". Unpublished Master of Arts thesis, Cornell University, 1971.

Rossi, Peter H. Why Families Move. Glencoe, Ill.: The Free Press, 1955.

Wilner, Daniel M., and Baer, William G. "Sociocultural Factors in Residential Space". Unpublished paper, University of California at Los Angeles, 1970.

The Warehouse Cooperative School

Jan Wampler

Do you think from an architectural standpoint
that it was helpful to you to participate in the process,
or would you have liked an architect to design the school?

Oh no! That would have been awful! Because then like
you'd come in one day and it's all there and you say "wait
a minute, I don't like this" and then you have to spend
about thirty zillion more dollars to get it all done over
again and this way you sort of know what's going to
happen and why you can't do this and why you can't
do this.

The Warehouse Cooperative School is located on Mt.
Auburn Street in the basement of the Armenian Church
Community Center in Watertown, Massachusetts. The
school was organized in the summer of 1969, that school
year being its first year. It is school for sixty-five people,
ranging in ages from five to seventeen years and represent-
ing a cross-section of the economic incomes of the greater
Boston area. Financially, the school is an economic cooperative, meaning that tuition is based on a percentage of
parental income.

The main room of the school is a sixty-by-ninety-foot
space located in the basement under the gymnasium.
Acoustically the space could not have been worse designed; the ceiling is approximately nine feet high, and
the room is completely enclosed in walls of concrete.
There are no windows for outside light or ventilation.
Many activities happen within this open space all day that
change its definition from a ballfield to a study space to a
classroom. The continual free interaction of ages and interests in this space is the visible result of the open chance
for learning in the Warehouse School.

The Warehouse School had been in session approximately four months when the older students announced
that something must be done to improve the environment
of their school, primarily to give it some self-control and
order and some acoustical treatment.

The general nature of the school, particularly in the
main room, was chaotic. Structurally the room contained
a six-by-three-foot space for each student and approximately six to eight work areas located in the center designated for carpentry, painting, pottery, science, and free
use.

Three negative qualities were immediately apparent.
The atmosphere of freedom, combined with the acoustical
problems of the space, made the school a noisy one, and
adding furniture and wall hangings did not reduce the
sound problem. The freedom also allowed the younger
students continually to invade the work spaces of the
older students, disrupting the learning process and sometimes destroying projects in inevitable accidents. Finally,
the layout of the individual spaces did not stimulate a
great deal of development. A few spaces became person-
alized, but most students knew that the disorder would
destroy whatever they created.

So I was asked as a parent and an architect to design
a space for the school to help with these problems. My
concern was twofold: to preserve or conserve the spontaneous quality and to design a school that would reflect
as many of the students' ideas as possible as well as the
parents' and the staff's, so that the actual design process
would include these people and not be the typical architect's idea of what a school should be.

From PROGRAMS OF PROMISE: ART IN THE SCHOOLS, edited by Al Hur-
witz, ©1972 by Harcourt Brace Jovanovich, Inc., and reprinted
with their permission.

Jan Wampler is an Assistant Professor of Architecture at M.I.T.

The project took six weeks from beginning to completion, on a budget of approximately one thousand dollars. I met with the students, parents, and staff usually two times each day, at lunchtime and after school, to discuss what they wanted in their room. We went on field trips to the M.I.T. architectural studios, where the students there have built their own environment within an already existing one, and to a used-lumber yard outside Boston, where we found inexpensive materials to use in the school.

We then started a series of design meetings. I drew a sketch and the students and the staff and whoever was around would make their corrections on the sketch with their ideas. The initial advice that I gave, only verbally, was to think of the room not as a dismal environment but as a potential reflection of the school, to think of the school as a village, and to think of the idea of frameworks in which the students could participate.

The first sketches simply indicated the basic idea that a village or small city has streets, parks, stores, public buildings, houses, neighborhoods, and even industries. We looked at these drawings and saw what the floor plan of the school had in common with the plan of a city, and we talked about the city problems that we wanted to avoid in the school. The sketches then incorporated these growing ideas.

A second sketch indicated the various neighborhoods, one each for the younger students, middle students, and older students. There also were four central areas designated for art, ceramics, and woodshop use. I asked the students if we should have something in the school similar to a city hall, if in fact we needed a city hall or wanted it, what was good about a city hall and what was bad. They decided that they did want a city hall and that it should contain the public telephone that they all use during the day, the refrigerator that keeps their lunches cold, a cubbyhole device for each student's mail, a mirror, a clock, and a place for making coffee. One boy wanted a place to hide, which I thought was a great idea of what a city hall should be. So their city hall was then drawn somewhat in the center of the school and surrounded by a green area that was called the park.

A later sketch pretty much firmed up the basic ideas of having the city hall and park in the center, with an animal farm at one end and a lounge or tree area at the other end. There were now two neighborhoods, for the young children and the older students, and each had a community center, small central areas in which people could eat lunch or have a discussion and some privacy. The public areas had increased from four to six. But in the process of the planning, the middle students had lost their neighborhood space. So they sent a representative to one of our luncheon meetings who said that he did not think that they were getting a fair shake. We then redesigned the whole scheme to include the middle neighborhood.

The younger students also had objections at this time. We had decided earlier that the younger students

Could you explain the process, what you saw?

Oh yeh! Well first you came in with a little piece of paper and you had a little drawing on it. And everyone started handing out ideas. You put the paper down in front of 'em and we all scribbled so you brought in a whole big roll of paper. Just doing and doing and doing and we just redesigned the room like many many times and figured out what we could do and from all kinds of standpoints.

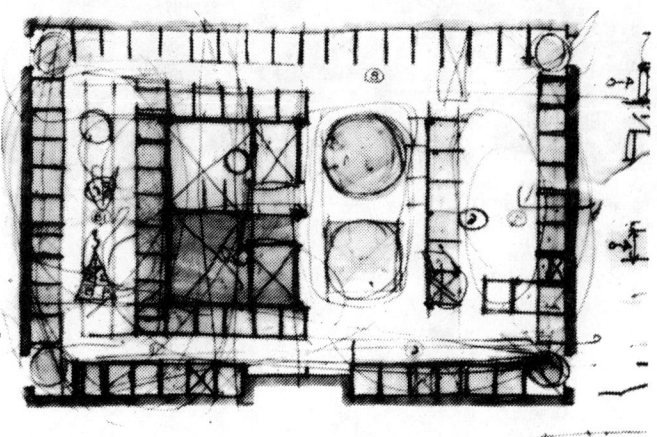

A working sketch of the layout for the Warehouse School.

would have three-foot spaces, the middle students four-foot spaces, and the older students, five-foot-spaces. The younger students objected, saying that they used as much space as the older students. The scheme again had to be changed, with every student having a four-foot space, which in fact turned out to be a definite improvement because the younger students do need just as much space as the older ones.

This process took about two weeks and produced a plan that accommodated almost everyone's ideas, the parents', the staff's, and the students'. The plan was then translated into a model, again by the students, the parents, and some of the staff. On the first evening, about ten people met in my office to build the model. My office holds about two people comfortably. During the evening, they managed to cut up all the Bristol board and their fingers, destroy other models in the office, and accomplish absolutely nothing. But we all had a good time.

The next evening was more productive. In about a week, the students built the model reflecting the idea of their last sketch. It showed the floor painted different colors to indicate different uses—brown for the streets, yellow for the community areas, green for the parks, and other bright colors for the public working areas.

The small students were to have a double space in their areas that anyone shorter than four feet could use because that was all the head room that existed there. The city hall was designed, and the community centers became small circles four feet in diameter with benches inside where anyone could eat lunch or have a meeting.

In addition to the ten people who were involved in building the model, everyone in the school participated by drawing a design they wanted to see on the floor and the walls of their house, in terms of either painting or interior decorating. They were asked to think of a well as their outside window. These designs were not to be the final schemes but were only to identify the spaces on the model and to realize visually the school's potential.

The sketches ranged from abstract paintings to detailed location plans of furniture, chairs, coatracks, benches, bookshelves, plants, weaving machines, a little bit of everything. Small students drew chairs, sometimes out of scale. Some drew American flags or great abstract personal messages. And the peace symbol was always a theme.

These sketches were then cut out and glued into the framework of the space, so that when the model finally was brought to the school, instead of seeing his name on a space, a student saw his own design. Then he could start moving his space to the neighborhood he liked. The space, delimited by four-foot high partitions, was four feet wide and five feet deep. It could be moved anywhere in the school as long as the former occupant of the area agreed to move somewhere else. In the corner, communes could be built, so in some cases four people utilized their spaces collectively.

What about building the model?

It was great because really I didn't realize what an architect had to go through—oy!

And then we decided to build a model to see what it would look like on a little scale so we just kept traipsing over, you know, to work on the model and then over Christmas vacation we just got a mess of wood and stuff and what we needed and everyone chipped in and built and painted and everything.

A working model.

After the model was built, and the students had decided on the location of their property and planned their neighborhoods, we started the monstrous building program. It took place over the two-week Christmas vacation, again involving a lot of the students, parents, and staff. The thousand-dollar budget purchased principally the half-inch plyscore framework items and the floor paint.

In about a week and a half, the framework was built. Then came the horrendous job of cleaning the floor to make it ready for painting. I practiced the principle of using available material and eventually found the yellow paint used by the Public Works Department to paint street lines at a cost of about two dollars a gallon. During this stage, the model was the only piece of architectural information, as most of the students knew exactly what they were doing, so it was not a matter of the architect telling them what to do. They understood it as well as anyone else.

During the last few days just before school started, people were up all night painting the floor. At two o'clock on the first day of school, the room was open and the students first saw how the room's character had changed. Someone said that when they let the students into the room, running around in their new spaces, it was like a thousand Christmases. The exciting thing was to see how this framework that I had worried might be too restrictive and destructive of all the beautiful chaos actually promoted more confusion in a very short time.

The students immediately began finding lumber and materials to build their own houses within their spaces. Small students quickly found that by putting wood on top of the four-foot-high partitions they could build a second level. Other students found that they could construct a second-level pedestrian way by interconnecting all the small spaces. Students began to paint their spaces. Some designs were elaborate. Some were close reproductions of the earlier sketches. Others were spontaneous. Chairs and other furniture were found and moved in and, in a very short time, each space had become a house.

Everyone in the school had a space, including the director and the staff.

Students also brought in photographs of their pets and their heroes, which they hung on their walls. Graffiti started to appear. The exciting thing was to see how the spaces changed each day; every day a new mood or a new idea would appear.

One student decided that he wanted nothing to do with the framework idea and built his own thing inside his space. He nailed together a great wooden tower. Each day it progressed. One day a set of steps appeared. Then, finally, he painted the whole thing black and placed a chair at the top, his sort of throne. Another student enclosed his space entirely and put a door on the front and a light inside, so that he could have complete privacy.

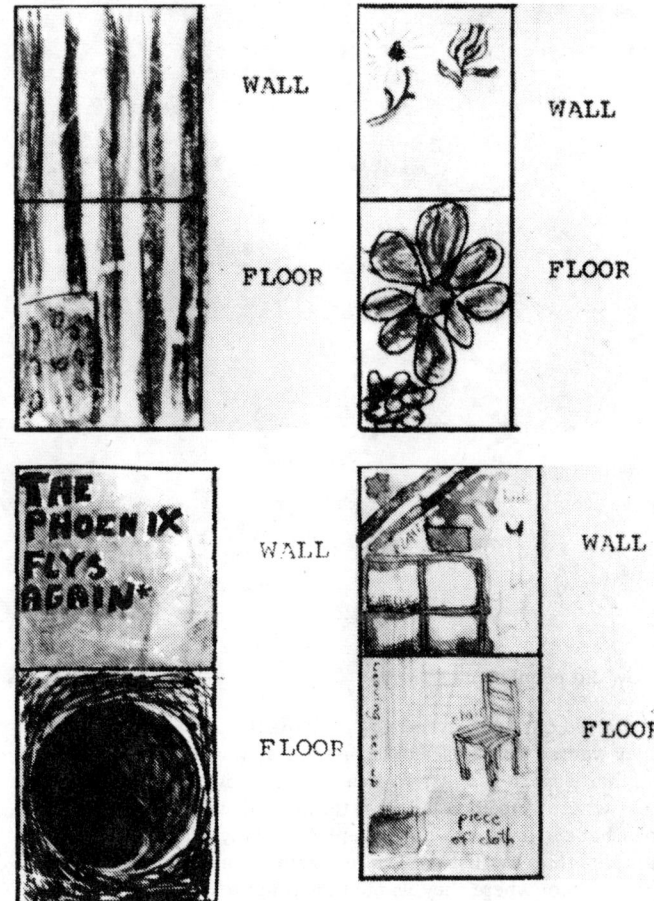

Children's designs for their individual spaces.

Quotations from Jan Wampler's oral presentation:

"These students protested--they said unless their environment was
improved, they'd quit the school.

"In six weeks, with about $800, and the cooperation of the kids,
the parents, and the teachers, we built a school.

"We decided to make the school a village, with a house for each
kid, streets, parks . . .

"One day a kid came in dressed up in a white shirt, and said,
'I've been elected by the middle-sized kids. We're not getting
a fair shake.'

"It didn't take long for the kids to realize that some spaces
were better than others, just like house lots . . . this one was
worth about one candy bar, while this other one was worth about
five candy bars. The kids started trying to trade spaces back
and forth . . . They couldn't work this out, so they decided to
elect a city council . . . They elected the council at about
two o'clock in the afternoon, and fired them about five. They discovered that the city council was getting all the candy bars."

How did people pick spaces?

It started out when everyone drew a picture of what
they wanted their space to be and it didn't have to
be that way. It was just an idea so we'd know whose
was whose. And then we went through the whole process
of what age level would be where and finally, when we
decided that, we just put 'em in together and later we
asked them where they wanted them approximately, and
then we said "if you don't want it there you can trade
with someone in another age bracket," and it worked
out that way.

How did it work when you traded with somebody?
Were there hassles?"

No. We had a few kids do it—like Debbie. She's done it
quite a few times. Like she got hers all fixed up and all
painted and all splashed all over everyplace and then
she decided that she didn't want to be there and so
she'd go and talk to somebody and with a little
persuasion and pressure she finally got them to change
with her, so then she changed and she'd redo her space
and splash everything all over again and then she'd decide
she didn't like it then, so she'd move into another place
and it just kept changing. But there wasn't much hassle.

Did the spaces create too much privacy—did it separate
you as a group or was this desired?

Mine's really good because it's in the corner and I can get
away and they can't see me and they can't find out
where I am.

Student spaces in the Warehouse School.

OR DON'T DO YOUR OWN THING

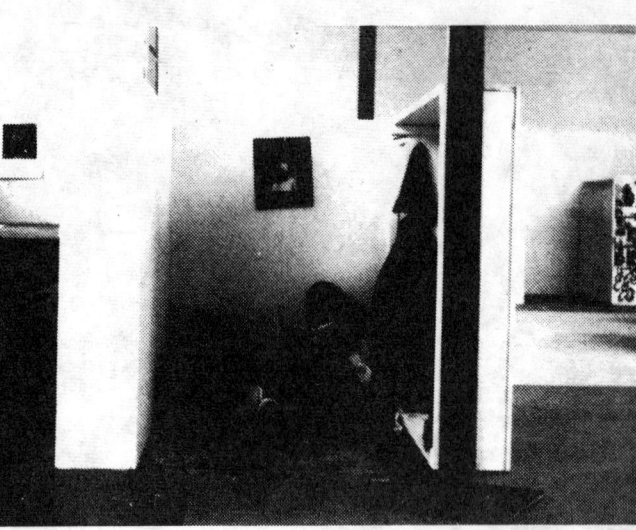

More student spaces in the Warehouse School

Students also named streets and numbered their houses so that they had addresses, such as 6 Country Terrace, Stillwater, Warehouse, Warehouse being the name of the city and Stillwater the name of the neighborhood.

Students are still changing their spaces, building in them, and using them as a house, a place to study, a place to sit. The Warehouse School is a changing collage of personal expression. The colors and order change within the necessary framework, reflecting the school as an interaction of students and teachers in a learning situation.

Obviously the Warehouse School is not going to change the architecture of educational institutions on a grand scale, but it has helped sixty-five students enjoy their school more than they did before.

What is perhaps most important, this project has helped to redefine art. More than a painting or a piece of sculpture, art is a personal and relevant extension of ourselves into the environment around us, our personalities and ideas made visual. All people should be involved in the making of their environment and changing it as they change. It is essential that we participate directly in the decision process so that it answers our needs.

This article has described one way of changing the environment in a specific situation. It need not be on this scale or in this form. It can begin with the design and construction of a small room or an outdoor space or, in the case of schools, the classrooms that generally are drab, sterile, and uninviting for learning. Instead of a conventional art course, students could work together with educators, study and correct their immediate environments, and improve the classroom, the corridors, and the space outside the school. Perhaps these small answers of positive action to realistic problems are more important and educational than all the grand plans for art education.

What do you think of the way it turned out?

We have improved its terribleness.

It's given those who have space more of a sense of possession. Now they have a space that's theirs and they can completely do what they want with it, whereas before it was just this great big open space.

It's restricted the little kids from running completely all over everywhere, over the big kids and jumping on them and all that stuff. (*Little kid:* "I still run." *Older kid:* "Yeh, but we don't notice it as much.")

END

Bertil Olsson
Rolf Nilsson

Flexible Apartment Units A Study of an Experimental Housing Project in Uppsala, Sweden*

English Summary**

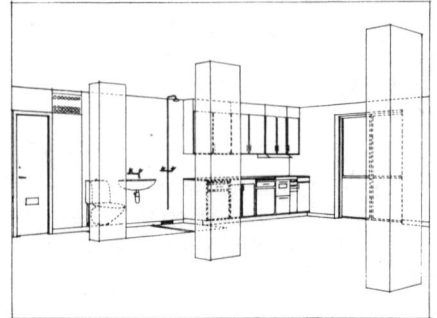

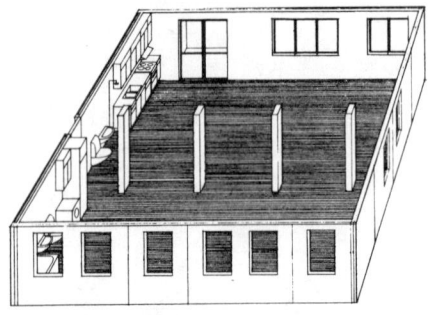

Two diagrams of the fixed construction of a typical apartment.

The Department of Building Function Study at the Lund Institute of Technology conducted this survey of tenants in a block of flats in Uppsala which had been specially outfitted by the building contractor with flexible partition walls and accessories.

In the 1960's Swedish consumers demanded housing units designed to comfortably accomodate a wide range of families with differing life styles, ages, and spatial requirements. Contemporary economic pressures, however, forced the industrialized housing industry to concentrate on a few standardized prototypes for the highly capitalized mass production processes. In search of a building system that would allow continuous production without modification for three years, the building firm of Ohlsson and Skarne abandoned attempts to provide a wide range of unit types, and instead selected a design which utilized the necessary standardized shell and interior components, but which allowed flexible adaptation of the interior space by the tenant himself, after moving into the completed unit.

The experimental apartment building consists of 16 one, two, and three bedroom units spanning the depth of the structure. A service wall between units contains the ventilation, water supply, and drainage systems for the kitchen, bathroom and laundry. The facade windows and entrance stairway occur in fixed positions and there are two to four fixed columns evenly spaced along the centerline between the exterior walls. All other interior components are designed to be movable by the tenants: wood stud and gypsum board room-height partitions in two and four foot widths; doorframe units in four foot widths; room

———— *Translation by Robert Allen Karasek

**Project Bb393:2 Statens Råd För Byggnadsforskning.
Distribution: Svensk Byggtjänst, Box 1403, 11184 Stockholm, 168 pages.

height closet units two feet square. A movable electrical service is attached to each door and wall element.

Tenants residing in this experimental apartment block were interviewed for this study in 1968, two years after they had moved in. Most tenants had taken advantage of the flexibility within that period and were generally positive about their apartment environment and the opportunities it provided.

The tenants ranged in age from twenty-five to fifty-five, with the majority between twenty-six and thirty-five years of age (oldest member of household). Eleven of the sixteen households had children, most of them between the ages of one and six years. In choosing an apartment, ten of the sixteen families had had several alternatives. The other six apparently had not seemed interested in an alternative apartment. Several households had actively sought out, or had been specially invited by the housing agency because of the "excitement" of participating in an experiment. Previous residence was an apartment for twelve of the sixteen families.

The rents seem comparable to other new apartments in the area where the experiment was conducted. A service man who took prospective tenants on an introductory tour commented:

> "Many were enthusiastic at first sight, and did not think
> the rents were too high.... All of the older people who had
> previously been living in less expensive apartments said
> "no" directly. Young tenants with decent incomes snapped
> them up. There was no difference between families with
> and without children."

The experimental project generated interest among the building industry and the general public and encouraged development of a "friendly community" among tenants.

Between houses with pre-school-age children, there was a spontaneous interest in informal contacts between families. Eleven of the sixteen families stated that they saw their neighbors socially at least once a week. They helped each other with care of the children and sometimes assisted with modification of the apartment components. Contact between households also occurred to discuss changes in the apartments and to look at the results. Half of the tenants felt that it was in this way that neighbors influenced or were influenced by each other.

I.
General Modifications

Residents changed their apartment plans in stages as they encountered shortages of space and realized the possibilities of the flexible system:

> "At first it was complicated to plan major changes because we really didn't what we wanted. We made sketches, but the whole time we felt restricted to the previously determined layout of the apartment."

Ultimately, a considerable number of modifications were made, and these led to a general reduction in the total number of separate rooms. Some young children's bedrooms were joined together. All small and deep rooms were combined, walls were removed that separated loosely defined functions (but not bedrooms or baths), and work places were moved from bedroom areas to larger spaces.

In a number of cases tenants made modifications that conflicted with the Swedish national housing regulations for "good housing:" several had bedrooms smaller than the minimum standard, and one had a living room without direct daylight. The basic function of a room was changed on six occasions. In no cases were bathrooms, laundries, or shower rooms significantly modified because of the permanent location of the plumbing.

The appendix of the Swedish report includes a drawing of each stage in apartment design for each of the sixteen apartments. the reasons cited for the modification are as numerous as the plans themselves. Like frames of a moving picture, these plans trace the families' utilization of space over time. Each pattern of development is quite different and forms a physical expression of that family's internal dynamic processes.

The following changes were made in the sixteen apartments over a two year period.

1. New living rooms were added on three occasions.

2. Living room walls were positioned to create new space on seven occasions.

3. Walls were moved to change room proportions on two occasions.

4. Walls separating corridors and living rooms were removed on three occasions.

5. Walls separating closets and the living room were removed on one occasion.

6. A cabinet or a wall was used to create a room partition on two occasions.

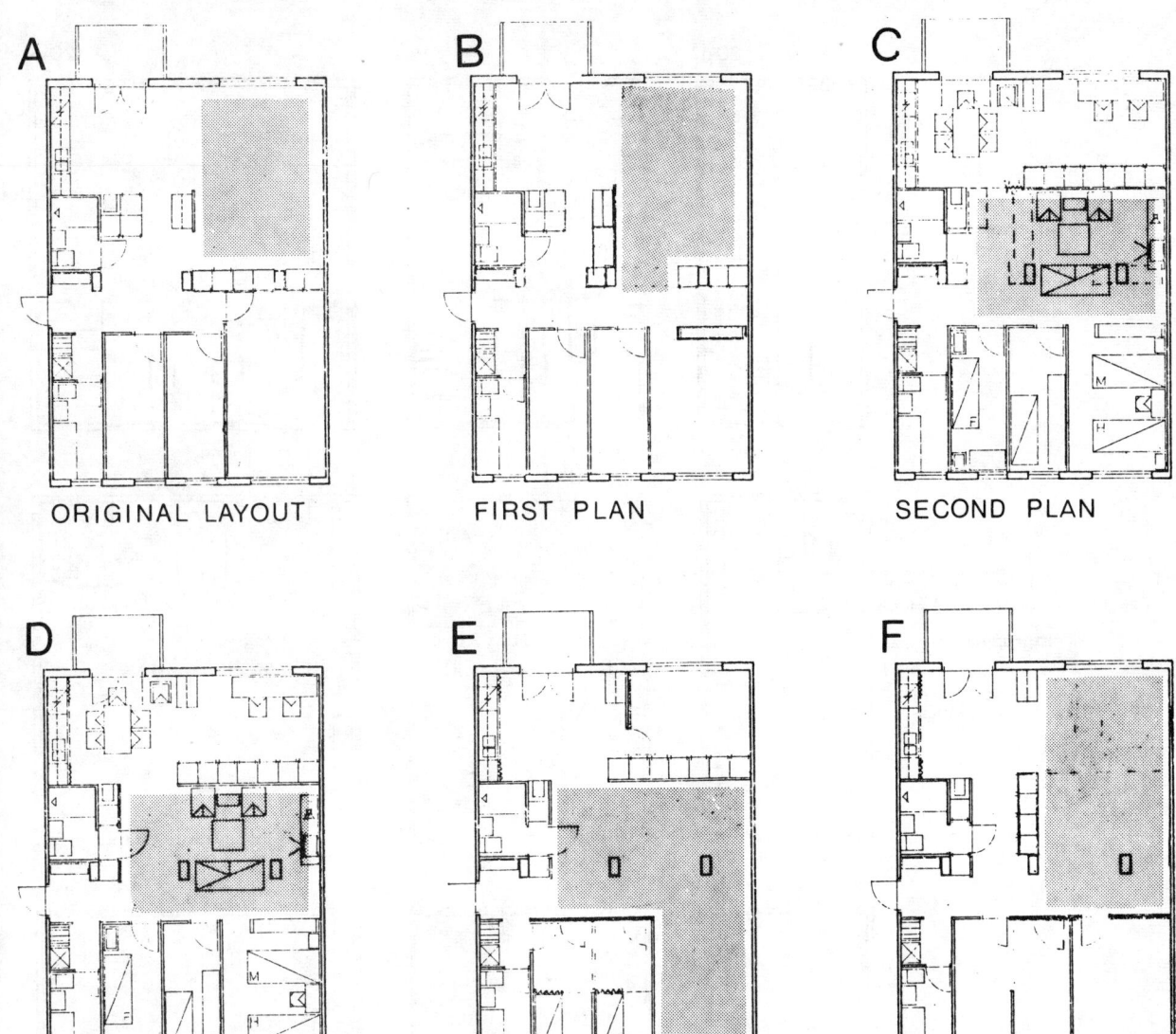

A	B	C
ORIGINAL LAYOUT	FIRST PLAN	SECOND PLAN

D	E	F
THIRD PLAN	FUTURE PLAN 1	FUTURE PLAN 2

Six versions of the same apartment.
The living room in each one is shaded
on the drawing.

7. Walls were set up without changing living room boundaries
on seven occasions.

8. A wall or a cabinet separating two rooms was removed on
two occasions.

A total of thirty-two room changes were made; including other
substantial modifications, a total of forty-two apartment plans
were generated for the sixteen apartment units.*

―――――*Three of the sixteen families had reserved their apartments
before the building was completed and the building company agreed
to arrange their apartments in accordance with those families' desires.

60 m² 90 m² 120 m²

sovplatser

sittgrupper

arbetsplatser

matplatser

Furniture groupings tried in
apartments of three different
sizes. "Sovplatser" = bed
locations; "sittgrupper" =
living room arrangements;
"arbetsplatser" = work desk
locations; "matplatser" =
dining furniture locations.

II.
Building Procedures
and Problems

Tenants desired a storage area for walls in the apartment, so that with the help of the landlord it would be easy to make new apartment arrangements. The building superintendent was a construction company employee and made suggestions to tenants about use of the system. He was not available as a full time "handy man" assistant, however.

1. "We were counting on moving (the walls) with help from a good friend; at that point economy also comes in; for example, the cost of apartment renovation with professional craftsmen would be on the order of $200 or more."

2. "There should be a central storage area, where one can get new sections (of walls) or exchange them for old sections. In conjunction with the warehouse, there should be some craftsmen who can help with the installation."

3. "I would like to move the walls myself, but it is fun and really goes faster if one has craftsmen's help. The economic benefits are less important."

4. "There should be a central storehouse, where one can turn in his own walls, buy and sell walls. There should be different kinds of walls that one can buy. There should be such an assortment (of walls) that one can pick up a teakwood wall or a real sound insulation wall, if one wants it."

5. "I would prefer to do the modifications myself, first from the economic standpoint, but also just because it is fun."

Complaints recorded by the interviewers were focused on the quality of finish of some of the interior components, and to poor sound insulation between rooms within the apartments.

Twelve of the sixteen tenants felt the apartments were "flexible," three felt they were "not completely flexible;" one said "inflexible." Tenants felt most restricted in their flexibility by the columns in the middle of the apartment.

Tenants made the following comments on the opportunities for flexibility:

1. "We have developed a plan and furniture arrangement from the starting point of column location. We have purchased bookshelves with the exact measurement to enclose the column closest to the wall (one-bedroom)."

2. "The column closest to the end wall was a barrier for the corridor and gave us a furnishing problem, but the size and placement of the windows was an even larger problem. The same can be said of the balcony door.(one-bedroom)."

3. "No major difficulties cropped up, after we took into consideration the location of the columns, windows, etc. But the column closest to the end wall together with the balcony door prevented many possibilities, and we had some difficulty furnishing this location (one-bedroom)."

4. "The position of the windows on the bedroom side prevented us from making four equal bedrooms, which we really wanted. Because of the position of the balcony door, it was impossible to change the kitchen as we had wanted with a daylit work area along the facade (three-bedroom)."

5. "The columns split the apartment into two halves. One tries to work them naturally in with the walls and cabinets, or make them blend into the background with furniture (two-bedroom)."

6. "Apartments should be narrower and longer. That would make it easier to divide into several small sections; if one wants many rooms one is forced to have very peculiar proportions."

7. "We tried one plan and the wall landed in the middle of a window (two bedroom)."

8. "We desire more window area and less apartment depth. It would be better to have windows instead of the existing apartment division walls, and a kitchen and apartment entry wall [present separation wall] instead of the existing facade panels (two-bedroom)."

The last section of the report is a comparison of the degree of flexibility possible with three current (1968) building systems in Sweden. The authors conclude that the design of components according to the criteria of successful tenant usage is still in early stages, and that more research is essential in this field.

III. Cost Evaluation of Flexibility

The majority (twelve of sixteen) responded that the rent should include the first apartment adjustment (in conjunction with moving in), and that all other changes should be paid for independently by the tenants.

In response to a question about how the cost of materials should be paid for, the tenants replied:

1. A specified number of components (doors, walls, closets) should be included in the rent, all additional components to be paid for by the tenants.(six respondents)

2. There should be extra charges only if components over and above normal needs are desired. (four respondents)

3. The apartment rent should be determined for empty space, with the cost of all interior components paid for by the tenants. (three respondents)

4. As many components as are desired should be made available at no additional charge. (two respondents)

In response to a question about what impact the flexible wall system should have on rents, tenants replied:

1. "I answer that the cost of movable walls must be determined in relation to the rental market. If the cost of moving walls is high, the rent should be the same, if the cost of moving the walls is low -- for example $10/month -- then the rent should be lower for apartments with fixed walls."

2. "The possibility of moving walls I think is worth $4 or $5 a month more in rent. Really, it should be less expensive to build an apartment with movable walls than with fixed ones."

3. "The hominess (congeniality) is greater because of movable walls, but one will try to get more than is possible out of an area. $10 a month lower rent for a comparable apartment with fixed walls."

4. "Flexible apartments should be cheaper to build than those with fixed walls, because it means a more efficient building process and thus results in lower rents. Flexibility such as this has an absolute value. One can stay in an apartment longer."

5. "I don't think that flexibility itself increases enjoyment. For me it is the opposite: I think too much about apartment plans; I am always searching for the 'ideal plan.'"

6. "Movable walls are clearly a fact in home enjoyment, but I don't think that should justify a difference in rents over a regular apartment."

7. "Without a doubt flexibility affects home pleasurability, even to the extent that I could accept up to $10 a month higher rent over a comparable apartment with fixed walls. If I build my own single family dwelling, it will definitely have flexible walls. Friends of ours, three families who are going to build their own houses, have decided to try to have movable walls for themselves after seeing this apartment."

8. "I think that the knowledge that one has the possibility to change and shape something oneself -- all have different friends and life-styles -- influence the pleasurability in a positive way. I don't like sameness."

9. "The same rent for a flexible as for a conventional apartment, not without regard to the fact that it isn't as simple to move walls as some might think. One has to take into consideration the electrical installation in the ceiling."

IV.
Conclusions

One factor which was studied and which gives a measure of the desirability of the movable wall system was the fact that of the eight families who were planning to move, none were doing so to find an apartment environment without flexible walls. Two were moving to find lower rent. Of the last six families, two -- with small children -- would like to have movable walls in their next dwelling, and two -- with teenage children -- would not.

Only two of the sixteen households stated that they did not feel flexibility was a significant factor in the comfort and enjoyment of the home. The relationship of flexibility and satisfaction with the apartment environment was clearly positive in this study.

Relationships between Man and Dwelling

N.J.Habraken

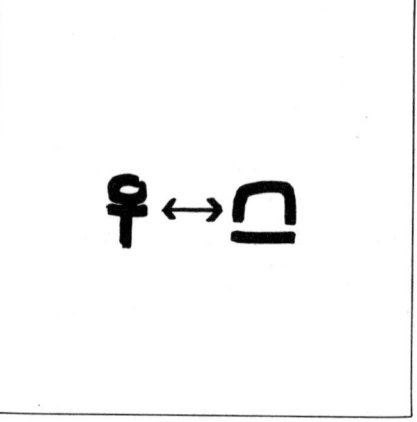

De mens leeft nu in een onnatuurlijke relatie met zijn leefgebied.
De onnatuurlijkheid blijkt als we weten welke natuurlijke relatievormen er bestaan.
Er zijn zes natuurlijke relatievormen.
De 'zevende relatievorm', die geldt in de massa-woningbouw, is een niet-relatie.
Deze zevende relatievorm doet niet-woningen ontstaan.

Nowadays man lives in an unnatural relationship with his domicile.
This artificiality becomes apparent when we know which types of natural relationships exist.
There are six natural types of relationship.
The 'seventh type of relationship', the one which holds good for mass housing, is a non-relationship.
This seventh form of relationship brings into being non-homes.

Waar houdt de gemeenschap op en waar begint de individuele verantwoordelijke mens?
Op welke manieren kan een individueel mens (bewoner) invloed uitoefenen op zijn huisvesting?
Hoe kan hij een rol spelen bij zijn huisvesting?
Het gaat hier om een relatievorm tussen mens en eigen leefgebied.
Dat is een fundamentele relatie; een natuurlijke relatie.
Kenmerk ervan is actie van de bewoner, van de leek.
Hoe kan die actie effectief zijn?
Er laten zich zes vormen beschrijven waarin die actie werken kan.
Drie 'individuele' relatievormen.
Drie 'collectieve' relatievormen.
In elk ervan is de bewoner actief bezig met zijn eigen woning.
Iedere classificatie is een hulpmiddel.
De werkelijkheid vertoont natuurlijk allerlei mengvormen.

Where does the community stop and where does the individual responsible human being start?
In which ways can an individual human being (occupant) exercise influence on his abode?
How can he take part in his housing?
We are dealing here with a relationship between man and his own domicile.
This is a fundamental relationship; a natural relationship.
Its characteristic is action by the occupant, by the layman.
How can this action be effective?
We can describe six types of relationship in which this action can work.
Three 'individual' types.
Three 'collective' types.
In each of these the occupant is actively busy with his own home.
Every classification is only makeshift; in reality all sorts of mixed types occur, naturally.

Wanneer men nu tracht productiewijze en individueel optreden in onderling verband te gaan zien, wordt spoedig duidelijk dat dan de hele kwestie van de relatie tussen mens en woning aan de orde komt. Welke is deze relatie? Hoe treedt hij op? Wat doet een mens met zijn woning? In zijn woning? Hoe kan een woning ontstaan? Welke rol kan de bewoner daarbij spelen? Hier is een studieterrein dat, zover mij bekend is, nog nooit werkelijk is onderzocht. Exploratie van dat terrein vraagt kennis van de geschiedenis van de huisvesting van de gewone man (op zichzelf een totaal verwaarloosd historisch gebied). Het vraagt kennis van sociale patronen en van de ontwikkeling van de bouwtechniek in het verleden. Ook hier dus koppeling van productie-mogelijkheden en mogelijkheden tot individueel optreden. De hier gegeven schema's willen niet meer zijn dan een aanduiding van wat de moeite van het onderzoeken waard zou zijn. Zij zijn ontstaan uit een poging om een eerste gedachtegang te ordenen, om het gesprek mogelijk te maken. Toch schetsen zij reeds hoezeer de mens van zijn huisvesting vervreemd is geraakt door de divergerende ontwikkeling van technisch organisatorische ontwikkelingen enerzijds en sociale ontwikkelingen anderzijds.

When trying now to view production methods and individual action as connected with one another, it soon becomes clear that the whole question of relationship between man and home is raised. What is this relationship? How does it become evident? What does a person do with his home? In his home? How does a home come into being? What part can the occupant play in this process? This is a field of study which, to my knowledge, has never really been explored. Its exploration demands knowledge of the history of the housing of ordinary people, (in itself a totally neglected field of history). It demands knowledge of social patterns and of the developments of building techniques in the past.
Thus, here too, the linking of production possibilities and the possibilities for individual action. The outlines given here are only meant as an indication of what might be worth examining. They derive from an attempt to bring order into an initial train of thought, to make discussion possible.
Nevertheless, they already illustrate how man has been alienated from his abode by.divergence in the developments of technique and organization on the one hand and of society on the other hand.

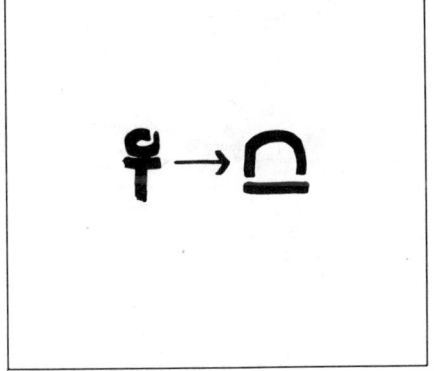

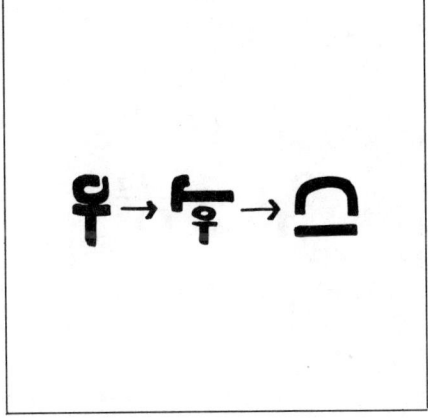

De eerste individuele relatie is de meest eenvoudige.
De bewoner bouwt zijn eigen huis met eigen handen.
Een vorm, die wij in de moderne culturen niet meer
tegenkomen, tenzij als resultaat van een noodtoestand
of op de camping.
Er zijn natuurlijk nog beschavingen waar dit gebeurt
zonder dat van een noodtoestand sprake is.

The first individual relationship is the simplest.
The occupant builds his own house with his own hands.
This is a type which we no longer meet in modern
cultures, unless it be the result of an emergency or
on a camping-site.
There are, of course, still civilizations where this
happens without any question of an emergency arising.

De tweede soort van de individuele relatievorm is
die waarin de vakman (bijv. de dorpstimmerman) zijn
diensten aanbiedt.
De vakman kan in deze figuur ook een groep
vaklieden zijn.
Deze vorm is ons uit de geschiedenis zeer vertrouwd.
Huisvesting ontstond zeer veel volgens deze relatie
in de westerse geschiedenis.
Ook de veel geprezen traditionele Japanse woningbouw
is er een mooi voorbeeld van hoe deze relatie tot een
prachtige harmonische wooncultuur kan leiden.

The second type of individual relationship is that in
which the craftsman (e.g., the village carpenter)
offers his services.
The craftsman mentioned here could also be a group
of craftsmen.
This form is familiar to us from history.
This relationship was very often responsible for
housing in western history.
Also the much-admired traditional Japanese housebuilding is a good example of how this relationship
can lead to a splendid, harmonious living culture.

De derde individuele relatievorm is die waarin tussen
bewoner en vakman de architect als tussenpersoon
optreedt. Dit is de vorm waarmee wij het meest
vertrouwd zijn. Die in ons denken eigenlijk normatief is.
Toch zien we deze relatievorm nog slechts in
uitzonderingsgevallen ontstaan in onze samenleving.
Er zijn maar heel weinig mensen die zich deze vorm
kunnen veroorloven, want daarvoor is nodig dat men
een architect opdracht geeft een huis te ontwerpen
op een stuk grond dat men zelf bezit.
Architecten denken het liefst in deze relatievorm.
Nu nog steeds.

The third type of individual relationship is that in
which the architect acts as intermediary between
occupant and craftsman. This is the type to which
we are most accustomed. The type which we tend to
consider as normal.
Yet we only see this type of relationship occur in
exceptional circumstances in our communal life.
There are only very few people who can afford this
type of relationship since for this it is necessary to
commission an architect to design a house for a
privately-owned piece of land.
Architects prefer to think in this type of relationship.
They still do nowadays.

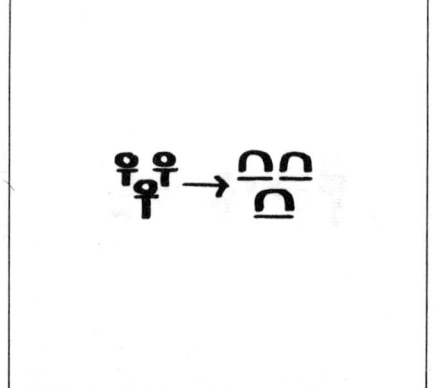

Het is voor architecten blijkbaar moeilijk om zich voor te stellen dat bouwwerken kunnen ontstaan zonder hun optreden. Nog moeilijker is het om te begrijpen dat door de eeuwen heen miljoenen zich gehuisvest hebben zonder tussenkomst van architecten, en dat de resultaten getuigen van een naar tijd en plaats optimaal bereikbare harmonie tussen mens en materiële omgeving. Is men eenmaal over deze schok heen, dan wordt de situatie duidelijk. Huisvesting is een sociaal proces in de eerste plaats en pas een technisch-organisatorische opgave in de tweede plaats. De architect, opgevoed in de gedachte dat hij moet bepalen hoe alles er uit moet zien, gepreoccupeerd met de gedachte aan het zelfstandige bouwwerk-als-werkstuk, heeft de illusie gehad dat huisvesting een kwestie van ontwerpen was. Nu niet een ontwerpen van één bouwwerk, maar van heel veel, en nog steeds wordt gewerkt onder deze dwanggedachte. Het resultaat is op zijn best architectuur. Nimmer huisvesting in de oorspronkelijke zin.
Wat kan de architect dan wel doen? Bouwtechnische voorstellen doen, die een harmonische huisvestingsproces in deze tijd met de industriële middelen, mogelijk maken. Moet hij daarvoor ontwerpen? Jawel, maar geen gebouwen, in ieder geval geen „woongebouwen". Wat een ouderwets woord „gebouw". Wat kan hij dan wel ontwerpen? Bouwwerken bijvoorbeeld, of industriële producten. Bouwwerken zoals straten en bruggen bouwwerken zijn. Producten die opgenomen kunnen worden in de stroom van het proces. Een architect is een vertaler van woonverlangens in bouwproductiemogelijkheden. De huisvesting kan beginnen als we ophouden met woningen te ontwerpen. Een woning is geen ding dat je ontwerpen kan. Een woning is een daad.

Apparently it is difficult for architects to realize that buildings can be made without their assistance. It is even more difficult for them to understand that through the ages millions have been housed without the mediation of architects, and that the results show that as much harmony as was possible under the circumstances of time and place was obtained between man and material surroundings. Once having got over this shock, the situation becomes clear: housing is first and foremost a social process and only secondarily a problem of technical organization. The architect, educated to think that it is he who has to determine the appearance of things, and preoccupied with the concept of an independent building-as-a-set-task, has been living in the illusion that housing was a question of design. Not, now, the design of one building, but of very many, and work is still being done under the constraint of this fixed idea. At its best, the result is architecture; but never housing in its original meaning.
What, then, can the architect do? Suggest building techniques which make possible the creation of harmony in the housing process in our times by industrial means. Must he design for this? Yes, certainly, but not buildings, at any rate not 'buildings to be lived in'. Such an old-fashioned word, 'building'. What, then, can he design? Building constructions, e.g., or industrial products. Building constructions such as streets and bridges. Products which can be assimilated into the stream of the process. An architect is the translator of longings for a home into productional feasability. Housing can start as soon as we stop designing homes. A home is not a thing that you can design. A home is an act.

De eerste collectieve relatievorm is die waarin- een gemeenschap gezamenlijk de woningen bouwt die hij nodig heeft en dit zelf doet zonder de arbeid te delegeren aan vaklieden.
Deze vorm vinden we in zogenaamde 'primitieve' culturen uit het verleden, maar ook nog in het heden. Hij valt (evenals de andere relatievormen) samen met een bepaalde samenlevingsvorm.
Hier, evenals in de volgende collectieve relatievormen is het individu nog wel degelijk rechtstreeks betrokken bij zijn huisvesting.

The first collective type of relationship is that in which the community builds collectively the houses it needs, and does this without delegating the labour to craftsmen. This type is found in the so-called 'primitive' culture of the past, but still in the present, too. This, like other types of relationship, coincides with a certain type of community life.
Here, as in the collective types of relationship described below, the individual is still very directly concerned in his housing.

De tweede collectieve relatievorm is die waarin een gemeenschap gezamenlijk de woningen doet bouwen die hij nodig heeft en daarbij de arbeid hoofdzakelijk of geheel doet verrichten door vaklieden. Ook deze vorm heeft zijn historische voorbeelden.
De eerste en de tweede collectieve relatievorm gaan natuurlijk snel in elkaar over en zijn niet altijd precies van elkaar te onderscheiden.

The second collective type of relationship is that in which a community as a whole has the type of houses constructed which it needs, thereby having the work wholely or partly done by craftsmen. This type, too, has its historic examples.
The first and the second collective types of relationship merge, of course, very quickly into one another and are not always clearly divisible from one another.

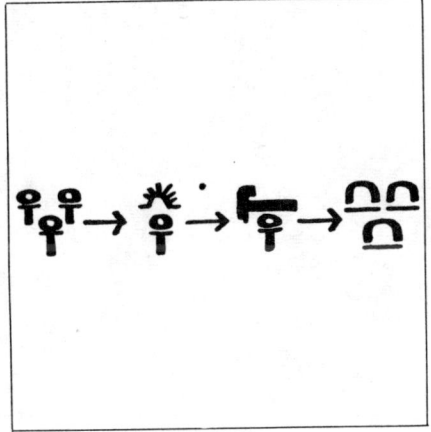

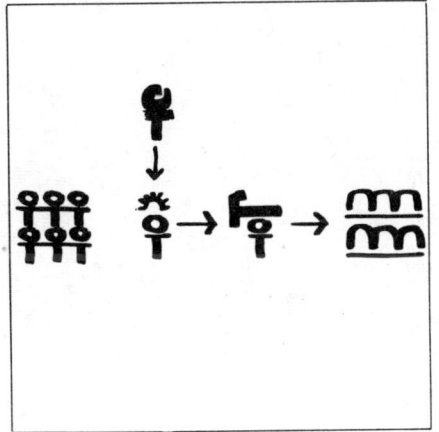

De derde collectieve relatievorm is die waarin de architect optreedt als gespecialiseerde tussenpersoon tussen de gemeenschap van bewoners en de vaklieden, die de bouwarbeid verrichten.
De woningbouwvereniging is oorspronkelijk vanuit deze relatievorm ontstaan, maar natuurlijk vindt men dit nu niet meer in de woningbouwverenigingen terug.

The third collective type of relationship is that in which between the community of inhabitants and the craftsmen who are doing the actual building
the architect acts as the specialized intermediary.
The building society originated from this type of relationship, but naturally it is no longer to be found in presentday building societies.

De 'zevende' relatievorm is een niet-relatie.
Geen van de voorgaande relatievormen vinden wij terug in de massawoningbouw.
Deze 'zevende' vorm van huisvesting kenmerkt zich door het feit dat daarin de bewoners eigenlijk niet voorkomen. Zij zijn onbekend tijdens het besluit-vormingsproces dat tot het ontstaan van woningen leidt. Zij zijn een anonieme massa.

Dit proces speelt zich uitsluitend af binnen de kringen van specialisten. De bewoner bestaat uit een abstract beeld dat deze specialisten zich vormen in onderlinge discussie om aan het werk te kunnen gaan. Zij observeren en onderzoeken daartoe 'de bewoner', die als zodanig natuurlijk niet bestaat.

Vandaar dat in het bovenstaand diagram van de groep 'anonieme massamensen' geen pijl is getrokken naar de architect. De architect krijgt zijn opdracht van een andere specialist, die evenmin als hij de bewoner is. Architect en opdrachtgevers doen hun best om dit dilemma op te lossen door 'de bewoner' te bestuderen. Oorzaak van veel semi wetenschap. Hier is dan ook eigenlijk geen sprake van een relatievorm waarin de bewoner aangegeven kan worden. Per jaar ontstaan nu volgens dit schema in ons land circa 100.000 woningen die eigenlijk niet-woningen zijn.

The 'seventh' type of relationship is a non-relationship. None of the previous types of relationship are found in mass production building.
This 'seventh' type of housing is characterized by the fact that the occupants really take no part in it. They are unknown during the process of decisions which leads to the production of dwellings.
They are an anonymous multitude.

This process takes place exclusively within circles of specialists. The occupant is represented by an abstract image shaped during discussion among these specialists so that they can begin work. To this end they observe and examine 'the occupant' who, of course, does not exist as such.

It is for this reason that in the above diagram nothing reaches the architect from the group of the 'anonymous multitude' of people. The architect is commissioned by another specialist who is no more the occupant than he is. Both architect and the people who give the commission do their best to solve this dilemma by studying 'the occupant' ... the cause of much half-knowledge. In this case there is really no question of a type of relationship in which the occupant can be recognized.
By planning in this way approximately 100,000 dwellings per annum are built in our country which are really non-dwellings.

Reprinted by permission from aap noot mies huis, by N. J. Habraken, Amsterdam, Scheltema & Holkema, 1970. Mr. Habraken is an architect in Eindhoven, Netherlands.

Section IV

"Heretofore we have thought of building in terms of the technology of today--the stamping machine, repetition. But the
technology of building will become all-capable . . . .
Ultimately, I would like to design a magic housing machine . . .
Conceive of a huge pipe behind which is a reservoir of magic
plastic. A range of air-pressure nozzles around the opening
control this material as it is forced through the edges of
the pipe. By varying the air pressure at each nozzle one
could theoretically extrude any conceivable shape, complex
free forms, mathematically non-defined forms. People could
go and push the buttons to design their own dwellings.

"This is a very exciting idea, indeed, because it suggests
that in the ultimate evolution of technology in the building
process, we may find that the highest form of organization
means the least standardization, that technology can make
industry as flexible as nature."

--Moshe Safdie, <u>Beyond Habitat</u>, Cambridge, Mass., M.I.T.
Press, 1970, p.244. **Reprinted by permission.**

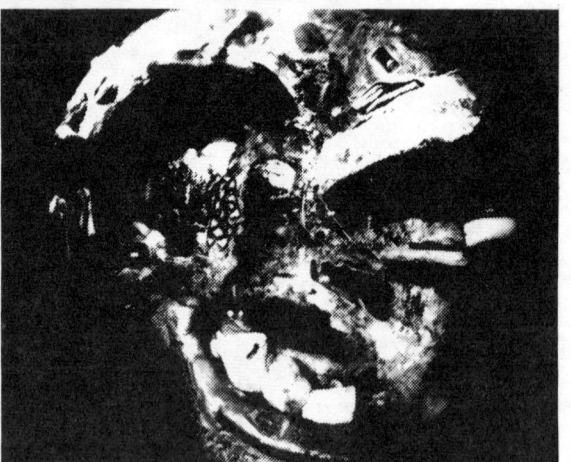

A sprayed-foam interior by Pierre Moffroid of Many Clouds, Inc., New York City. "Bathroom is entirely made of foam--tub and sink are high-density foam, some of it dyed; alongside of bathtub is a salt-water aquarium. Sink faucets are sea shells, no spigot, water gushes out like a spring. . ."

System-Pre-Design

Winslow Elliott Wedin

System-Pre-Design: A Definition

System-pre-design is a process of constructing adaptive shelters. It comprises the following features:

A. All aspects of the structure are pre-engineered. Thickness-to-span ratios, thermal insulation factors, aperture detailing, mechanical services, acoustical properties, and surface finishes are selected, solved, and specified in detail before construction begins.

B. A limited selection of simple materials and simple methods of construction provides a minimum-joint system. One basic material (polyurethane foam plastic) and one process (spray gun application) are used for all walls, floors, roofs, and ceilings, in one continuous, seamless surface. Built-in counters, beds, tables, seats, and cabinets are also part of the system. Coatings and other finishes are spray applied as well.

C. Maximum freedom of form is possible, allowing the user to create natural and organic expressions relating the building to the site and the requirements.

This approach produces sculptured environments incorporating an endless variety of form and surface texture. The process is prefabricated, but not the finished form. The factory is brought to the site and is user-directed. Individual expression leads to an environmental continuity emulating natural organic growth within the landscape.

By binding, molding, stacking, and carving, the building rises. Domes and vaults are built directly without recourse to fashion or style. Shapes are self-supporting, and ornament is an indigenous product of the needs of construction and structure. Spray in place technology allows custom-built, individualized environments to be constructed more rapidly than present mass production permits, without having to invest in fixed plant and expensive tooling.

Methods

The System-Pre-Design concept utilizes polyurethane foam sprayed-in-place. Fifty-five gallon containers of liquid plastic resin become, by chemical reaction, a solid volume 30 times that of the liquid. The drums weighing only 500 pounds, are easily handled and transported. With the system-pre-design approach, form and desired function can be manipulated at will. Form flows from the fingertips of the imagination in much the same way as a potter works his clay, feeling intuitively the strength and weakness of the design. The material is plastic, as the need is plastic.

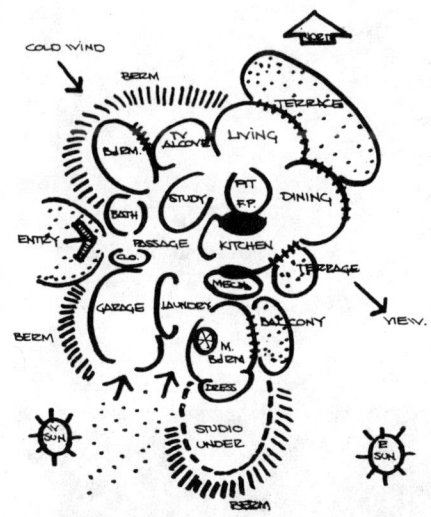

Function diagram for the Ensculptic III house.

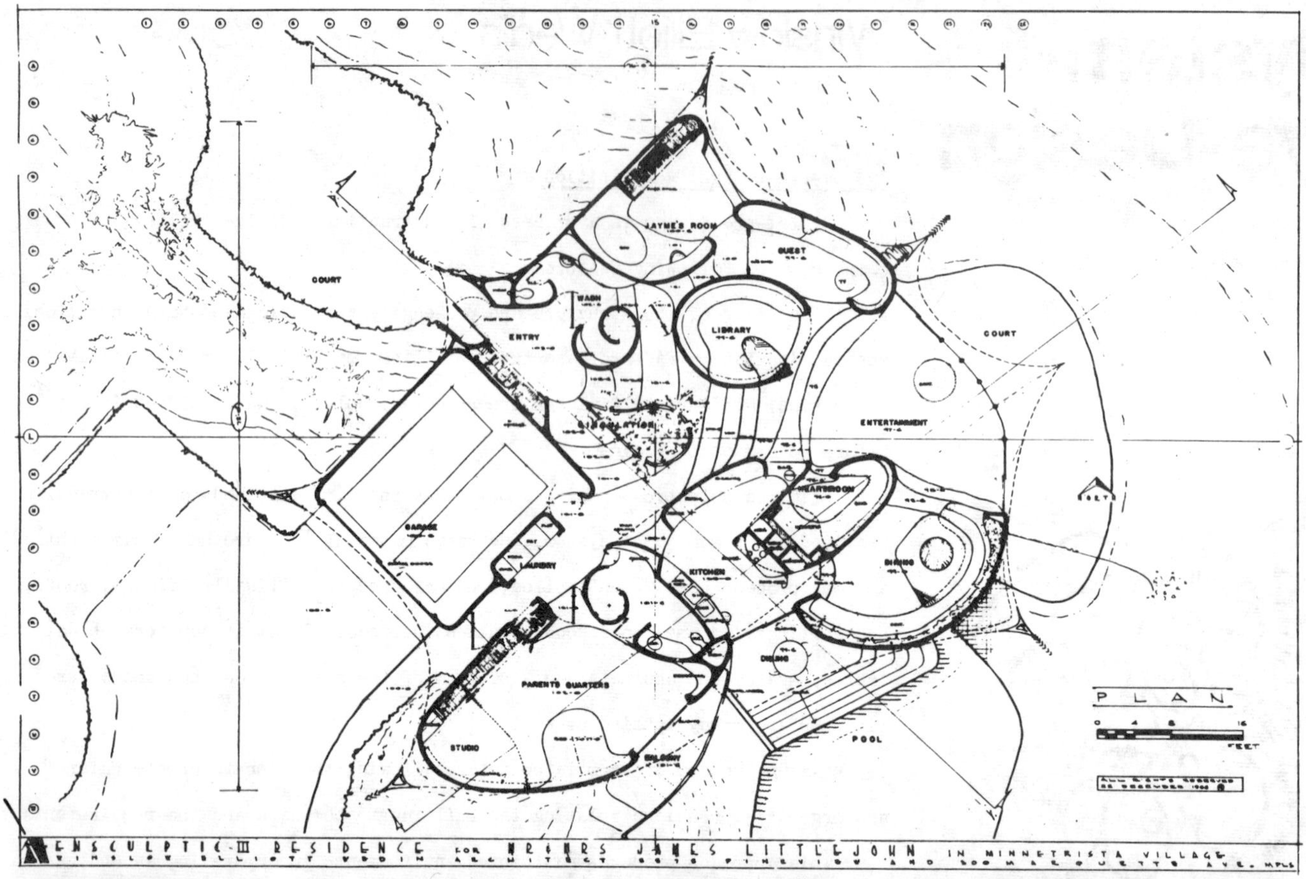

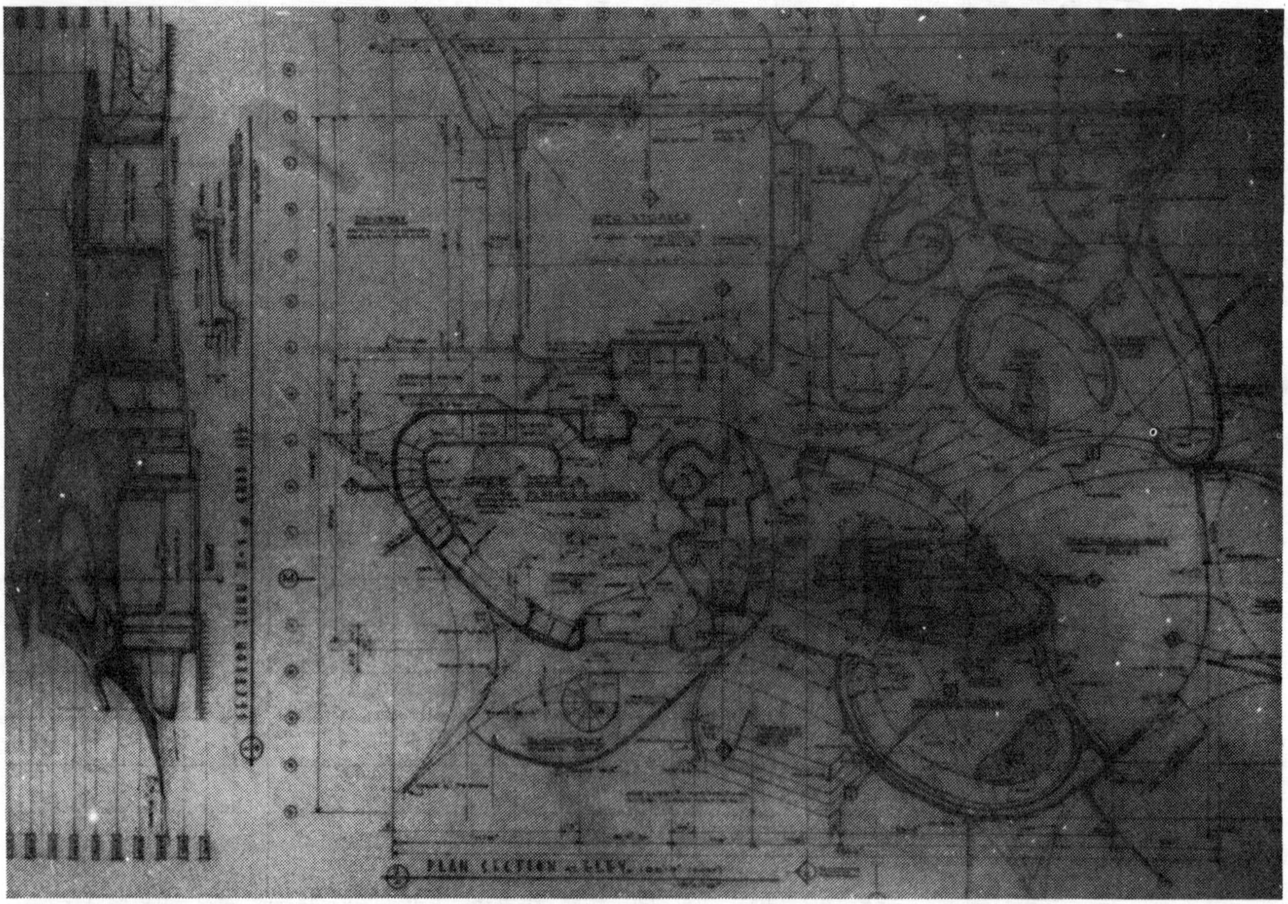

Design drawing for the Ensculptic III house.
Working drawing plan for the Ensculptic III house.

Construction of the
Ensculptic III house. The
mast and cables.

A cable anchorage.

The result is a synergetic unity of shelter, need, and individual creative
fulfillment.

Due to the slight bearing weight of the structure, the totally utilized
construction, and the inherent material flexibility, no footings or foundation
walls are required. Suitable bearing soil need not exist. The building
could float. A unit, for example, could be floated on swamp land, peat beds,
or fuller's earth (expandable clay), with no greater problems than those
which exist with conventional buildings on stable soil. Tie-downs, similar to
utility pole guy wire anchors, are employed as anti-wind-lift devices.

The technique utilizes on-site construction. The system is portable,
and not limited by highway width or weight limitations. The system is ideal
for developing nations, being simple, rapid, requiring only a low level of
technological skills, and utilizing simple equipment. The system can be air
dropped (both equipment and materials), and is therefore suitable to
inaccessible sites.

The resulting form is the consequence of circumventing most of the
conventional constraints of cultural tradition, structural stability,
available materials, and acquired technology. Nothing dictates fitting walls
to a grid, or the avoidance of compound surfaces. There is no need to build
accurate geometric figures. Each shelter can be produced as a direct
response to the needs of its occupants, in its required form, dimension, and
disposition.

For the first time in the history of mass-produced, machine technology, we have a material and method capable of reflecting and expressing human needs in any shape or form, as defined by one's imagination. Mankind can again carve and sculpture the environment. Formwork is very simple, due to the light weight and plastic nature of the material. Many lightweight, easily workable, locally obtainable, non-permanent materials can be used for shaping the basic shelter. Canvas, netting, tobacco shade cloth, bamboo, thatch, cardboard, even haystacks can be sprayed with foam plastic to become permanent shelters.

Field construction of several demonstration projects has utilized nearly all of the techniques described to this point. Onsite experience has indicated that doors and windows can be easily cut into the surface after initial foaming, simplifying installation procedures and avoiding on-job conflicts. Openings are sawed to required sizes and the "cut out" used as a pattern for the window insert. Mechanical devices are attached to the formwork prior to spraying, and become integral to the shell. Since construction methods are simple, close dimensional tolerances are not required (as in prefabrication); unskilled labor can perform most of the tasks. The exception to this is the spray gun operator. The production of a smooth surface requires experience. But formwork, cabinet installation, floor leveling, plastic plumbing, glazing, and 12-volt lighting can all be accomplished with little training.

Burlap stretched over the cables.

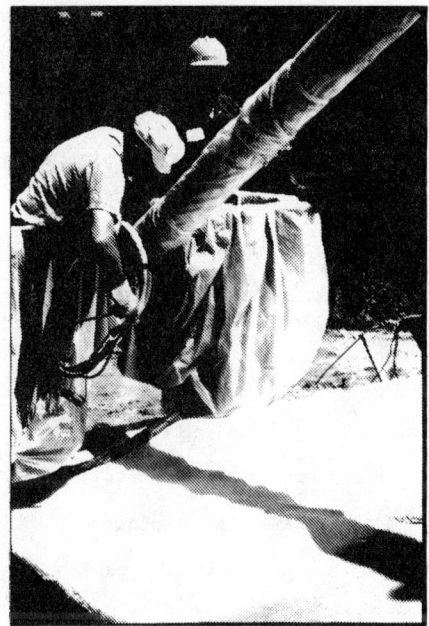

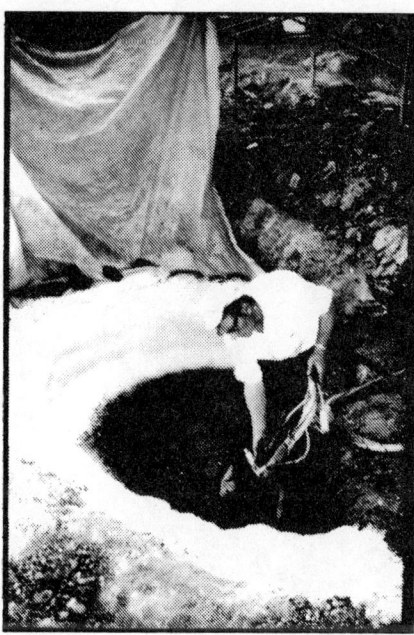

Polyurethane foam being
sprayed over the burlap
from a "cherry picker" crane.

The architect prepares a
window opening for spraying.

Spraying a foam foundation
directly against the earth.

Technical Innovations

        The following is a list of some innovations which have been incorporated
into the demonstration projects:

Floors on Earth: Three inches of high-density foam are applied directly to
smooth and tamped non-organic earth, providing an insulating, moisture-proof,
non-rotting barrier between man and the ground.

Glazing Gaskets: Hollow neoprene tubing is used for all fixed glass heads,
sills, and jambs, providing a flexible stop which is also temperature- and
sound-insulating, and which also allows for easy replacement.

Pivoting Doors: All doors pivot, a less costly and more durable solution
than conventional hinges. The pivot is spring-loaded, with the panel
closing noiselessly on foam plastic gaskets, sound-insulating adjacent spaces.

Finishes: Polyurethane seamless flooring material is used on all major floor
surfaces, forming a continuous, jointless, flexible surface between floor and
walls. Polyester gel coat is used on counter tops, forming also the sink
wells and backsplash.

Mechanical: Plastic plumbing is used for its simplicity and flexibility.
Since connections are easily joined with solvents and glue, a semi-skilled
worker can be employed.

Electrical Heating: Continuous electrical resistance cables are buried in all
surfaces in contact with the exterior (walls, floors, and ceiling). A surface
temperature of 78° maintains a room air temperature of 72°.

Lighting: A 12-volt lighting system operating on light-gauge wire, similar to
an automotive electrical system, is incorporated, giving greater simplicity
and flexibility to location, switching, and wiring, while at the same time
incorporating a "no shock" safety feature. Flat stick-on conductors are now
becoming available from the aerospace industry.

Demonstration Projects

    "Ensculptic III," the first example of system-pre-design, was built in
Minnesota, where yearly temperature variation may exceed 140°, from -30° to
+110° F. The Ensculptic III experience has confirmed that substantial
cost savings (from 1/2 to 2/3) can be expected from this type of technique.
Material costs for the shell (enclosure) were about one third of those for a
conventional wooden house of the same floor area.

    In the second demonstration project, "Hollyhouse," the architect used the
knowledge obtained in Ensculptic III to design a prototype low-cost dwelling.
The system-pre-design technique was combined with a pre-fabricated mechanical
core, solving in part the cost deficiencies of Ensculptic III in the area of
mechanical integration. Based on the "Hollyhouse" experience, the following
projections are made:

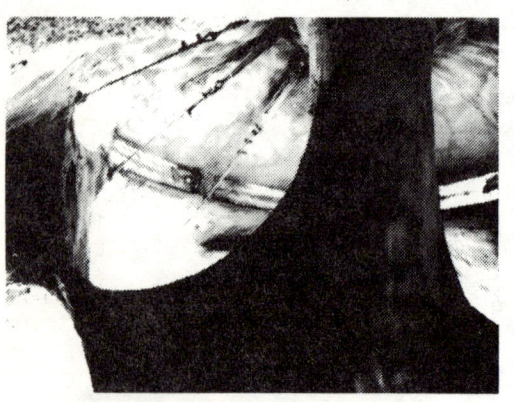

The interior of the house
before interior finishing
operations.

Completed exterior of the
Ensculptic III house.

1. A home of 1,000 square feet can be built for $8,000.00.

2. Budgeting a house at 2½ times yearly income, an annual salary of $3,200 would be sufficient for purchase.

3. Using a figure of 20% of yearly salary allocated to housing, a monthly payment of $53.50 would make it possible for a family to acquire ownership in a low maintenance, well-constructed dwelling.

It is important to note that this solution, while low in cost and utilizing local unskilled labor and local forming materials, provides a finished product of vastly superior quality and considerably reduced maintenance problems.

Further demonstration projects are currently in construction in Colorado and Florida.

Directions for the Future

The Ensculptic projects have demonstrated the potential of low-cost spray-in-place technology. But numerous investigations are yet to be conducted, if a truly adaptive environment is to become readily available. Among these must be studies on the re-use of discarded foam, and the addition, subtraction, and alteration of portions of the house after initial construction is complete.

Winslow Elliott Wedin is an architect practicing in Tallahassee, Florida.

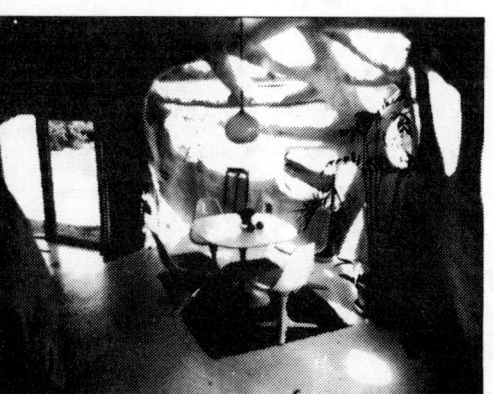

The dining area.

The System-Pre-Design concept.

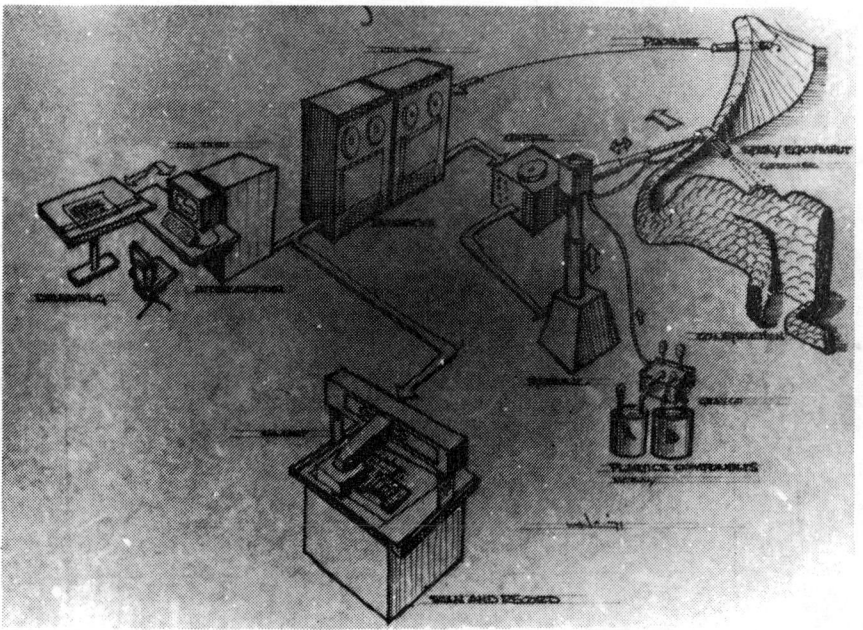

Pneumatic Structures, Cybernetics and Ecology

Toward Ecostructures for Habitation of People and other Lively Systems

Blair Hamilton

Pneumatic structures have a technology that is important in its potential for development of a responsive architecture. As with each advance in ~~materials,~~ the field of structu[re] ~~the first use,~~ of pneumatic structures has been to imitate previous structures. Now that we are beginning to become skilled at using lightweight structural materials, we can pursue their unique possibilities.

Present building materials and their technologies fill our environment with architectural forms that tend naturally toward the freezing of time, the destruction of their environment, and the de-humanization of their inhabitants. This architecture encourages the designer to shut out and screen out the environment so that he can reasonably create a controllable sub-environment to inhabit. We have learned to value this isolation from our surround, unaware that it, like most of the rest of our technology, does not include any sense of ecology.[1] Architecture as shelter has been built of hard materials, dependent on their rigidity or insularity (structural, thermal, acoustical, visual) to separate our living spaces from their surrounding environment. I visit too many modern buildings where I am isolated in such a conditioned space that I am completely unaware of what is going on outside the structure. Like the automobile driver who doesn't have to breath the fumes he leaves behind outside his shell, I am unaware of how my building, or my activities inside it, are affecting the outside environment. This may seem the extreme case, but in most of our architecture almost all the information we perceive inside about what is going on outside is what we see through a few windows, hear because the "walls are to thin," or feel because the "insulation isn't good enough." You see what I mean by isolation being highly valued—and worthy of cost. The loss of any sense of continuousness with the environment is just beginning to be valued differently.

This sounds like general theorizing, and I'm going to do more of it, but I'm also going to talk about our work at finding ways to create the innate technology of soft materials by real building. If we want to create pneumatic structures that will be valued by people, our work cannot be just in our heads. Words, drawings, and models are not sufficient when our object is something experiential. We must build our ideas. We must make our

Blair Hamilton is working at Ecology Tool and Toy in Milford, New Hampshire.

fantasies something we can feel. We must build them before we know if they will work. We have to interact with our prototypes by inhabiting them, evaluating that interaction, and re-designing to solve problems and explore further. Our prime design consideration is no longer material, structure, or product, but the total experiential process. Our language about it must either be a metalanguage or we must start doing it and talk about what we are doing.

The need for doing is reflected in the pneumatic structure project at Antioch College, in Columbia, Maryland. An environmental design program was started so that students and faculty can be primarily responsible for developing their own facility. After building several smaller prototypes, we are now ~~currently~~ constructing a 30,000 square-foot structure to house all the administrative and academic facilities required by the programs in Columbia, one of several units in the Antioch system.

The facilities problems faced by not only most of our schools, but by most social institutions caught in an environment of accelerating change, has been extreme at Antioch-Columbia. Anticipating that the campus might choose to relocate as the focus of academic programs shifted, heavy investment in a conventional campus facility was ruled out. The rental space available proved to be expensive, cramped, and inappropriately designed. This campus of the college has, in its three years of existence, been in a continual process of radical change in institutional structure, programs, and related facility needs. The inability to anticipate further inevitable and dramatic changes compounds this problem and exemplifies how it is becoming increasingly difficult to design conventional buildings which will be useful for any reasonable amount of time.[2] A new kind of capability—the ability to accomodate unanticipated requirements — is called for now, and will be required for the future if we want the changes of our personal and social growth to be encouraged.

A precondition to this responsiveness is a process of informal convergence upon an architectural embodiment which includes the user as a central figure in the design, building, and later modification of his architecture in a real-time and on-line mode.[3,4] Recognizing the importance of this at Antioch, the Environmental Design Program has as its major focus ~~on~~ the development of this kind of process. Many of the users have no prior training in design, but this has the advantage of easy acceptance of fresh approaches. Work on the building goes more slowly, and there is much frustration of well-intended effort, but there is also an incredible amount of education and a high level of involvement with their architecture in a process of mutual evolution. The process is learned by involvement rather than academic explaining. The ease of learning about, building, and modifying pneumatic architecture makes it very appropriate for this process. Heavy construction equipment is avoidable.

The ability of pneumatic structures to enclose large open spaces provides a further level of adaptability. Lightweight materials and portability now become reasonable for interior architecture, since the need for load bearing and weather resistance has been eliminated. At Antioch we have begun developing an interior system of furnishing, spatial definition, and services which is compatible with the potential portability and softness of the surrounding architecture.[5] But there is still great research needed in this neglected area, particularly in the context of real use and evaluation. Many of the traditional problem solutions become ridiculous in this new context. A high level of invention is needed.

All this flexibility is trivial, however, until we shift our inventive energies from trying to make pneumatic structures like hard structures with objectives like longer-lasting membranes and single-state high stability, and instead begin to value developing the innate softness and potential responsiveness of this form.

We might begin to approach this by going from the notion of architecture as shelter to architecture as interface. Now we are talking about a system which recognizes that there is a relationship between the architecture and its surround as well as ~~between~~ the architecture and its inhabitant(s), and that these relationships are loops where short- and long-term mutual adaptive change is taking place. The construction of this kind of sophisticated pneumatics takes us into the realm of living things and ecology.

If this is difficult to recognize, it may have to do with the magnitude of these changes and their unfamiliarity in this context. A stone building will crumble as a response to its environment. All of our structures and materials have perishability, just as all living things. Our problem with hard architecture is not that it cannot change, but that the changes are not usually of a scale which has adequate meaning for people. The changes are not of a human scale (human effort will little affect the rate of change), and this is like the two parties of a conversation speaking different languages. The information transfer which is necessary to the process of adaptive change has been lost; participation by the perceiver is made impossible or irrelevant. But not all the changes should be of this very particular time-grain. The architecture we are developing is in larger forms (city as architecture) and smaller forms (clothing as architecture), and the multiplicity of its time-grains of response should be as rich as that of its environment. In each situation, the magnitude and rate of change ought to be compatible with what is meaningful in that environment. For example, at Antioch this means that there should be some changes which correspond to the moment to moment changes of the individuals and their groupings, some of the rhythm of the weather, some which reflect the academic schedule and calendar, some of the scale of the sun and moon. The participants derive meaning from these rhythms because of correlated personal changes which never before had an external referent.

With this level of responsiveness, then, we can see that one of the things the architecture can do is to absorb or feed back the information or energy it comes into contact with as its environment. In the context of biological-like behavior, this is simple. As an interface, however, the architecture also has the ability to transmit. The two loops become related and there is a transaction (between architecture-surround and architecture-inhabitants) which is (controlled by) the architecture. In hard architecture, the control which is built in to the architecture is a screening out, sometimes selectively, of environmental information and energy. Another possibility, however, is that the interface may act as a lens to enhance aspects of the relationship it separates. It is an active interface and its functions are necessarily selective. This architecture has a behavior of its own which draws on both sides of the interface. It is like the selectively permeable membrane of a simplistic micro-organism. The notion, then, becomes one of an architecture which has, as a result of its structure, a capacity to change its own basic units of self-organization. It is self-referent. It modifies itself in response to changes in its env-

238

ironment, both inside and out, and the interactions which take place in real time.

These words will seem far-fetched to some, but obvious to others who are familiar with the development of self-organizing, adaptive and self-referent systems in the mechanical and electrical modes as a high technology. I believe if we value the task, we can reduce this technology to a less abstract and more usable form, particularly with our architecture.[6]

As we increase adaptability, we can change the present over-design of structures necessitated by the need for one state to respond to varying environmental conditions.

To develop this responsiveness in pneumatic structures requires the adequate application of control technology which, for the most part, has already been developed, particularly for aerospace applications. Pneumatic structures are ideally suited to this kind of development because of their control potential. This suitability is largely because instead of relying on so much mass to build our architecture, we are substituting energy.[7] Hard architecture requires great energy input to begin to approach the responsiveness we are looking for, but when energy becomes identical with the greater part of our architecture, control becomes relatively easy. Instead of brute-force muscle, we're using ju-jitsu. We have added to idiot muscles of a hard system a beginning of artistry in control.

The parameters and mechanics of control in this architecture, since they operate in a number of time-grains, are evolutionary.[8] The problem is the orchestration of a number of environmental qualities, each in its own time scale[9], to create a courteous and playful environment.[10] The control system functions, like the energy it controls, as an integral part of the architecture. It explores parameters, modes, and rates of change to create relevant responsiveness. Ideally, such a control system is distributed in the material of the architecture, combining structural with other forms of computation.[11]

The functioning of such a control system might be as follows:

Sensors which are providing information on the results of behavioral changes in the architecture (skin and cable tensions [tendon sensors], internal climate [body temperature], lighting characteristics, condition of mechanical equipment [pain when malfunction begins], acoustics [hearing], power supply, interior structures, etc...), act through a self-organizing controller which cross-couples non-linear effects by continuously experimenting for achievement of high-level intentions. The controller is modifying the conditions of the architecture within the parameters set by a meta-control function. The fragments of sensed behavior are not responded to by corrective changes. Rather the system continually explores its own behavioral effects. Input and output are looped rather than separated by a controller. The loops and network of loops are provided more capacity to explore as they stabilize in an unsatisfactory way. The exploration which results from the addition of carefully constrained noise (so it will experiment in a non-random region) provides for not only magnitude of change, but also change in the basic units of response-time, complexity of cross-coupling, influence of any one behavioral sensor, rates of change, rates of rates of change, etc. This is the primary meta-conrol function.

The control loop we have described here enables the architecture to have a behavior of its own in exploring various internal relationships in the architecture. A behavior it effects interacts with the inhabitants and the surrounding environment as they loop back with new behaviors, and the new condition of the architectural loop is then sensed and modified again by the self-organizing controller.[12] The importance of cascading shorter and longer control intervals, behaviors that change hour by hour being nested within controls that operate quite independently day by day, will be evident. This interactive exploration of environmental behaviors enables a playful relationship between the architecture, its inhabitants and the surrounding environment (either of which includes people, plants, animals, etc.). Information on how well the system is doing loops back to be modified at the local, structural computation level, or by more general regional control, or by meta-control causing re-inforcement on the whole state as it tends to settle into a meta-stable condition. This might take the form of indicating to the meta-controller that the discrete set of behaviors should be labeled and remembered, increasing the long term frequency of its repetition and making it available for future recall and use in special situations.

Reinforcement of behavior also takes place within the control loop as some fine-grain behaviors are re-inforced by those of larger grain.[13] These processes

239

tend to insure the gradual self-tuning and acquisition of courteous behavior by the system. This tuning is towards the creation of the information level that insures the variety necessary for the survival of the total system.[14,15]

All of this responsiveness is, thus far, within certain parameters which change only in an explorative, evolutionary manner as the system learns when it is doing better at being actively supportive of inside and outside healthy environmental functioning. There are, however, times when a rapid change of these parameters is appropriate.

There are times when the control system should move into an emergency mode. This is far simpler and closer to systems with which we are familiar than the dialogue mode described above, and these controls can easily be put into a facility such as the Antioch structure. The function here is that of a decision in the context of information from sensors in the surrounding environment and monitoring of the primary control loop network, that indicates the necessity of shifting to certain emergency procedures through overiding control of the effectors in the architecture. For example, strain guages on the cables and wind speed information may indicate the need, within a certain response time, to vary combinations of blower use and speed, ventilation, and interior temperature to quikly change the inflation pressure and corresponding cable tension to maintain structural stability. For gross emergencies, this kind of overiding control could be achieved my manual control by inhabitants who recognize and want to deal with an emergency situation, but often can be more effectively provided for by setting limits on loop behaviors, extending our level of prediction as our control skill warrants. When this function is built in to the system, we have an analogue to the control exercised in a crisis situation by the reticular core of the central nervous system, where certain sensory information results in abrupt behavioral change which assumes priority over all other behavioral tendencies.[16]

Another need for abrupt change in the architectural environment comes when the function of the space it encloses (or its role in the surrounding environment) changes abruptly and requires a whole new set of environmental behaviors. Control of this decision might require direct indication to the meta-controller that a new mode of behavior should be called from memory or synthesized. However controlled, the decision of the system involves knowledge of learned modes of successful behavior.[17] This ability to acquire, recall, and explore when the system is not critical, new modes of successful operation means that the system becomes richer and more responsive as its experience increases over time.

It is important to note that I am not describing a general system of control here, but rather a particular evolution for a first structure which need not be duplicated. As soon as there is interaction and skill is acquired by the inhabitants and surround, the behavioral model has already changed. Loops might be added into the surround where the architecture would be nourishing a vegetable garden. The system should suggest possibilities we could never imagine.

This childhood of a control system is a poor beginning which calls for extention, re-thinking, and modification based on full-scale building and experience. Some of this can happen with the Antioch project, where we are trying to implement parts of it now, but still within the rather major constraints of a simple structural design and available materials. We need to move toward designs which allow for far greater adaptability of form, shape, size and other controllable characteristics.[18] We need membranes of more ecologically sound chemical composition that are more desirable for people to interact with; membranes of controllable elasticity and controllable transmission; membranes which function as beautifully and complexly as human skin.

My repeated use of the biological analogy is not merely illustrative. I am talking about a wedding of biology and architecture in the context of ecology to create what Rudolf Doernach calls "biotecture."[19] The use of pneumatic principles is found throughout nature and its use in this task is obvious. Our architecture, and the technology it is an extention of, must become as living organisms if we are to go on living.

This is still only descriptive. The task of doing is enormous. We are working with a whole new technology; a post-industrial technology. The hard mechanical technology of the industrial revolution is useless in this task, and the inability of this old technology to deal with our present problems is becoming increasingly critical. We need help.

It is time to move into the new space.

REFERENCES

(1) Reference is made here to the people whose thinking, revealed to me mostly in dialogue rather than through their listed publications, has been synthesized with my own in pursuing this task and describing it here

(1) Rossman, Michael, "The Technology of Technology and Social Reconstruction", part of On Learning and Social Change, unpublished manuscript

(2) Hamilton, B., and Whitaker, W., Bubble Primer, Antioch College Environmental Design Program, Columbia, Md., 1971

(3) Wellesley-Miller, S., "Real Time Design and Self-Organized Environments," Participatory Architecture (conference proc.), Manchester, Sept. 1971

(4) Storm, H.O., "Eolithism and Design," Colorado Quarterly, 1, (3), Winter 1953.

(5) Development of a system of low-cost, highly manipulable interior components for large space-enclosing structures has been a major focus of the work of the Research and Design Institute of Providence, R.I.

(6) Pask, G., "My Prediction for 1984," Prospect, Hutchinson of London, 1962 - (a particularly clear and delightful exploration of how these principles might enter into our lives with such examples as a "self-organizing dam")

(7) Goering, P.L.E., "Energy Structures," Canadian Architect, 1971

(8) Brodey, W.M., "Soft Architecture: The Design of Intelligent Environments," Landscape, Vol. 7, No. 1

(9) Brodey, W.M., "The Clock Manifesto," Ann. New York Acad. Sci., vol. 138, pp. 895-899, 1967.

(10) Johnson, A.R., "The Three Little Pigs Revisited," in Eleven Views, Student Publications of the School of Design, vol. 20-1, North Carolina State University, April 1972.

(11) Johnson, A.R., "Dialogue and the Exploration of Context: Properties of Adequate Interface," Proc., ASC, October 1970.

(12) Johnson, A.R., "Self-Organizing Control in Prosthetics," in Advances in External Control of Human Extremeties, Yugoslav Committee for Electronics and Automation, Belgrade, 1970.

(13) Brodey, W.M., "Information Exchange in the Time Domain," General Systems Theory and Psychiatry by Gray, Dahl and Rizzo, Little, Brown and Co., 1969

(14) Brodey, W.M., "Biotopology 1972," Radical Software, no. 4, Raindance Corporation, N.Y.

(15) Herbert, F., Dune, Ace Books, N.Y. 1965. (see Appendix I on ecology, diversity and survival).

(16) McCulloch, W.S., and Kilmer, W.L., "Some Mechanisms for a Theory of the Reticular Formation," in Systems Theory and Biology, Springer-Verlag, New York 1968.

(17) Johnson, A.R., "Organization, Perception and Control in Living Systems," Industrial Management Review, vol. 10, no. 2, Massachusettes Institute of Technology, Winter 1969.

(18) This work is being pursued particularly by the Eventstructures Research Group in Amsterdam, the Chrysalis group in Los Angeles (Mike Davies, Alan Stanton, Chris Dawson), Simon Conally and Mark Fisher (Air Structures Design) in London, Charlie Tilford in Columbia, Maryland, Ant Farm in San Francisco, and probably several others, as well as here at Ecology Tool and Toy in New Hampshire and with present projects of Sean Wellesley-Miller in Boston.

(19) Doernach, R., "Biotecture," Architectural Design, February 1966

ALSO:

Brodey, W.M., and Lindgren, N., "Human Enhancement:
Beyond the Machine Age," IEEE Spectrum, Sept. 1967
and Feb. 1968.

Bateson, G., Steps to an Ecology of Mind, Chandler
Press 1971

Bateson, G., "The Role of Somatic Change in Evolution," Evolution, vol. 17, no. 4, Dec. 1963

Iberall, A.S., and Cardon, S.Z., Analysis of the
Dynamic Systems Response of some Internal Human
Systems, NASA Contractor Report CR-141, NASA, Wash.,
D.C., Jan. 1965

Negroponte, N., The Architecture Machine, MIT Press,
1971

Ekstrom, R., Technical Report (on the feasability
of the Antioch structure), Antioch College Env. Des.
Program, Columbia, Md.,April, 1971

Doernach, R., "Bauen auf dem Meer" (Maritime Structures), Kunststoffe im Bau, KiB Oct., 1968, Verlag
Chemie, Heidelberg, 1968

241

Discussion following Blair Hamilton's presentation:

Hamilton:

"In a pneumatic structure, you begin to move from <u>mass</u> toward <u>energy</u> as the main component of the structural system. The building doesn't exist without energy. Once energy is an integral part of the structural system, then you have the beginnings of <u>control</u> . . . . There's a loop that goes on between architecture and its surroundings. There's another inside, between architecture and its interior. Architecture is the interface. In hard architecture, the architecture is a screen between the inside and the environment. What I'm looking for is a means of control so that this transaction between inside and outside takes on more the quality of a <u>lens</u>."

Van der Ryn:

"It seems to me there are two things we need to consider in all such systems: What's the capital that's required to do it (like if you have foam equipment, once you have it, things go very fast, but we all know that there's a technostructure and money required;) and the other issue is the ecological issue. Those are the two criteria.

"If you need high tech and a lot of money to do a simple building, it limits the access, it limits who can do it.

"We were up at Goddard College, where fifteen unskilled architecture students are building quite large spaces . . . I think it's possible to do quite large spaces with simple tools and simple technologies."

Hamilton:

"I'm in no position where I want to defend pneumatic structures in general. I think they have a particular potential. But I also have some reservations similar to yours."

Richard Britain:

"I'd like to say something about the buildings at Goddard . . . they didn't even manage to get the bloody things enclosed in time for the winter. It was incredible--there were four feet of snow lying inside. This is a real reflection on the way they were built and the type of technology they were using."

Negroponte:

"Most inflatables are what I would call 'Miesian Mushies.' Less is more, and a bit cheaper. What <u>you're</u> talking about is that the inflatable is a highly controllable organism which now does a great deal of responsing in a <u>physical</u> sense."

Hamilton:

"I'm trying to put myself out on as many limbs as I can to try to get more people involved."

Negroponte:

"You're doing it."

Hamilton:

"Clearly the game that Daria Bolton presented before is
no game, it's the state of the art. It's time that architecture
started responding to its inhabitants."

Several people:

"But how does pneumatic architecture respond?"

Hamilton:

"I'm looking for the kind of relationships that happen in
the biological world. I'm looking for an architecture that's going
to relate to me . . ."

Hans Harms:

"Responsive to what?"

Wellesley-Miller:

"The answer is, to people."

Negroponte:

"I want to know, how does it respond to me,
by knowing me? . . . the whole issue of responsive environments is a very, very suspect one, because we don't know how
they should respond. We all feel they ought to respond, but
the only examples are the most banal, second-rate light shows
. . . Responsiveness implies a very definite, overt behavior
on the part of the environment."

Van der Ryn:

"What I hear from Negroponte is that we're
going to interpose machines, and that those machines are going
to program our buildings and monitor our response to buildings
so the buildings can change. That kind of thing is respondent
architecture, not responsive architecture.

"Any building situation is responsive. It's a matter of
how much energy you're putting in to create responsiveness in
a shrunken time."

Hans Harms:

"All this machinery that you build in may have
meaning to the people who build the machines. Whether it has
meaning to the people who live in it is another question."

Richard Britain:

"I come from a culture where things are very
largely built in response to a very long time scale, where
buildings increase in value through time, and people very much
depend on materials that are very long-lasting . . . there's
a time scale way beyond their own lives which is built . . .
For my money, we should never be thinking of making an environment totally of inflatables or totally of stone. We need to
find out what the use of each type of construction is, and
there should be no conflict. I'm interested in building
buildings out of both."

Hamilton:

"What I would like to be going on in this room is that the environment should be going through a set of behaviors . . . exploring different combinations."

Watson (interrupting):

"I think the people should do it first. If the people can't do it, why bother about technology?"

Hamilton:

"Because our imaginations are limited, and there are certain combinations which the architecture might suggest to us; then we need to be able to indicate to it when it's doing better."

Hilbertz:

"Perhaps we could take a very crude example of a responsive system; let's take lovemaking . . . (Laughter)"

Rob Freedman:

"A simpler example would be that a building would know when people want light inside, and would open up a window."

Hamilton:

"I'm saying you don't know that. [Let's say] I'm playing with a whole different kind of acoustics, exploring with the architecture how I can change them, and the phone rings, and I've got to change the acoustics real fast. Another thing that I want to be able to indicate to the architecture is that it should revert to a previously-learned mode of behavior."

Hans Harms:

" . . . you have to pay a very high price; this requires high technology . . . you're hunting sparrows with cannons."

Lucas:

"I don't understand why everybody's hung up on high technology. It's not necessarily a bad thing. It's a neutral extension of man himself."

Harms:

"It costs a lot of money."

Lucas:

"Your brain is high technology."

Voice:

"What if there are several people in various corners of a pneumatic structure?"

Hamilton:

"Part of the response of the architecture might be that it could divide, acoustically, an interior space . . . "

Voice:

"Why leave it all to the architecture? Why can't you just speak more softly?"

Britain:

"Yes, why is that any better than just going into another room and shutting the door?"

Negroponte:

"I'm sorry, somebody at this point should just say 'bullshit' . . . The two kinds of questions we are hearing here are trivial, but relevant, in a very important sense: One, 'Why bother responding? What will it do?' The idea of its turning into two spaces instead of one so you can answer your bloody telephone is just not convincing. The other question is, 'Why stop there? To hell with inflatables! Let's just put little nodes into our heads and plug ourselves into a machine, and simulate every environment and every possible experience.' "

Van der Ryn:

"You can't ask rational questions of our technology any more (applause)."

Freedman:

"Why do people choose the metaphor 'biological' to describe what they want in buildings? . . . What is it about the quality of a living thing that you want to get into your buildings?"

Voice:

"Whatever people are interested in doing, they want to be able to _do_ it. It's that simple. There are incredibly few areas of life in which you can do that . . . Someday children will be taught in third grade that that's what responsive environment is; it's an environment that enables you to do what you want to do."

Huck Rorick:

" . . . the implication of these things is that lots of people end up using them, but the guy who made it has had all the excitement of manipulating it . . . I think that's one of the problems."

Kahn:

"I keep getting the feeling that you who are working on these pneumatic structures and/or computers are flacked out on the things you are working on. You're trying to justify what you're doing by all these absurd analogies. I can't relate them to shelter, and I was thinking for a minute that what you're talking about is art, and in that respect it's fascinating, but it doesn't have anything to do with housing people that I can see."

Negroponte:

"I'm sure that in the 1890's people got up and said that kind of thing about airplanes . . . most of what was written about aerodynamics in the 1890's was wrong. I'm sure that most of what _we're_ talking about is wrong."

Van der Ryn:

"There _is_ a technology of responsiveness. It's called _human awareness._"

Negroponte:

"Seymour Papert . . . built teaching systems for children which were nothing more than computing systems to enable little kids to build teaching machines for their class-mates. There are now about fifteen grammar schools throughout the country that have this system called LOGO . . .and in the process, their students have all the fun, and they do, in their classrooms, just what the guys at RCA in Palo Alto do . . . our model for the responsive house may not be the old teaching machine syndrome, but perhaps the LOGO model. We're after the kind of machine that a person builds for himself."

Strategies for Evolutionary Environments

Wolf Hilbertz

"Not form, but forming, not form as a final appearance, but form in the process
of becoming, as genesis."
                                    Paul Klee

1. INTRODUCTION

If we ever will be able to discuss what has been called interactive, self-organizing, adaptive, intelligent, responsive, cybernetic, or even evolutionary environments in a sense other than utopian, large scale integration of the arts, architecture, engineering, and the hard and soft sciences has to occur. In order to
promote the forming of the necessary substrate of integrated knowledge and creativity, unusual hypotheses will have to be set forth and fresh questions will have to
be posed. The approach is heuristic.

The development of environments within the previously mentioned framework is in
its embryonic state. Notions like flexibility, changeability, design participation, fit, increased choice and decision making, accomodation of continuous change,
man/environment interaction, etc., are abundant, but the meaning is not clear.
Goals are not stated and the underlying philosophy is obscure.

On the other hand we have to guard against a scientoid reductionism which certainly would prove to be detrimental to this emerging field. Only holistic approaches
promise to successfully cope with the problems of increased complexity that we
encounter.

2. BACKGROUND

In 1966 we began defining the potential role of cybernetics and evolution in a
genuinely progressive architecture. Since then, principal research interests
center around:

1. Evolutionary, self-organizing environmental open systems capable of
    forming higher orders of organization; dynamic morphological and psy-
    chological manifestations in transactional symbiotic response to contin-
    ually changing interior and exterior forces.

2. Man-animal-plant-technology-nature symbiosis including the interpretation
    and effectuation of behavioral, social, and other information from animate
    and inanimate sources for environmental solution generation and processes.

3. The environment as an evolutionary code and the interfacing of information
    and morphogenetic systems.

4. Exploration of man's inner and outer self in a rapidly evolving synergistic setting with the prospect of enhancing and complementing organic and socio-cultural evolution, both being the result of organism-environment interaction.

5. The development of morphogenetic (material distribution, manipulation, and reclamation) systems on land, in air and water, under the earth, on polar icecaps, and in extraterrestrial space.

6. Energy requirements, energy-harnessing developments and energy conservation.

7. The socio-political and biological implications of proposed complex symbiotic environmental systems.

3. RESPONSIVE ENVIRONMENTS

To chart the historical development of building technology it is useful to compare the flexibility of use with the degree of industrialization. (Figure 1). Beginning with existing and modified cave volumes such a progression eventually leads to the evolution of a cybernetic technology which leads to responsive environmental systems. This implies that the user becomes the stimulus to which the environment responds.

Conceptually, such systems eliminate interpretive linear skills of the building profession which have so successfully adapted architecture to past circumstances, while placing the user on Procrustes' bed and ignoring him thereafter.

The potential implications of responsive environments are far reaching. First, the development of diverse life-styles within varying social contexts would accelerate rapidly.

Second, the dynamics of such environmental systems would serve to weaken considerably outdated beliefs and petrified institutions; and would ultimately generate previously unpredictable solutions based on the currency of the moment. Conceptual activity occurring in this and related fields seems to have already laid the groundwork for these changes.

Third, the development of cybernetic environments requires new "yardsticks" by which to gauge and reflect changing goal structures. The use of measures of flexibility and industrialization would no longer be adequate.

Fourth, the development of these environments appears to be merely a requisite stepping stone on the way to achieving what I refer to as <u>evolutionary environments</u>. (Figure 2).

Since, in this context, the morphism of the environment to meet changing needs and stimulate further development demands increased plasticity, the traditional separation of morphogenetic capability and material has to be abandoned. Thereby the morphostatic component is weakened. A detailed description of proposed morphogenetic hardware and reversible materials with sensing properties is given elsewhere (Hilbertz, 1967-72).

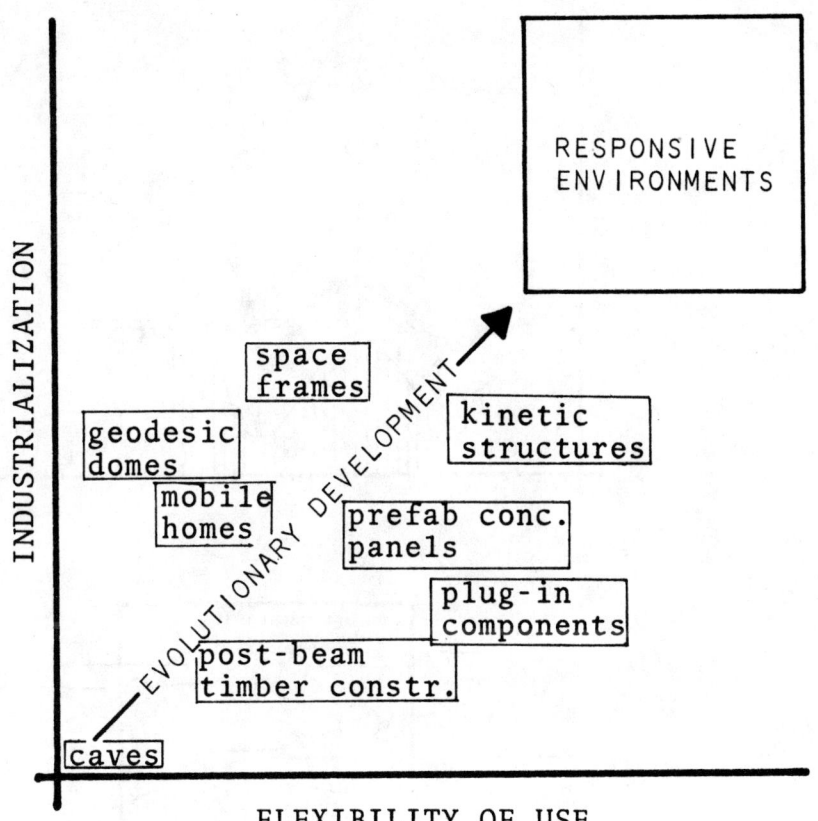

FIG. 1

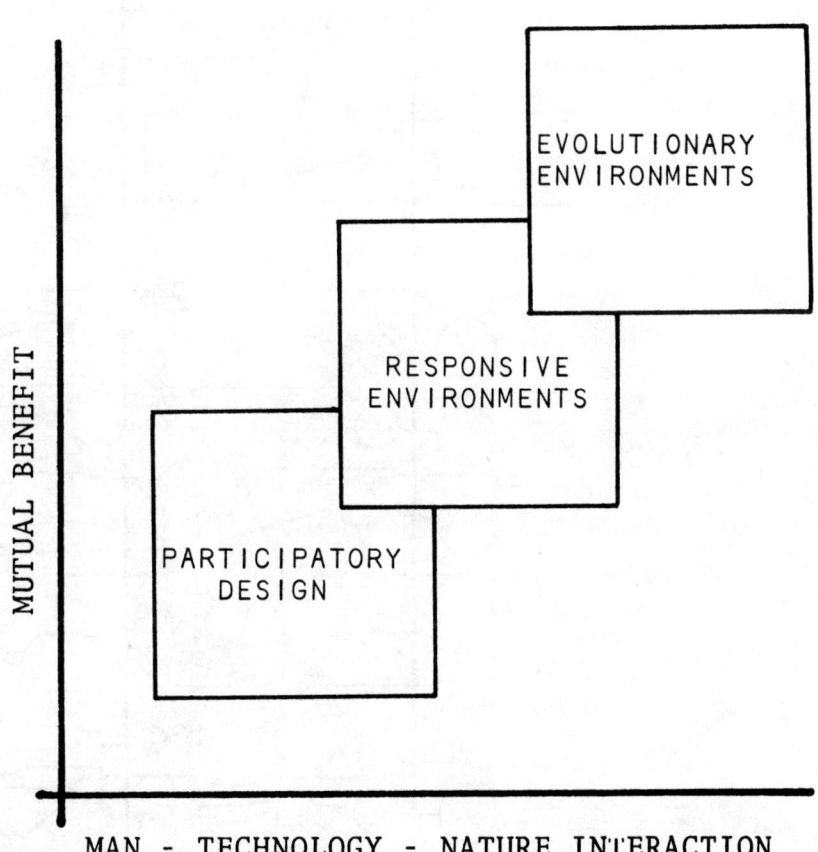

FIG. 2

249

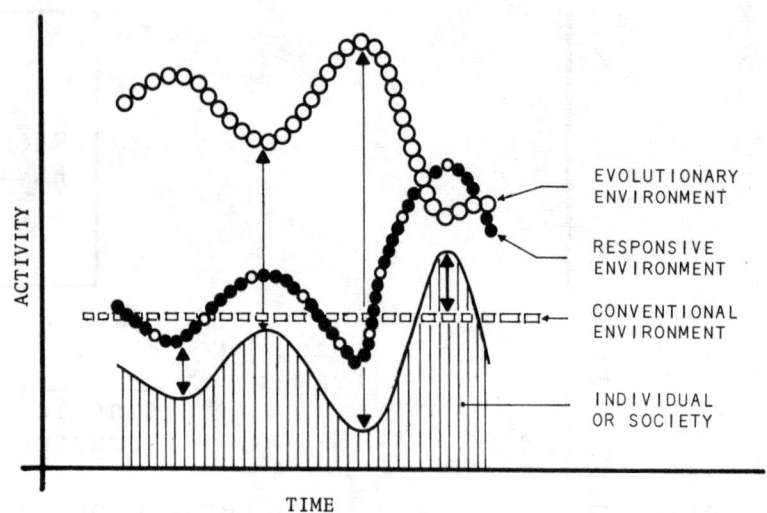

EVOLUTIONARY
ENVIRONMENT

RESPONSIVE
ENVIRONMENT

CONVENTIONAL
ENVIRONMENT

INDIVIDUAL
OR SOCIETY

FIG. 3

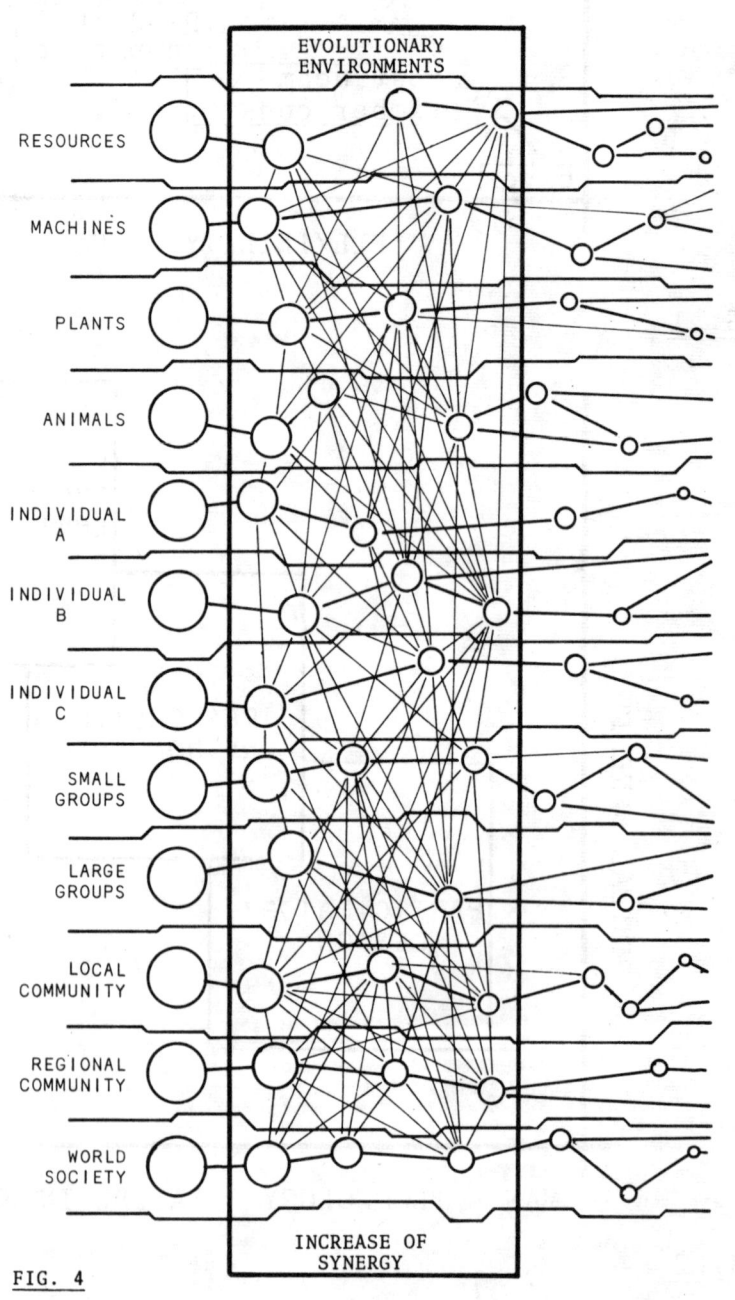

EVOLUTIONARY
ENVIRONMENTS

RESOURCES

MACHINES

PLANTS

ANIMALS

INDIVIDUAL
A

INDIVIDUAL
B

INDIVIDUAL
C

SMALL
GROUPS

LARGE
GROUPS

LOCAL
COMMUNITY

REGIONAL
COMMUNITY

WORLD
SOCIETY

INCREASE OF
SYNERGY

FIG. 4

4. EVOLUTIONARY ENVIRONMENTS

The differences between traditional, responsive and evolutionary systems are obvious. (Figure 3). The conceptual separation of the user (stimulus) and the physical environment (response) in the responsive system implies that at best only one-sided evolution or a superficial fit between the two can be achieved. In an evolutionary environment, however, this cause-and-effect dualism is replaced by dynamic interrelationships. (Figure 4). The richness of connections between components determines the system's performance. Whereas the responsive system produces a "mindless fit," the evolutionary system accelerates both socio-cultural and biological evolution through purposeful stimulation. The evolutionary system is comprised of man, his extensions and nature; being simultaneously beginning and end, originator and result, producer and user. (Figure 5).

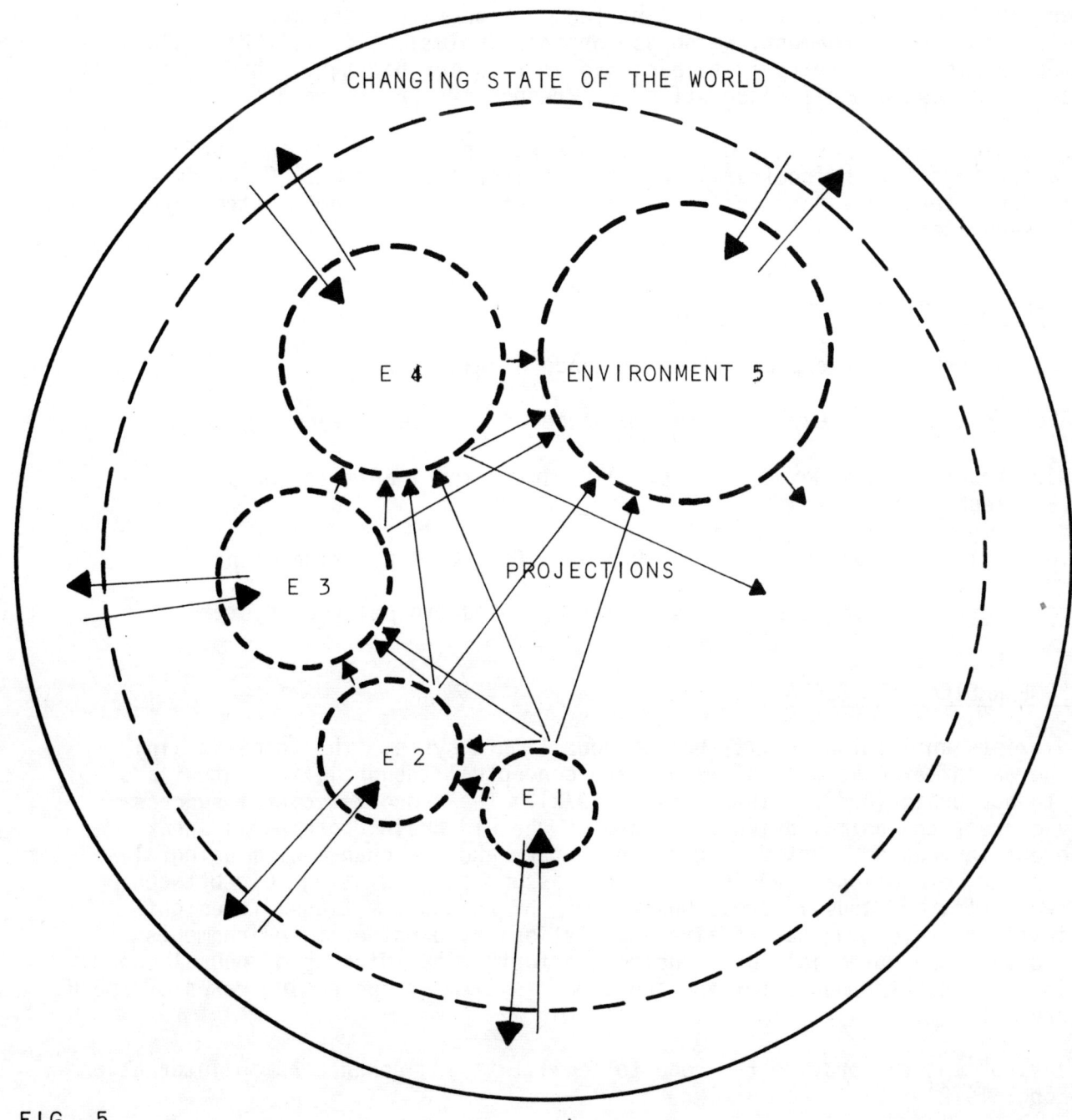

FIG. 5

251

Since the early Greeks (Aristotle, Herodot, Hippocrates) the influence of the environment on man has been studied. During the 18th century, Montesquieu developed a 'theory of the milieu,' and around the turn of the century the scientific exploration of this field began (Malacarne). Although it was generally assumed that brain changes as a result of organism - environment interaction would occur, only recently scientific proof of this hypothesis was given.

Animals were placed in impoverished and enriched physical environments. After varying exposure times their brains were examined. It was shown that both environments caused distinct and measurable changes in brain anatomy and chemistry. Animals with enriched experience had a greater weight and thickness of the cerebral cortex and greater activity of certain enzymes. They developed larger nerve cells and nuclei, more basal dendrites and larger synaptic junctions. (Bennett, Diamond, Krech, Rosenzweig, 1964; Rosenzweig, Bennett, Diamond, 1972).

Walter (1969) contends that by adapting the environment to our models rather than ourselves to the environment, we oppose organic evolution, i.e., that evolutionary mechanisms no longer apply to ourselves but to our habitat. This is perhaps the earliest most general description of a responsive environment.

But evolutionary environments will be structured in a way as to insure further and faster organic, sociocultural, and environmental evolution. Organism - environment processes and consequences are mutually dependent systems and cannot be separated.

5. GOALS OF EVOLUTIONARY ENVIRONMENTS

1. Abolishment of the exploitative dominance of man over man.

2. Abolishment of the exploitative dominance of man over nature.

3. Directed potential development of all animate and inanimate forms. (Entelechy, Evolution).

4. Conciliation of all animate and inanimate forms. A new kind of nature.

Figure 6 describes some characteristics of proposed evolutionary systems.

6. SOME RADICAL PROPOSALS

There exists an incompatibility between our limbic system (the animal brain) and the neocortex (the seat of reason and conceptual thought), the latest addition to our brain (Ardrey, 1970; Esser, 1972). The neocortex cannot successfully correct the animal drive functions of the old brain; both parts speak different languages. Considering human history and the chance of meaningful development, evolutionary environments can assume the mediating role between the two parts, and thus insure a healthy mix of reason, emotions, foresight and instincts governing our affairs. Evolutionary, prosthetic environments also possess the potential for inducing the further building-up of neural connections between the neocortex and the limbic system at increasing rates of speed. (Figure 7).

The environment can provide the code to facilitate and enhance human interaction (De Long, 1972).

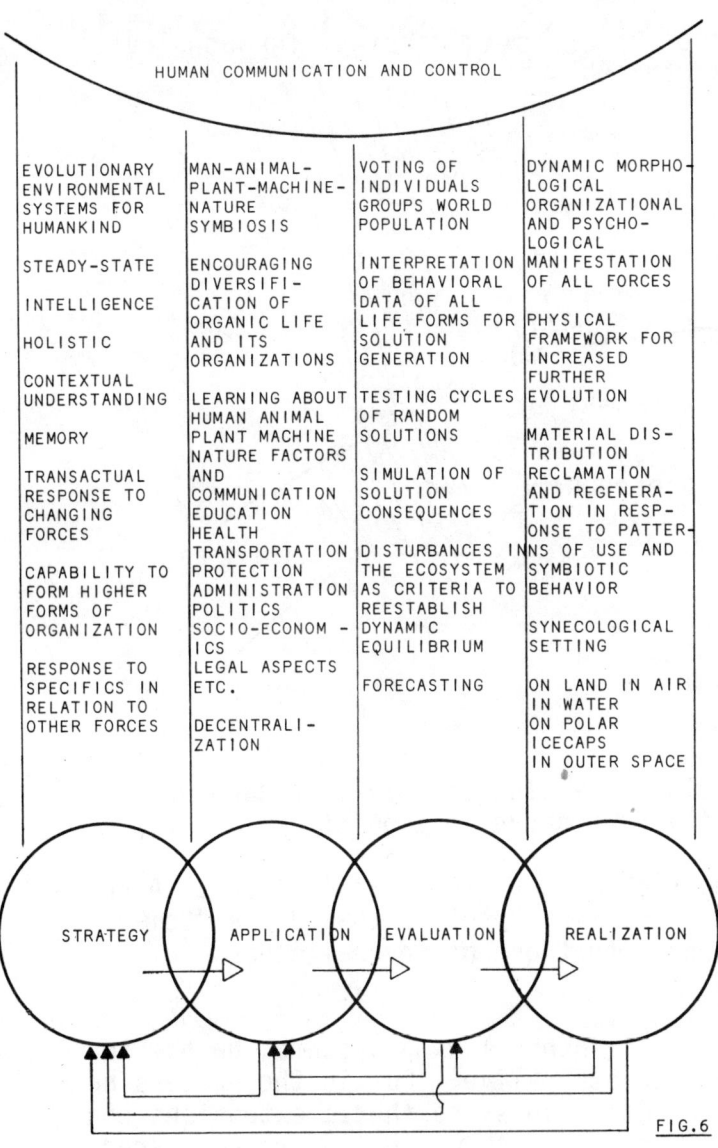

FIG.6

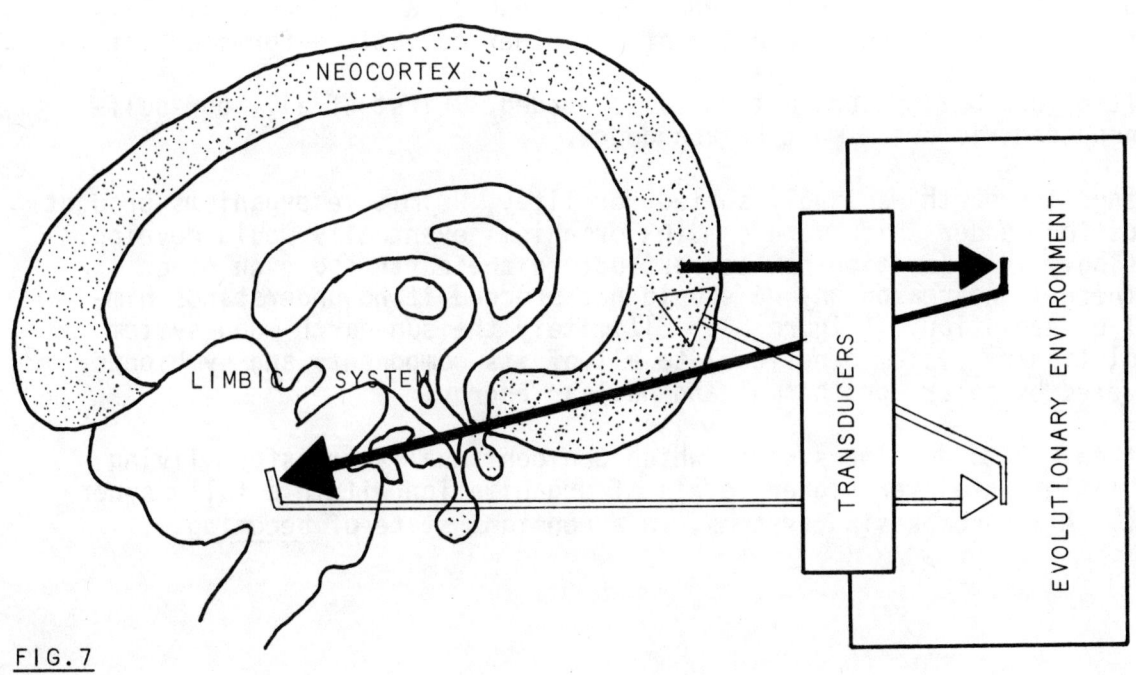

FIG.7

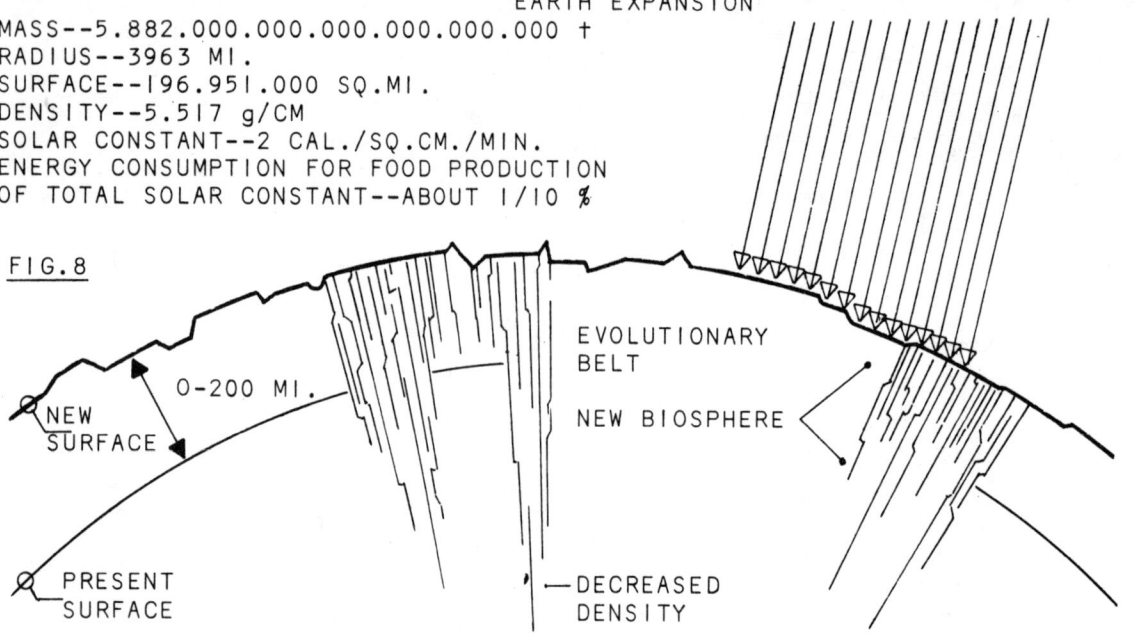

EARTH AS EXISTING: INCREASED USE OF SOLAR CONSTANT THROUGH
                                 EARTH EXPANSION

MASS--5.882.000.000.000.000.000.000 †
RADIUS--3963 MI.
SURFACE--196.951.000 SQ.MI.
DENSITY--5.517 g/CM
SOLAR CONSTANT--2 CAL./SQ.CM./MIN.
ENERGY CONSUMPTION FOR FOOD PRODUCTION
OF TOTAL SOLAR CONSTANT--ABOUT 1/10 %

FIG.8

0-200 MI.

NEW
SURFACE

EVOLUTIONARY
BELT

NEW BIOSPHERE

PRESENT
SURFACE

DECREASED
DENSITY

To trust blind and capricious organic evolution would not only be hazardous
but outright deadly. History and the daily news prove this point.

Today, the human species is self-evolving (Dobzhansky, 1962; Hall, 1966). According to J. B. Calhoun (1971) we will have to evolve ourselves toward the "Compassionate Revolution", a notion consistent with those of several others (e.g.
Maslow, 1964).

Man is an integral part of nature, and like all other living systems, he has to
draw negative entropy from his environment to stay alive. But in the process he
increases positive entropy in his surroundings. Man's prosthetic extensions con-
form to the same law. Within the present context of shortsighted, man-centered
technology we live at the ever-increasing expense of the rest of nature. We,
thereby, not only endanger our own existence but refuse to accept responsibility
for the conservation and further evolution of all other compatible forms of life.

A radical shift of our beliefs and attitudes is needed. First of all, our self-
deceiving anthropocentric dream must be abandoned.

In its beginnings the earth was badly suited for life. Primitive organisms brought
about the conditions under which more complex organisms eventually would develop.
Man is increasingly in a position now to restructure the earth and even other
planets, and there is no reason why he should not proceed if he understands himself as a tool of evolution. (Figure 8). Ultimately the sun-earth-moon system
will be an evolutionary system beneficial to all of its components and symbionts.
It will be powered by solar, geothermal and nuclear energy.

A beginning, then, is to develop systems which can continually transform living
and nonliving matter into ever higher levels of organization within a fully synergistic setting. All-encompassing systems, in a constant state of <u>becoming</u>...

7. REFERENCES

Allan, J., Hilbertz, W.H., De Long, A. J., "The Symbiotic Processes Laboratory," Man-Environment Systems, S 60, September, 1971.

Ardrey, R., The Social Contract. New York: Delta, 1970.

Bennett, E. L., Diamond, M. C., Krech, D., Rosenzweig, M. R., "Chemical and Anatomical Plasticity of Brain," Science, Vol. 146, No. 3644, PP. 610-619, October 30, 1964.

Calhoun, J. B., "Space and the Strategy of Life," in: A. H. Esser (ed.) Behavior and environment. New York: Plenum Press, PP. 329-387, 1971.

De Long, A.J., "Environment as code," AAAS Meetings, Washington, D.C.: December, 1972.

Dobzhansky, T. Mankind evolving. New Haven: Yale University Press, 1962.

Esser, A. H., "Evolving neurologic substrates of essential forms." General systems, XVII, PP. 33-41, 1972.

Hall, E. T. The hidden dimension. Garden City, N. Y.: Doubleday & Co., 1966.

Hilbertz, W. H. "Cybertecture, an evolutionary environmental system," Working paper , The University of Michigan, Ann Arbor: February, 1967.

Hilbertz, W. H. "Toward Cybertecture," Progressive Architecture, PP. 98-103, May, 1970.

Hilbertz, W. H. "Cybernetic Architecture, a teleological process" in: Michael Kennedy (ed.) Proceedings of the Kentucky Workshop on computer applications to environmental design. University of Kentucky, PP. 95-99, 1971.

Maldonado, T. Unwelt and Revolte . Hamburg: Rowohlt: 1972.

Maslow, A. H., "Synergy in the society and the individual," Journal of individual Psychology, 20, PP. 153-164, 1964.

Rosenzweig, M. R., Bennett, E. L. Diamond, M. C., "Brain changes in response to experience," Scientific American, Vol. 226, No. 2, February, PP. 22-29, 1972.

Urbanitat 2002, Modern American Architecture. Traveling exhibit in Europe (Fuller, Hilbertz, Knowles, Safdie, Soleri, Vivrett), May 1972 - May 1973.

Walker, W. G., Observations on man, his frame, his duty and his expectations. Cambridge (G.B.): Cambridge University Press, PP. 37-38, 1969.

Evolution of Future Environments

Robert E. Lucas

It is apparent that some new, more flexible building system is necessary; when we see the rot of the urban core, the abandonment of whole city areas, the immediate obsolescence of a new structure, and the constant demolition and rebuilding going on, we know that the design and construction professions are not keeping up with the rapidly changing needs and conditions of the world. It should be apparent that what we need far more than a new system is an entirely new philosophy of environmental design.

On the simplest level, a "responsive" building technology would be comprised of a means to distribute material in space according to some "design" or directing factors and, to provide flexibility, a means to reclaim and/or change existing structures to fit changing needs and conditions. Implicit in such a technology is the desire to increase efficiency through reclamation of materials and a built-in sensing structure to detect changes in needs (structural, through changing loads, etc.). However, if this is the limit of the system, it is only a flexible technology, and flexibility alone is merely a mindless fit. What does a responsive building system respond to?

Most work in this field indicates that the user controls the system, i.e., the system responds to the "needs" and "desires" (whims?) of the user. Presumably there is a control panel (whether remote or not) at which the user "punches in" how he wants the environment. Depending upon the sophistication of the system, the complexity, frequency, and speed with which the system "responds" may vary. Somehow, such a system seems to be merely a "super-thermostat;" let's call it "Dial-a-Space."

If we imagine such a system building a dwelling or group of dwellings, we can see certain immediate advantages. The initial form potential would be almost limitless and the required labor would be extremely small. Certainly the boredom inherent in "mass production" systems today would be alleviated; individual "custom" design would (or should) be automatically provided. However, if this is a representative description of systems now envisioned, there are two fallacies -- internal and external -- that need to be dealt with. Moreover, the system described is still no more than highly flexible; there are many "flexible" building systems at present, and they don't work either.

Again we must ask -- what does such a system respond to? The needs of the user? Form follows function? The structured environment will be changing accordingly. What changes -- size, shape, color, temperature, furniture, lighting? I assume that we have the necessary technological skill to change any or all of these attributes at will; the crucial question is --Why? At some point the user knows what he wants; he exhibits this knowledge every time he turns on a light switch. But other than such direct needs, what else is there to respond to? When his family grows, the user needs more space; the need is obvious and simple, but is the response? Does the user know how much space, where, and what shape? I think most designers would agree that, although the user professes to know exactly what he wants, his knowledge is superficial, to say the least. By definition, the designer is the one trained to translate user needs into spatial solutions. Thus, we have defined another necessary characteristic of the system; it no longer "responds" to needs, it translates them.

It is highly desirable to give the user the ability to structure/restructure his own environment. Purely from the psychological point of view, notions of identity, creativity, stimulation, and spontaneity can be attained. One restriction is, though, that the system must have built-in parameters and techniques to act as the medium through which the user acts on his environment.

Robert E. Lucas works in the Symbiotic Processes Laboratory at the University of Texas, Austin, Texas.

Are these "designing" parameters the only restrictions the
system must have? If we stop at this point, all we have given
the user is a glorified, automated erector set with a fancy
set of instructions. The internal fallacy lies in the fact
that no system can respond or react to only its user(s).
It is time to recognize that all manipulations of the environment,
large or small, depend directly on the system's integration
into the total environment. Obviously, an individual user
cannot be allowed total freedom in determining what his
environment is like. What happens when two separate systems
want or need the same thing or space? Decisions in such cases
must be handled by the environmental system as a whole,
correlating physical, social, political, psychological ... factors.
If the environmental system is not fully integrated in
making such decisions, we will make the same fatal mistake of
putting the technological cart before the environmental horse.

The recognition of the need to fully integrate environmental
systems is fairly simple; carrying out such a system is much
less so. For one thing, it involves a change in attitude.
Contrary to popular theories, man does not and cannot control
the environment or nature; if we could we would certainly
not be facing the present ecological crisis. According to
Robert Ardrey in The Social Contract, this "philosophy of
the impossible rests on an article of faith, that man is
sovereign." Ardrey fairly sums up the situation:

The philosophy of the impossible has been the dominant
motive in human affairs for the past two centuries.
We have pursued the mastery of nature as if we ourselves
are not a portion of that nature. We have boasted
of our command over our physical environment while we
ourselves have done our urgent best to destroy it.

Although man lives in an almost entirely synthetic environment,
the natural and social environments not only are beyond his
control but also are reacting strongly against man because
of the blind application of technology to reap short-term benefits.
In the final analysis, the world is the ultimate responsive
environment -- a vast interdependent system. Thus, in reality
we don't have to "create" a responsive environment -- we've
already got one that works. There can be no improvement over
present trends, even with a new "super-building" technology,
unless we realize that man and his structured environments
have an affect on and in turn are affected by the natural
environment.

To this point, I have briefly described what a new "responsive"
building technology should entail. The basic idea is a result
of continuing research at the Symbiotic Processes Laboratory
at the University of Texas. The work of this group was
originally directed toward creating a new "super technology"
based on on-site, computer-controlled material distribution
and reclamation. After frustrating years of Rube Goldberg
attempts, we adopted a different approach. We are not scientists
nor do we claim to be; we leave technological refinement
to others. We feel that our best efforts should go toward
the formulation and conceptualization not only of environmental
techniques but also of guiding philosophies and goals.
A summary of the direction, goals, and philosophies behind a
responsive system is in order.

The linear thinking behind the rapid growth of scientific
discovery and technological achievement results in the total
exploitation of the world for man's "benefit," with no concern
for the environment in general or the future basis of
man's existence. I do not condemn advanced technology itself;
it is merely a neutral tool created by man. It is the mis-
directed application of technology for material wealth and
security that is wrong; technological application has not been
preceded by a well thought out philosophy or direction.
The increasingly apparent error in such one-dimensional
thinking has led us to establish what we somewhat loosely
call "responsive environments."

Ultimately, we are working for the realization and promotion
of the interdependence of all things. The guiding moral
principle behind the system we envision has been loosely stated
by others as "democracy without majority rule." We may say
that all things in the system have a "conscience" and thus a fish
or an amoeba has a "vote" in the system. The implied morality
of such a system asserts that any "individual" in the system
is allowed to enjoy total freedom of action as long as it
imposes no restrictions upon the freedom of any other "individual."
By definition, this morality is remarkably close to anarchy;
we suggest that herein lies a paradox which must be resolved.

This paradox seems to be the result of our human point of
view; in part, the idea of self-liberty and independence of
action is based on being outside the system. Humans can
speak of liberty among themselves only because we stand above
the environmental system and exploit it; such a moral and
political philosophy has in part led to the wanton destruction
of the environment by man. A totally interdependent system
including man is not possible until we reach what J. B. Calhoun
calls the "compassionate revolution."

It would seem that an environmental system based on the idea of
individual freedom of action is no system at all. If a large
fish cannot eat a smaller fish because he is impinging upon
the smaller fish's freedom, the large fish (and, by extrapolation,
everything) will die. The exploitation implicit in the food
chain seems cruel to us only because of our viewpoint from
outside the system. It is apparent that we must redefine our
system of morality. B. F. Skinner rightly notes that
freedom is an illusory concept. We are, in reality, controlled
by our environment. Can the concept of the individual's
freedom be integrated into the morality of the food chain?

A facetious example points out the arrogance of man in assuming
dominance over and ownership of the environment. It is
apparent that man is near (if not at) the top of the food chain.
Theoretically man has a place in the chain and must contribute
equally as he receives. Although there is no super-creature
around that eats man like beefsteak, at the very least we
should become the fertilizer for next year's crops. And yet
man has the arrogance to seal himself in an iron box (in fact,
he is legally required to do so) so that even the grass
cannot benefit from him. To correct this situation, we
propose a new ecology action drive: BE CALM: DON'T EMBALM.

To summarize briefly, the ultimate goal of our efforts at
creating responsive environments should be the promotion of
the realization of the compassionate revolution. In effect,
we should put man in his place.

We must ask -- What do we do now? It's obvious the environment
is screwed up and it is man's fault. Jumping ahead conjecturally,
we can theorize that until man learned to create extensions
of himself he was part and parcel of the system and was subject
to the response of that system. We can also conjecture that
the environment has suffered to an extent directly related
to the degree of conceptual space man has discovered. Thus, to
save the environment our task is to interrelate physical
and conceptual space.

It is probable that man's use of conceptual space (e.g.,
technology, etc.) has now overbalanced the environment to the
point that, if left alone, it could not "recover" (i.e., with
man still surviving). Everything negative man has ever done
affects the entire world; everything positive affects only man.
We must therefore create, from conceptual space, systems on
the positive side to remedy the overbalance. As users of
conceptual space it is our responsibility.

Since man has not been able to detect the needs of the
environment, we must make a system that can. Our mid-range
goal, then, is the creation of ever larger environmental
systems that are aware of the existence and needs of plants,
animals, the atmosphere, etc. It is here that our "new
morality," as yet undefined, is involved in deciding between
possibly conflicting needs within the system.

We may now be able to drop the label "responsive" and substitute the title "evolutionary." If we have an interdependent environmental system that can adjust according to changing needs, we will actually increase the evolutionary selective rate. The direction (or goal) of evolution has been previously described as the increasing complexity of organization and the increasingly efficient utilization of energy. According to Skinner, we can speed up the evolution of a culture by making man more susceptible to the consequences of his behavior. An interdependent environmental system will do just that.

It is not necessary to conceive of an environmental system as a gargantuan all-encompassing control mechanism. The same basic principles apply to small-scale efforts. Even an "evolutionary" dwelling or dwelling complex should retain in principle the ability to take an interdependent "look" at all relevant factors.

Even though the individual would have little control over the system in the traditional sense, largely because there are many interrelated factors enforcing parameters on his decision, man himself would greatly benefit from any application of such a system. A popular cliche today is that the only constant is change (see Future Shock, et al); it does not take much effort to reinterpret the word "change" as "evolution." As I have already noted, one effect of the system would be the increase in the rate of evolutionary change. One result of this fact may ultimately be the realization of the discredited Lamarckian theory of evolution. If our response turn-around is shortened sufficiently by technological advancement, we may be able to do "on-line" editing of evolutionary selection.

On the purely individual basis, such a system should be able to satisfy all of man's needs. On the lowest end, the system could certainly provide the physiological and security-oriented needs, while the sense of recognizable environmental response to his behavior should provide man with stimulation and identity, the high-order needs according to Ardrey. Ultimately, through a necessarily closer integration with the world and increasingly less time devoted to providing for security, man should be able to approach the possibility of Maslow's "self-actualization."

At this point, one caution must be noted. Given all relevant environmental factors, the system working for or responding to any two individuals must necessarily respond differently. By definition, contrary to Thomas Jefferson, all men are not created equal, and the system should not try to make them so. The system should promote diversity as an evolutionarily advantageous attribute. At the same time, the system must provide equality of opportunity to all individuals; thus we recognize the two basic needs of society, according to Ardrey.

I have tried to outline briefly the goals and objectives for an evolutionary environment as developed by the SPL. We feel that no more definite goals or directions can or should be stated; only strategies can be developed. If at any time the structure or "growth" of such a system is pre-defined or pre-set, either by its creators or operators, the true evolutionary aspects will be lost. In this lies the beauty of the system -- the ultimate flexibility. There is no one who can predict what the social, political, etc., needs of man will be 5, 10, 100, or even 500 years from now; we should not try. What we should try to establish is an interdependent environmental system with sufficient capability to "learn," if you will, to adapt, and to evolve as the unpredictable future presents unforeseen circumstances.

To the authors and authorities I have quoted here I apologize for a possible misapplication of what they have stated. However, in closing I will compound the possible error in extrapolating an author's ideas; it is the newest and possibly the most exciting and far-reaching concept stemming from our research in evolutionary environments.

It is the opinion of at least one authority on proxemics, linguistics, kinesics, psychology, anthropology, sociology, and architecture, Prof. Alton DeLong of the University of Texas, that the environment is and should be a code. (In passing, let me emphasize the growing need for the inter-disciplinary approach to environmental problems.) Recognizing the critical lack of communication between men as a result of different and specialized language and culture (particularly in non-verbal communication), the changing environment itself might become the coding and translating device for breaking down cultural barriers.

Man to man communication is definitely deficient and a serious problem. But an even greater problem is the communication between man and himself. The rational/irrational duality of man has long been unexplained, since man is supposedly the "rational animal." We may find the answer in a physiological look inside man's head.

It is a fact that in man's evolution, about 700,000 years ago, the hominid became Homo Sapiens; the brain of the pre-human primate more than doubled in size. It all happened at once; evolutionary selection generally happens gradually. More than this, we find that the new "big-brain" man was not behaviorally or socially much different from the hominid of 12 million years ago. In other words, when man got the big brain, he didn't use it. Not until the invention of the bow and arrow 25,000 years ago did man substantially change. This whole incident is contrary to evolutionary progression; when a trait is beneficial to a species, it is gradually selected for. But the brain expansion was not used; it had no benefit. The only explanation is that it was an accident -- either through freak world-wide climatic conditions or widespread inter-breeding of isolated gene pools, a far-reaching mutation occurred in brain development.

This may all seem totally out of order, but to clear it up let's take a look at the brain itself. Man's brain is comprised of three parts: the reptilian, mammalian, and the "human" (or neocortex). According to Paul MacLean, director of limbic research at the National Institute of Mental Health, the reptilian brain, or brainstem, the oldest portion, plays a primary role in instinctually determined functions: territory, shelter, hunting, homing, mating, etc. The second oldest, the mammalian brain, evolved over more than 100 million more years; the oldest brain was too clumsy when confronted with new situations. The second cortex provided greater ability to learn, adapt, and feel, etc. Over 100 million years, the two brains evolved and inter-connected, making strong neural connections, particularly in the areas dealing with emotions.

It was at this point that the human brain "exploded." The neocortical development has only had roughly 500,000 years to evolve and interconnect with the older brains; it has not done so. Few connections exist between the neocortex and the limbic system and, worse yet, they seem to be only unidirectional. To quote Ardrey:

The new brain speaks in a language that the old brain does not understand. With the gargantuan neuronal resources of the human neocortex, extensions of foresight and memory, symbolic language, conceptual thought and self-awareness became possible. But the animal brain does not know the language. Through moods and emotions the old brain can communicate with the new. But only with the most difficulty can we talk back, for it is precisely the equivalent of talking to animals. And that is why, for example, we may understand perfectly the cause of a psychosomatic affliction and just as perfectly be unable to do anything about it. We "act against our better judgment," "let our worse impulses get the better of us," plead that "somehow or other we could not control ourselves."

There is an animal within us, put there by millions of years
of evolution, and we must control it, or at least communicate
with it, or we will destroy ourselves in an irrational,
violent, animal mob.

We can't hope to stimulate overnight the evolutionary growth
of 100 million years. But our "rational" neocortex may
save us yet; we might either simulate or bypass that evolution.
If we can create an environment that can act as a code between
man and himself, we might have the answer. Presumably the
"animal brain," mostly dealing with emotions, can respond to
the environment -- space, color, etc.(For example, this
might be why blue depresses us, red excites, etc.) If we can
rationally shape the environment around us, we might be able
to communicate with the animal within us. The environment
is the common language.

I have proposed many different ideas -- the realization of the
compassionate revolution, the speeding up of cultural evolution,
the interdependence of man and environment, the environment
as code. All are part of the same concept: an evolutionary
environmental system. Obviously, such a system won't be
realized overnight -- or even 100 years. And yet we do have
the wealth and the scientific knowledge to carry it out: as
I write this, two men are on the moon. If we can stem the
tide of unthinking technological advancement, we can solve
the most crucial problem of man's history -- his relationship
to the environment. All it takes is an idea.

Structuring an Adaptive Environmental System

Joseph Mathis

An adaptive environment can be defined as a self-organizing, self-maintaining environmental system capable of soliciting information about user needs and preferences and adjusting its behavior and physical configuration according to those needs and preferences. It is the purpose of this paper to present a workable structure and organization of the artifacts necessary to facilitate an adaptive environmental system.

The concept of adaptive systems is by no means a new one. Indeed, in a mechanical sense the basic principle involved has been known since the invention of the governor (some time before the Industrial Revolution). In the biological sense adaptation is the foundation of evolution itself. However, it is only recently that man has applied these concepts to environmental control. As a result of the wide-spread use of analog and digital logic devices, process engineers have been able to make much progress in this area and have developed adaptive systems which rival the human intellect. Unfortunately, these systems have been used primarily to control process-oriented functions and large-scale plant environmental control, too complex for human operators. Inspection of these elaborate adaptive systems reveal that they are, for the most part, applicable to man's physical environment.

In an environmental context the relationship between observed behavioral phenomena and desired environmental change can be considered the performance criterion. In man's environmental context no such yardstick is possible as yet. This, then, must be the first and primary concern of all "adaptive environmentalists"; to develop a satisfactory and reliable method for determining the "success" of any environmental state.

The key part of any system is the control unit which governs its performance. In order to even begin to discuss structuring an adaptive system, one must have some basic understanding of the control system which would make it possible.

There are two basic concepts of control: predictive and exploratory. Predictive control is more commonly referred to as "feed-forward," anticipatory, or open-loop control. It is very linear in character because all information is passed in one direction; i.e., fed forward without any information being returned or looped back for verification. As a result of its linearity the designer must have sufficient prior knowledge of the processes as to be able to predict or anticipate the influence of all possible inputs. As a result of its aforementioned characteristics predictive control is extremely inflexible and totally unsuited for application in an adaptive environmental system.

Exploratory control is commonly regarded as a feedback or closed-loop control. In this concept measurements are made on the process and then feedback closes the information loop in order to verify and modify control actions. This feedback process gives an adaptive control system the capability of changing its actions in order to compensate for wide variations in environmental or operating conditions. In other words, an adaptive control system responds to any new input by modifying its actions to compensate. It then not only checks to see if the modifications have been carried out but also continues to check to make sure that they are maintained, while waiting for new inputs.

The great flexibility of an adaptive control system is a direct result of its three basic elements. These elementa are identification, decision, and modification. In order to facilitate an adaptive environmental system the present environment must be restructured into the following:
1. The Sensing Structure (identification)
2. The Adaptive Controller (decision making)
3. The Output Mechanisms (modification)

Joseph Mathis works in the Symbiotic Processes Laboratory at the University of Texas, Austin, Texas.

The function of the sensing structure is to provide the
system with a link to the outside world. It provides the
system with not only information about the state of the
"world", but also the present state of itself. The sensing
structure to a large dagree determines the flexibility of
the system, in that both the quality and quantity of
information which the controller has to work with is directly
related to the sophistication of the sensing structure
Indeed, the larger the quantity of information provided by
the sensing structure the larger will be the data base
upon which decisions can be made. The quality of the information
provided will directly effect the flexibility of response
the system is able to make. It is necessary that this
information pertain not only to the physical state of the
environment but also to the user preferences, as shown through
his behavior. This aspect is very important if the environment
is to be at all responsive to the user.
There are any number of sensors in the present environment
which register user preference, such as switches, thermostats,
rheostats and the like. If we add to these the tremendous
list of more sophisticated sensors available (hodo-meters,
video scanners, etc.) it becomes evident that a sensing
structure could be constructed having the ability to solicit
all relevant information about user needs. A sensing
structure of this sophistication is necessary in an adaptive
environmental system because without it the controller has
no means of relating its decision to user needs.

The ability for adaptive or responsive behavior resides
primarily in the adaptive control element. The traditional
adaptive controller functions as follows: It receives
information about the state of the world from the sensing
structure. It then constructs a model of this information
and compares this with another model, which represents the
optimum state of the system. This optimum model has been
constructed from information provided to the system by the
designer and is modifyable by the system at any time according
to parameters established by the designer. After comparing
the two models, the controller calculates the differences
between them and the action necessary to eliminate these
differences. It then generates the necessary siginal to
the output mechanisms in order to bring about the necessary
change.
The controller for an adaptive environmental system should
function in basically the same way, with the only exception being
the manner in which the optimum model is constructed. The
main reason why a different method is needed for modelling is
that, with human behavior as an input, the input possibilites
become so large and complex that neither the designer nor
the system could possibly anticipate them all. Therefore it
is obvious that the controller must possess the ability to
recognize patterns, learn from past experience, generate new
solutions and adjust its strategies accordingly.

The output mechanisms are those elements of the environment
which to a great extent have control over its physical state.
They are responsible for implementing the decisions made
by the adaptive controller. The kind, amount, and resolution
of adaptive behavior which the environment can display is
restricted only by the sophistication of the output mechanisms.
Theoretically this behavior could range from simple lighting
changes all the way to complete reorganization of physical space.

This paper has presented only the skeletal structure of an
adaptive environmental system. However, it should be clear
that the implementation of such a system would completely
change traditional man-environment relationships. For the
first time man would no longer need to adjust his behavior
in order to survive in a sometimes evan hostile environment.
An environment capable of determining user needs and adjusting
its behavior accordingly would for the first time place man
and environment in a true symbiotic relationship.

The Subtraction Method of Producing Structures with Robots

Forrest Higgs

The building process as it is presently known is characterized by intricate materials manipulation processes which form a wide spectrum of materials into a completed structure. Because of the complexity of this process attempts to mechanize the present day building process have produced marginal results.

An entirely different approach to producing structures can be taken. High quality structures could be produced by entirely robotic manipulation of building materials. With the utilization of robots the construction tasks could be performed on the site where the structure was to stand. Reclaimable building materials and robotic construction processes would enable new structures to be constructed from the thermoplastics of existing obsolete structures on or near the site. This technique would largely eliminate the need for the massive transportation systems now required for non-reclaimable building materials.

Removal of most of the labor factor from the building process and on-site reuse of building materials can result in a building technology which can meet the needs of our society in a much better manner than those presently in use.

OVERVIEW

We presently possess a housing delivery system which has reached the point of diminishing returns with respect to the amount of resources invested versus the quality and quantity of structures produced. The major point of stability within the present system is the dependence on skilled laborers employed on-site in a manner ensuring a relatively fixed level of worker productivity.

Attempts by the laborers within the construction industry to increase their economic and social stature within the society at large while maintaining relatively fixed productivity has had the tendency to price individuals in lower income brackets out of the traditional housing market.

Government at all levels has responded to the developing trends (i.e.: less housing for more money versus the increasingly popular attitude that good housing is a necessity) with a wide spectrum of short and medium range programs. In the short range the federal government has introduced measures to make the use of non-traditional housing types such as mobile homes more attractive to potential buyers. In the medium range the government has sought to reduce the cost of housing by encouraging the industrialization of the building process.

The chief operating concept behind the present attempt at industrialization is that of the early first industrial revolution. This technique utilizes assembly lines manned by workers performing highly repetitive tasks. The technique usually results in an increase in worker productivity great enough to justify the initial outlay for setting up the production line. As the individual worker operations are understood the workers begin to be replaced by automatons capable of performing the operations more productively than the human laborers.

The unique nature of housing has produced problems which have discouraged the efforts over the last half-century to subject homebuilding to industrialization. The desire of the public for individuality in housing, however marginal that individuality may be, has proved one of the greatest deterrents to the success of industrialization. Many persons now purchasing non-traditional housing types tend to view their living situation within such housing as a transient condition. Their ultimate goal with respect to housing usually remains a individually designed home.

Forrest Higgs is a graduate student in the School of Architecture of the University of Texas, Austin, Texas.

This feeling and related feelings within the housing industry has made the mobile home industry among others a largely throw-away form of housing. The notion that our society can afford throw-away housing and the enormous problems associated with it is at best questionable.

PROBLEM ASSESSMENT

In general much of the problem with the housing types produced today is that the structure and support systems are intricately bound together. To compound this difficulty the structure alone often consists of as many as thirty distinct types of materials. This complex of materials is both hard to assemble and extremely resistant to economical reclaimation. Due to this fact most structures are reclaimed by what amounts to a degenerative process. Materials reclaimed from a structure are generally not suitable for use in another structure in the same function. Materials committed to structures usually become trash after one to three use cycles.

The committment of our society to rapid change and mobility is the basic reason that the characteristics described above have become critical The massive population shifts which have become a commonplace in the last quarter-century have had the effect of making structures built for one purpose serve completely unexpected functions within the service life of the building. This development has encouraged the generation of general purpose structures which serve no purpose excellently but instead serve many functions with at least a tolerable efficiency with minor changes.

The tendency toward general purpose architecture has had a greying effect on our society. The sameness of living experience brought on by the bad fit between this sort of architecture and the society which it serves has greatly decreased the richness of the living experience within this country.

At this point two major courses of action are open. Societal mobility could be discouraged or a building technology synthesized which would allow for efficient change of structural form. Societal mobility could be reduced somewhat without radical change. Unfortunately mobility within our culture is intertwined with the concept of social and economic advancement. Increasing ones stature within the society almost assuredly requires a number of physical moves in all except the individual sorts of professions such as law and medicine. Even if we take the course of reducing social mobility the society will find itself with an inventory of generalized design traditions which would probably linger for several generations.

A more acceptable course could be the development of an architectural vernacular more closely fitted to our society.

Several sorts of things are implied when we speak of handling change in structural form efficiently. Obviously this change can ill afford to cost the society the same level of resource committment and labor investment that the present building technology requires. We are, at this point, only barely able to afford the costs of providing our population with adequate shelter and activity space. One of the major factors of the high cost of construction is the labor intensive nature of present day construction techniques. This factor directly ties the cost of structures to the standard of living of the workers involved in the construction process.

GUIDELINES FOR SOLUTIONS

Mechanization of the building process is the only visible alternative at present to the problem of allowing for more efficient change in our architectural inventory. The present method of assembly line production is one method of achieving mechanization. The assembly line has received more than fifty years of field testing in the housing industry with results to date being very discouraging.

Another method of mechanizing the building process
remains. This method produces structures with robotic devices.
A sort of localized assembly line is created with the end
product being many times larger than the assembly line. The
assembly line rather than the product is, therefore, moved
from place to place. Adding reclaimable building materials
to this system reduces the transport system required to support
construction activity and makes cheap, efficient change a
definite possiblilty.

SYSTEM OPERATION TECHNIQUE

The method of automated construction which will be
used for the project is known as the subtraction technique.
When the construction project commences a robot begins to
extrude layer on layer of foamed thermoplastic the density
of which varies from point to point according to the structural
and aesthetic requirements of the master plan residing in the
memory of the computer controlling the robots on the project.
The foamed plastic is differentiated into two parts; structure
and scaffolding.

The structure consists of foamed plastic of densities
determined by the master plan. The scaffolding consists of
foamed plastic of a sufficient density to bear the weight
of the robot and the structure which will be put upon the
scaffolding in subsequent layers. The scaffolding is built
up with the structure. Ramps enabling the robots to move easily
from the ground to the level of current construction are extruded as an integral part of the scaffolding.

When the topmost portion of the structure is completed
the extrusion device is removed and cutting and cleaning
robots are sent in on the job. The robots are then put to
the task of removing the scaffolding layer by layer and putting
final finishes on the exposed faces of the structure as it
is exposed from the scaffolding by the cutting robots. Shards
of scaffolding and excess structural thermoplastics are collected
and saved for future use by robots mounted with vacuum cleaning
devices. When all of the scaffolding is removed the process
of construction of the structure is complete.

The structure can be reclaimed when it is no longer
useful by what amounts to the reverse of the above process.
A useless structure is enveloped in foamed thermoplastic
scaffolding and reclaimed by the cutting and cleaning robots
on the same layer by layer basis until the ground level is
again reached and the structure no longer exists.

MAJOR ISSUES GENERATED

At the end of the project period which we intend to
pursue a firm groundwork for the introduction fo robotic building techniques into our society will have been laid. When
later development and research by both the public and private
sectors produce production models of our prototype system
housing can become a cheaper and higher quality commodity. The
presently existant delivery system for housing and the materials
supply systems for the housing industry will become increasingly
less important as the new system becomes accepted. Much land
devoted to tree farms will be allowed to revert to wilderness
areas.

The chemical industries supplying the thermoplastics
for the structures need not become extensive. Their production
capacity should be keyed to the rate at which conversion to
the robotic construction systems is to take place. As the
conversion proceeds the nation will begin to possess a resource
pool of thermoplastic building materials in the inventory of
structures required by the society. This resource pool will
need additions only to cover deterioration losses and whatever
growth factor is required by societal imperatives.

Great savings in energy expenditures required by housing can be made if the superior insulating qualities of thermoplastic foams are correctly utilized. Comfort conditioning loads from preliminary calculations indicate that solar radiation utilization to control internal climate would be much more feasable with thermoplastic structures than with traditional structures. A general utilization of thermoplastic structures could greatly alleviate energy generation requirements if solar energy utilization for climate control was combined with the insulating qualities of thermoplastic structures in a new housing technology.

"It's crucial to the whole idea of responsive environments...
Architects have <u>not</u> been able to determine what a user
really wants in a space."

"We feel that there should be a biological architecture,
in the truest sense of the word..."

"All this is based on biological development of environmental systems, a true organic building."

"Through a more advanced, humane technology, we can provide
individualization of housing in real time."

"We asked ourselves a couple of years ago, 'What do you
respond <u>to</u>?' The immediate answer was,'people'... We have
a feeling that people <u>should</u> create their own environments.
We also feel that there is not a great deal of awareness
of the environment in the person...'"

"By and large, a person can't define exactly what he wants
in his environment. So we went to psychologists and asked
'How can we find out what these people want? We're supposed
to be architects, the mediator that does that, and we're
not. Now is there some way we can get closer to this man,
and find out what he wants and what he needs?' The psychologists said they didn't know."

"We thought that maybe instead of extruding things, we could
project an image of an environment in space, and somehow
or another give it substance... We found out that you can
build, using interference patterns, three-dimensional
volumes of light. We were also lucky enough to stumble
across several materials that solidify under light. So
we started fooling around with them."

"We've been working at two ends, both trying to conceptualize
how one makes environments responsive and what responsiveness
is; and we're also looking at ways in which the environment
can change."

"These systems are an extension of man, a part of man, and
man is a part of them, and they would evolve through time,

testing subtle variations in the environment, and according
to the user needs and emotions and reactions to the system,
would selectively evolve the most successful system for that
purpose."

"If this building system can decide on a certain spatial
need for a person, and put it up, [first of all, it can
change; and second of all,] if the person doesn't like it,
it evolves until the user does like it."

"Now we can talk about the environment... as a rational
extension of our brain..."

"We feel that a building should have a spirit"

Discussion following Texans' presentation

The Texans, working as a group, offered a ninety-minute triple-
screen color slide presentation of their ideas. Several hundred
images were projected, only a handful of which can be reproduced
here. Midway in their presentation, other participants began asking questions and offering comments, and a very heated discussion
broke out as the last slides crossed the screens. It lasted for
more than an hour, ending only when hunger pains drove us, still
arguing, to the local eateries. A number of participants have
mentioned in the ensuing months that this presentation was among the
most thought-provoking that we witnessed, and that they recall
the debate which followed as being bloody, but extremely useful
in delineating the problems that we all face in searching for
a responsive architecture.

A review of the tape recording of the discussion reveals that
a majority of it was devoted to arguing whether the right number
and right kinds of slides were shown, which is perhaps symptomatic
of the problems of any group that consists predominantly of architects:

"Why are you showing us this stuff?"

"There's one thing that can be said for all this: The slides
are beautiful."

"You've got a magic carpet here, but you're trying to nail
it to the floor."

"So many of your images of the future are obviously unpleasant."

A couple of the conferencegoers seemed particularly incensed that images of various forms in nature were being used as justification for a machine-built architecture, where they had used the same images to indicate the organic quality which they felt could be achieved only through a long evolutionary process between an owner-builder and his handcrafted house. The Texans were also attacked because they often seemed unsure whether their images, despite their strength, represented the answers they were searching for. This helped to turn the discussion into a more useful path:

Voice: "The important thing at this conference is that everyone should help each other, because everybody's having trouble, I think. One of the issues that we're trying to get into, and are having just a hell of a time with, is responsiveness . . . what do we mean by responsiveness, anyhow?"

Jacobson: " . . . our approach at the Center for Environmental Structure is that we talk about responsiveness, but we try to discover, through the memory stored in our buildings and in our culture, those aspects of the environment which seem to be the most worthwhile. We look at an example which seems to lead to a good experience . . . we try to extrapolate . . . we try to say, 'that's what's worthwhile, that's what we want to preserve.' "

Hamilton: "The kind of process that goes on--the joy that happens sometimes when you do experience this thing we're calling responsivemess, is something that we're evolving pieces of, and . . . you go through the dome thing for a while, and it's a great trip . . . it's something that you value. Then you move on into a different one. We're fishing around for something that's based on the qualities of that process."

Kahn:"The process I like is doing things with your hands. The people who are building things with their hands don't want to be released from doing that. They don't dislike putting in windows. To build a house takes an awful long time, a half a year, but they don't want to be released from that."

Allen: "The one thing so far that's really brought this group together, the one thing everybody here can identify with, is Daria Bolton's game. We could work out a lot of the things we're hassling over here, if we could all sit down and play that game. It would put us all through a common experience, in which we would find ourselves reacting to what's happening, and as we began discussing our reactions, we might find ourselves beginning to get someplace."

Wellesley-Miller:"The one thing about the game was the immediate response you had, the ease of manipulability. But if it was played in real time instead of as a model of the situation, it would be much longer and slower, and I wonder if, in a situation where you're pumping concrete and banging in nails, if the game would be as enjoyable. In the game, in matter of hours, you go through things that in real time would take years."

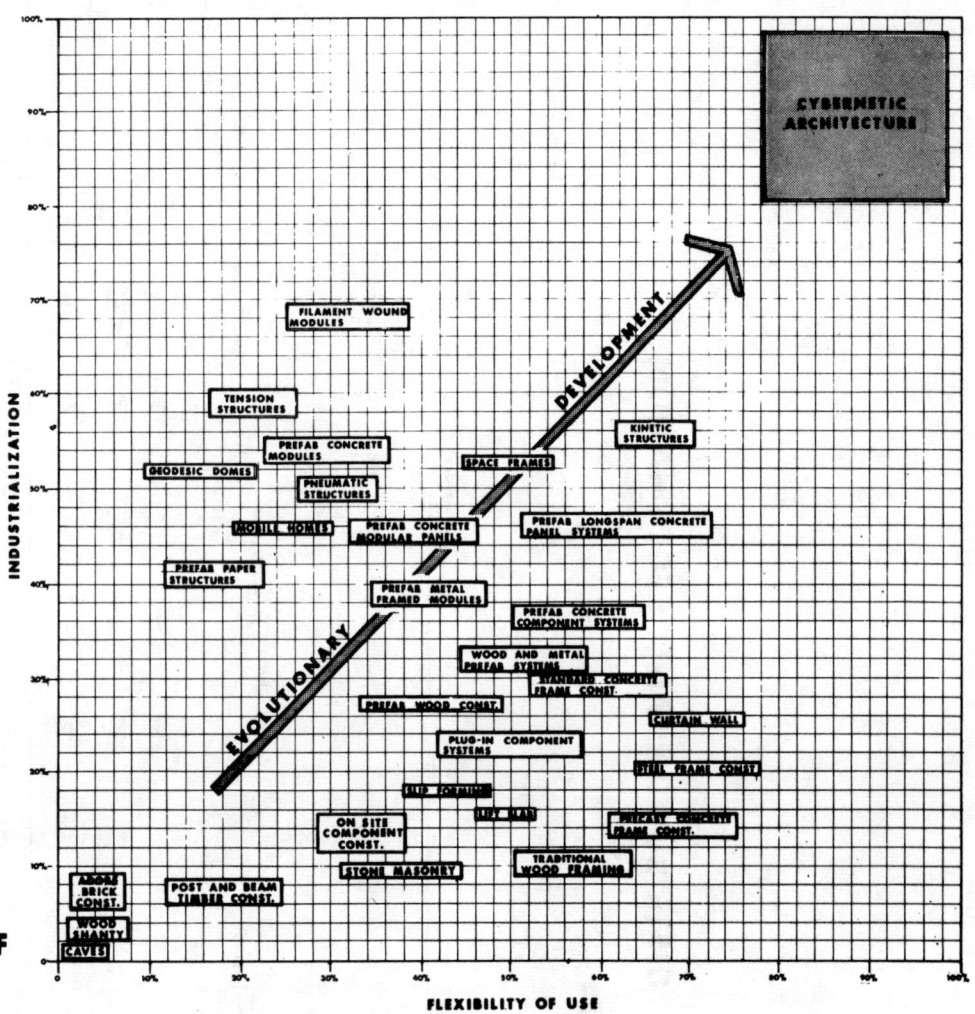

COMPARATIVE STUDY OF BUILDING TECHNIQUES AND SYSTEMS

Cybernetic architecture as it relates to known building systems.

MATERIAL DISTRIBUTION AND RECLAMATION

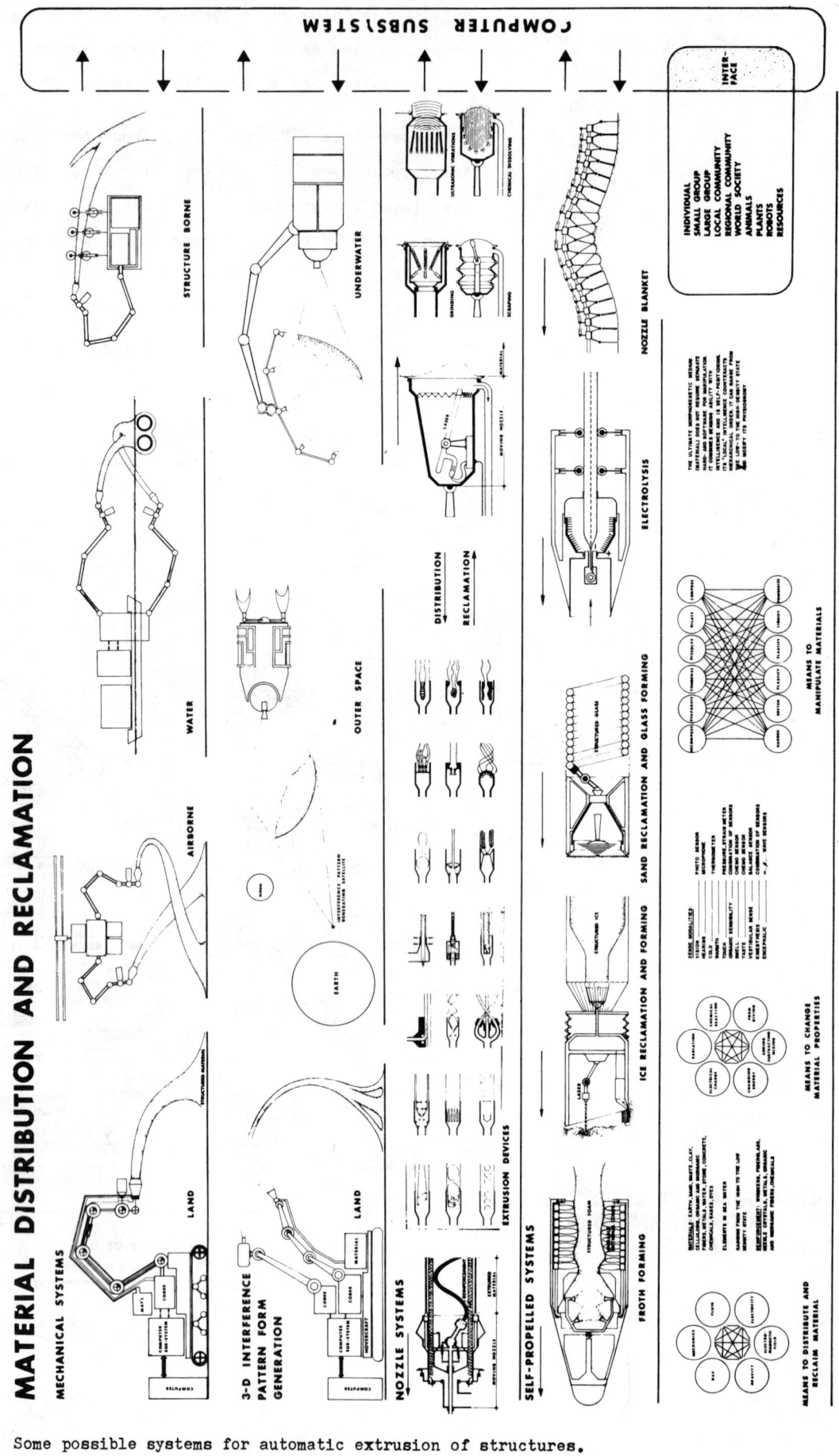

Some possible systems for automatic extrusion of structures.

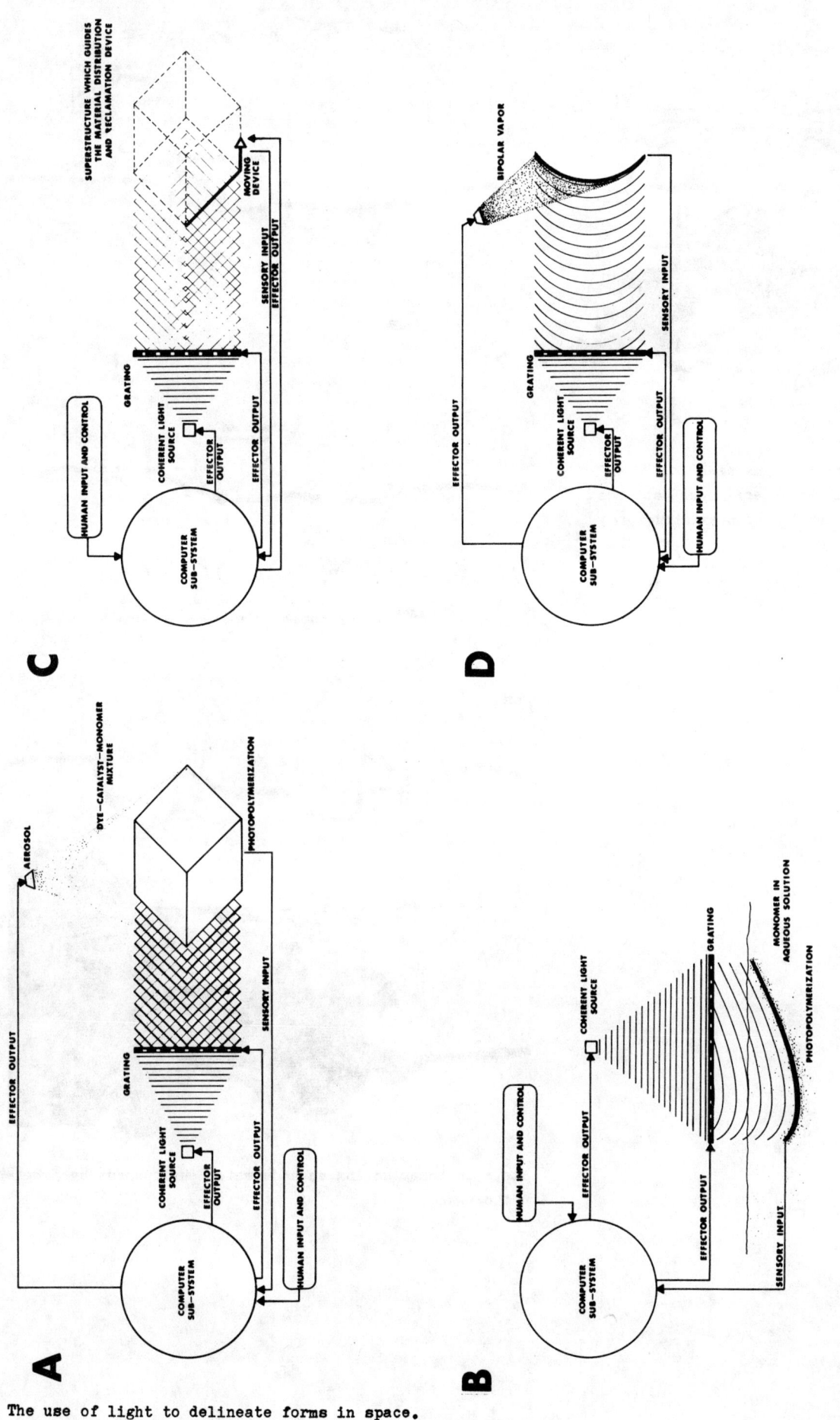

The use of light to delineate forms in space.

3-D INTERFERENCE PATTERNS FOR FORM GENERATION

A hypothetical machine
using light interference patterns
to construct a complex structure.

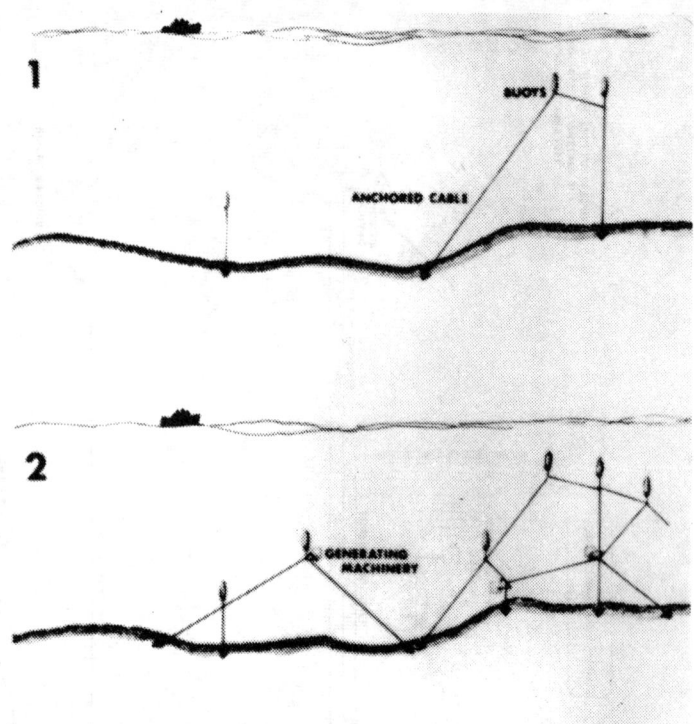

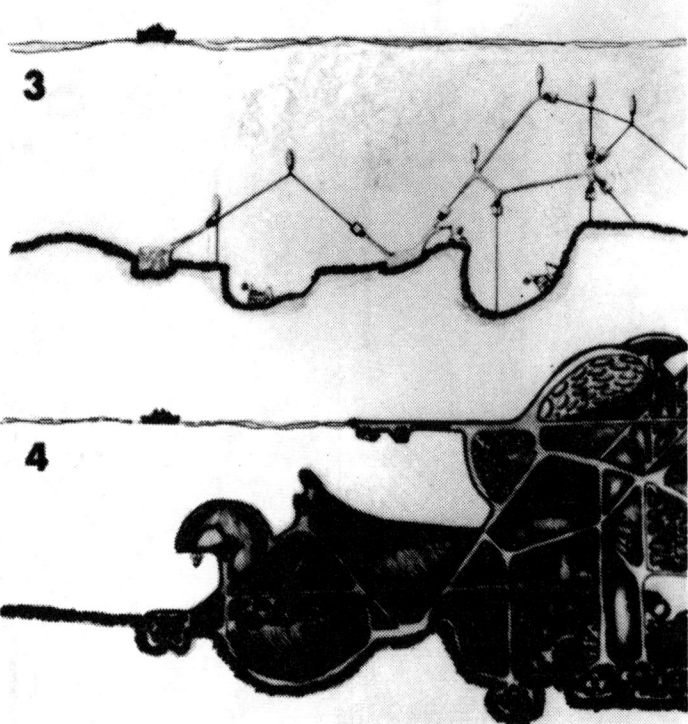

Steps in constructing an underwater environment by cybernetic means.

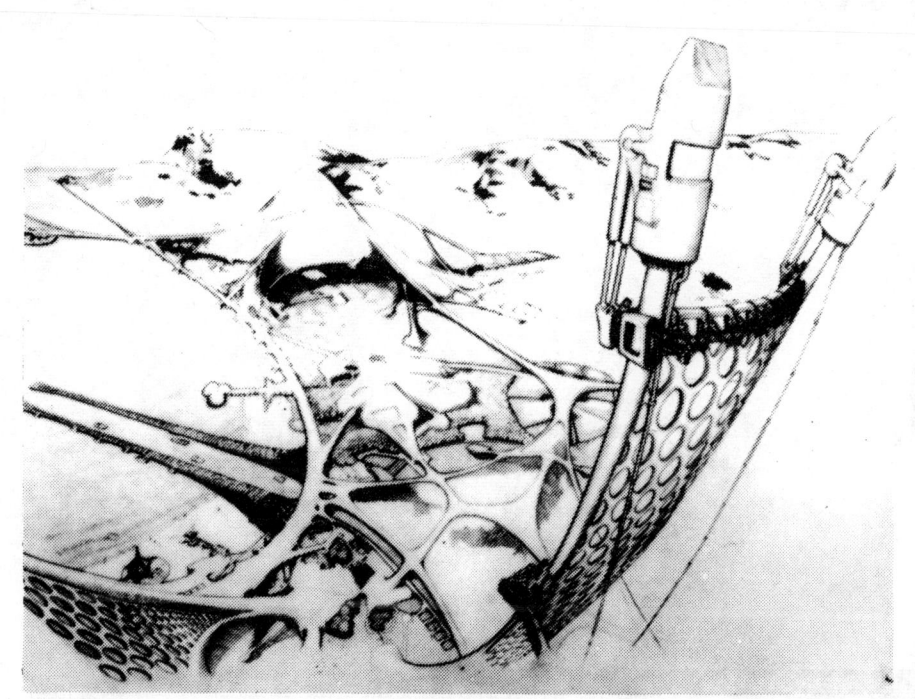

Machines building
terrestrial structures under automatic control.

A laboratory version of an automatic
building machine.

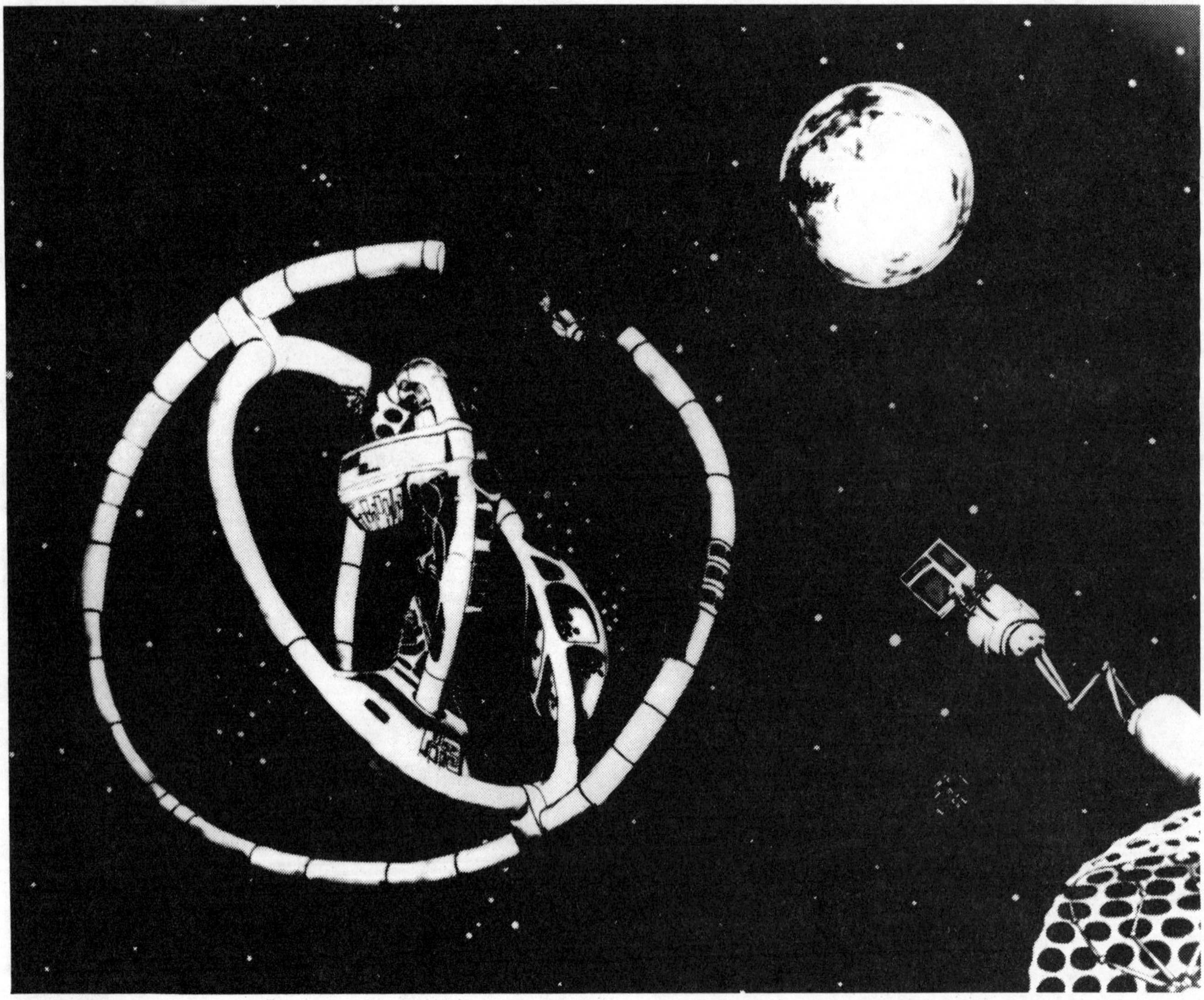

Machines building structures in earth orbit.

The Three Little Pigs Revisited

Avery R. Johnson

Then along came the wolf and said:
    "Little pig, little pig, let me come in."
"No no, by the hair of my chinny chin chin."
    "Then I'll huff and I'll puff and I'll
    blow your house in."
So he huffed and he puffed and he puffed and
he huffed and at last he blew the house
down, and he ate up the little pig.

                 Ubiquitous

      Y'see, kid, ya can't make houses of straw or sticks; they aren't any protection against the wolf! Ya wouldn't want to be eaten up by the wolf would you? Ya gotta have a brick house: solid, strong (yeah, and lumpish and square and unchangeable and one that will isolate you from the out-of-doors as much as possible). Houses of straw and sticks are play houses: y'know, for kids. Y'don't live in houses like that! It's not safe. It's not sanitary. It's........

SECURITY IS WHAT YOU WANT, RIGHT?

      Sure, I know it's _fun_, but what's that got to do with it? Living is a serious matter! Ya gotta make up your mind; decide what you want to be; commit yourself; don't take chances; live in a stable neighborhood; acquire the right friends; get a secure job, _make_ something of yourself (and be able to name what you are if anybody asks). Phooey!

DAYDREAM FOR A MOMENT

      Imagine your own rendering of "The City of the Future". Take a minute to envision the whole thing........now look at it. Does it show one building crane, or a torn-up street, or any other kind of change in process? Be honest, now. Don't cheat by putting those things in now just because I asked for them. Did your utopian vision allow any room for activities or subsequent changes that you haven't thought up yet? Did you focus on a snapshot which was to exist for all time?

First published in Eleven Views, Vol. 20-1, Student Publications of the School of Design, North Carolina State University at Raleigh.

Were you _in_ the city itself, or outside looking at it? Was it a living, growing, evolving creature? Was it uniform or did it offer high variety?

> Were you trying to establish some kind of "truth"?
> Have you thought of giving relevance a chance?
> What might be involved in that?

A CYBERNETICIST'S VIEW

I really didn't want to insult you but rather to start your participation with me in dealing with some problems. All of us have a large stake these days in improving our environment and our communications. Within the field of cybernetics we think we've hit upon some approaches that stand a chance of meeting the criteria of complexity, relevance, and responsiveness that have been plaguing everyone for answers. I would like to address myself particularly to the problems of urban planning, and within that context especially to the relation of the individual participant to the community within which his energies are spent. If that community fails to metabolize his energies in such a way as to enhance both him and its relationship to him, then it fails to fulfill itself. Cybernetics concerns itself with the viability of complex organisms.

What do you want a community to be? If you think that question is too broad, try another one which I consider to be equivalent: How are you going to get it across to the members of that community the fact that it has those properties you have wished for it?

For me, the entity that is a community has its existence for the individual within the process by which he explores its responsiveness to him. This is true of good politics, good education, good transportation, and it should also be true of good architecture. How often do we ourselves have the opportunity to experience it that way? Buildings, urban environments, landscapes, and even traffic plans all seem to have an immutability akin to a stone wall. Each acquires its character for the potential user upon first exposure within the time it takes for him to discover which limited subset gives him the least grief in going about his survival within a larger system. Thereafter all other topologies of the remainder are simply ignored (as much as possible) because they have become redundant. The overwhelming irrelevance of objects and events around us is attributable largely to their inability to show us any reflection of our passing: to ackowledge some unformulated wish on our part that they might somehow, just for once, be different.

--- Different because we willed it so.

--- Different because it just happened. We didn't will it, but we didn't quite guess it either and so it raised our information level and introduced us to new alternatives.

--- Different because the environment has some interesting behavior of its own and is _exploring_ _us_ for our responses of approval or delight.

--- Different because if it doesn't change sometimes we will remind ourselves of its existence by writing obscenities on it or burning it down to make it respond. Or if we are well schooled in the science of labelling, we can _name_ _it_ and let it thereafter come to life on picture postcards as an art form --- not for living in but for visiting. There's relevance, baby, you can write on it!

Humans cannot tolerate total low-variety environments for long without becoming sick. We need just the right amount of good guessing to produce the difference that makes a difference. If _our_ behavior makes a difference, then change takes on meaning for us.

CHANGE

We hear a lot about change:

the necessity for it

the inexorable fact of it

the need to be ready for it

the inevitable discomfort of its coming.

Then from another side we hear that CHANGE --- any change --- is inherently good. The Change Makers believe the outworn maxim that "more is better" whereupon they throw a heap of garbage in our faces and expect us to groove on it. The conceit of the expert: "do unto others what you believe is right for them", virtually guarantees irrelevance. Their magical belief is the notion that moving a particular quality of an object or event, such as its size or number or CHANGE from its prior status, is itself inherently good (or bad). That is simply expert nonsense.

I repeat: if _our_ behavior makes a difference, then
change takes on meaning for us. We have been too well
schooled by the truth-sayers who proclaim that meaning can
be pointed at: observed. That, too, is nonsense. Meaning
can only be discovered _in_ _context_ --- and upon that barb hangs
not only the modern dilemma of mathematics but also the crux
of this article.

CONTEXT AND WHERE TO FIND IT

Do you know what I mean by CONTEXT? It is the essential ingredient left out in the truths of the "objective"
sciences. The context of an object or event is not something
you can point to and say: "There it is!" We have no calculus
for it (yet). It is generated in your active processes of
perception and can never be fully shared by another. That
is why it is left out by the truth-sayers. Context acts
_as_ _an_ _operator_ (9) to assign meaning to the metaphorical
signals we receive from the world, but it is not to be found
in those signals. It is to be found, rather, in the consequences of our response to those meanings in that environment.
"Get undressed" does not convey the same meaning in a doctor's
office as it does in the back seat of an automobile --- but
it would be a mistake to identify the background setting in
either case as the context. Look to the consequences that
are implied and to the relationships that point to them.
Keep a running account of the infinite recursion of those
relationships and consequences and you may have a handle on
the context. It is difficult to simplify further; impossible
if what you want is a formula that can be applied in another
case.   Fine..........O.K...........sure..........but...
what does all of this tell us about urban environments?
It tells us quite a lot if we can first leave behind those
habits of thought which are based upon context-free premises
which promise to lead to easily printable, transmittable
"truths" in conclusion. The primary premise that we must
drop is the one which removes the untrained community member
from participation in the changes occurring in his environment --- because he cannot play the game of "all people
universals", is interested in what he personally wants and
in his capacity to produce the level of difference he wants
when he wants it and so he knows it. What we should be

seeking, then, is the means by which members of a community may affect their environment in informal ways which are adequate to return to them a sense of active participation.

There is a payoff which commences almost immediately but is seldom recognized as arising from a common source since it always appears in a garb closely associated with the particular activity in question. Participation is the wellspring of appetite and feeds back upon itself to deepen and enhance the individual's involvement with the broadening of his own contexts. Some day our psychologists may recognize that <u>appetite</u> is the fount of motivation; hunger is not.

THE COMMUNICATION PROBLEM

I talk to people trained in other ways of thought. They believe, for example, that one must be prepared with a lot of knowledge before one can <u>decide</u> upon action to be taken. They fail to recognize the circuitous, digressive way that thought proceeds to recontext prior experience and thereby imply new responses and new consequences. These latter then concatenate to generate new doings and new seemings and thence to wholly new anticipations which in their turn beget our sudden recognition of "aha!" I wish more people would read "Eolithism and Design" by Hans Otto Storm (16) for it would convince them of the value of whimsical game playing. Undirected exploration --- real-time groping --- seem to have no place in what is taught as real. But what questions, what play is possible if one knows that soon the concrete footings will be hardened and the mortgage money loaned for a specific bank-approved design? The box is cast. Students, be still! <u>I</u> <u>will</u> <u>not</u> <u>accept</u> <u>the</u> <u>common(non)sense</u>. Call me childish if you will.

TOOLS AND TOYS

I want toys as well as tools.

What is a tool to you? To me a tool is an extension of my hand or eye or whatever which allows me to manipulate some part of my world in a way that would otherwise be unreachable or at least more difficult. That manipulation permits me to express some intended change upon the world around me. The scope of my expressions of intention may be severely limited by the tool and my use of it. You give up possibilities for difference that the tool cannot cope with at least while you are conditioned by it: cars produce fat bellies in place of strong legs.

What is a toy? I think of a toy as something which embodies relationships which are otherwise not available for exploration. Its modelling of ordinarily unrealized relationships is based on a revealing shift of size, or of time frame, or of material........or in any event a shift of context which is recognized implicitly. A TOY IS SIMPLY A TOOL TO THINK WITH. It renders inconsequential any "errors" of exploration on your part and allows you to place into juxtaposition many relationships which would be either unlikely or be passed off as irrelevant in a less playful context. Toys invite exploration of what was taken for granted or was otherwise unknown.

A child --- or a childlike adult --- acquires knowledge of how things work and of how to change their workings through participation in making changes happen, and in simultaneously observing consequences. I emphasize simultaneity. Sending off the Wheaties boxtop and later seeing the prize return in the mail requires mama's word that something of consequence happened. In real participation results cannot be grossly separated from their instigations. Identification of causes and effects by name cannot lead to a meaningful description of the experience. Purpose and rates-of-change in the direction of that purpose are a more apt statement. Furthermore, the recursiveness of purposive systems is far easier to experience in a playful setting than in an analytical one.

Remember elementary school? On your own time you developed skill in the whole-body movement of game playing. You knew objects in terms of their fun potential. Names were for identification, not for explanation.

Then you were taken into class and told to sit straight. Don't fidget! Look, listen, speak when called upon. The names were made over into energy consumers and thereby began to acquire a reality of their own. You became a budding expert at taxonomy for that is the major tool of the scholar --- and then like any other tool it imposed the conditions upon what it could make. Take a hard look around you to see where you can find a trace of playfulness: one, that is, which still invites you to play; not some frozen metaphor of someone's long-lost toy.

A BRIEF RECAPITULATION

If you want someone to grasp complex relationships and to identify with their processes then you must not only allow him to experience them, but also to have some effect upon them which may be observed first-hand. The involvement that you offer need not be total --- the ordinary citizen doesn't want the responsibility of redesigning the city, nor does the hospital patient want to be his own surgeon --- but somewhere there must be an interaction between the individual and his surrounding which admits of his existence because it responds to his use of whatever skills he has.

AN UNCOMFORTABLE PARALLEL

Architecture and medicine share a common professional fault which at this time in history is doing them both a great disservice. They are conducted as priesthoods: the services they provide are performed upon the recipient but he himself is not allowed participation. Playfulness is taboo. When the modern planners of "health-care delivery services" finally come to recognize that the patient himself is a well-intentioned and highly motivated self-organizing system and can be trusted with information about the meaning of his own physiological signals, then they will begin to achieve some success both in reaching patients and in modernizing their own concepts. The parallel of the medical priestly attitude with the urban design "expert syndrome" is not so immediately obvious because the client in the latter case is not so aware of the source of the pain to be remedied. Nevertheless, his ignorance of the processes of change and of improvement is similarly based: he has never been afforded the chance to participate in the changes which are imposed upon him. Some people learn the rudiments of painting walls and fixing leaky plumbing but these efforts are akin to the application of a bandaid on a superficial cut. Nor am I suggesting that they must learn the "expert" task of setting goals and specifications. Not at all! We must

build environments that invite their playful participation
so that their self-referent knowledge of their community will
grow with their appetitive involvement.

COURTEOUS ENVIRONMENTS and NONROBOTICS

What I have been leading up to is the notion that the
environments we provide for people must have some intelligence
built into them so that mutual explorations can commence at
an informal, unskilled, elementary level. People must be
allowed to discover for themselves that it is not difficult
--- and may be quite enjoyable --- to attempt expression of
their intentions. A few of us have spent some years toy-ing
with such environments and have made inroads into the
techniques which can produce behavior that is "courteous"
to the participants. I will attempt to set forth a best-to-
date description of their properties which will allow you to
start toy-ing for yourself. If you do not wish to become
involved with real things but prefer simulations, then stop
reading here to save time for whatever _you_ think is important.

I am most certainly _not_ intending to raise the spectre
of a mechanical or electronic Big Brother Robot which is
hyper-attentive to you, watching your every movement and
every change of heart-rate or respiration or alpha-rhythm
as if to quiz you constantly and surreptitiously to find out
what you want. No, that sort of behavior is not at all
courteous. That way of imagining "intelligence" assumes
that the data which the robot is collecting can somehow be
made meaningful (decoded, interpreted, translated) so as to
tell it what to do next. It's the old "decision model"
which we have already laid to rest. For example: a robot
armchair programmed to play soft music every time you get
restless, to dim the lights when you rest your head back,
or to keep the temperature of the space surrounding you at
some preset level. No, and again NO! Therein would irrelevance soon be guaranteed. It happens when a mechanism is
preprogrammed (therefore acontextually) to do for us what
we will want.

Machines and machine-like human systems that people
propose suffer from the decision concepts that theorists find
easy to manage. In my opinion a gross misunderstanding

prevails about how wants and meanings arise for us. It
produces the fallacy which leads people to believe that our
brains process sensory data and decode it into a description
of the world around us. I would state the rule I use as
follows: In order for us to elicit meaning from any data
entering our sensorium, it must either have arisen as the
consequence of our effector (outgoing, active) interaction
with the source of the information, or at least imply an
interaction in which we might engage with some other
sensorimotor combination in our perceptual apparatus.

The notion of a necessary participation in the events
and objects which we wish to make meaningful to us cannot be
overemphasized.

Let's see if we can arrange matters so that an
architectural environment will be able to follow the same
rules in dealing with us.

MACHINES AND THEIR CONTROLS

So as to avoid an obvious omission, let me say a few
words about the ways in which we now are accustomed to control various "bits" of our environment. For the most part
it is by way of switches, valves, control knobs, levers, and
other manipulables. We do not communicate with our fellow
man in such an arms-length manner which somehow has seemed
appropriate for mechanisms or environments and even homes
which have no "life" of their own. The problem has always
been --- and it remains to this day --- that we have not as
yet been able to teach our machines to grasp our intentions.
Why not? Because those machines have been denied exploratory
behavior of their own through which they could establish, in
terms of their own self-referent responses, the CONTEXT of our
gestures toward them. The alternative that has always been
chosen has been to limit the context of those intentions so
severely that they are in no danger whatever of being misinter-
preted. Turn a valve, push a button, flip a switch: ON ---
OFF --- UP --- DOWN; easily understood because the context
of the gesture is absolutely explicit.

The advent of computers has clouded the issue lamentably
because they seem to be able to engage in highly complex

exploratory behaviors of which we are incapable unassisted.
Overlooked in the wonderment at these feats is the very clear
fact that those machines have interpreted the instructions
given them as explicit and as meaningful in the extremely
narrow contexts of the language in which they are stated.
Our fellow man, on the other hand, shares a commonality of
experience with us and therefore can identify with us so as
to be able to discern the meaning of every word, gesture, or
change of timing as they convey our meaning richly in its
full context. He can literally (almost) put himself in our
place well enough to make the interpretation of our meaning
a _self-referent_ act implying _his_ _own_ interaction with our
world.

If you don't believe me, try engaging in a deeply
meaningful conversation with a stranger from a culture
foreign to yours the first time you meet. Watch him struggle
to find self-referent material in what you place before him.
If you can't find a stranger handy, try a child --- or do you
doubt that a child has deeply meaningful experiences he would
like to be able to convey? Help the child to find a means of
interaction common to you: use a toy and watch his understanding of intentionality deepen.

The problems we face in trying to make an environment
communicate with us are not so difficult as you might at first
imagine. The reason that this is considered by custom difficult is related to the fact that the object of the communication is so intimately bound up with the communicants themselves.

Let me put it another way.

Every dialogue is _about_ something, but the manner of
the transaction will depend upon the "distance" between the
referent and the place and time of the dialogue itself. I
have written elsewhere (_13_) at some length about this, but
let me summarize with an example or two here.

In human experience the most intimate dialogues ---
as between lovers or mother and child --- are carried on in
a physical mode of touch and movement which involves each
person totally, but which demands very little of his capacity

for decisionary behavior during the encounter. One experiences more a sense of flow and this is real in spite of its unaccountability. We use little or no symbolic language in these affairs. We can carry on the dialogue with infants, people of other cultures, animals, or idiots because the referent of the dialogue is each other and the relationship between.

There is a broad spectrum of dialogue situations which shows the "distance" of the referent (what the communicants are refering to) gradually increasing and there is a consequent requirement placed upon the communicants for more elaborate and refined behavior. At the far end of the spectrum one finds symbolic language where the referent need not even exist in fact. Try talking of mathematics with an infant or an animal!

Between the limits of immediacy at one end and symbol at the other we see varieties of situations where people can share a communications medium, but where the direct interaction of each with that medium may itself fall within the awareness of the respondent to a greater or lesser degree. The more immediate that awareness is, the less complicated and metaphorical the language has to be.

I won't reargue the paper here. I only want to indicate that the kind of dialogue which an environment can be made to engage in with us can and must be pushed as far to the lower, intimate end of the scale as one can manage. There the design and fabrication of courteous environments becomes astonishingly simpler than of ones that manipulate symbols. The materials and techniques may be unfamiliar to most architects, but perhaps that is because flexibility is in the domain of the first little pig.

SOME PIECES OF THE PARADIGM OF COURTEOUS ENVIRONMENTS

What follows are some principles or rules of thumb which have simplified themselves out of a number of years of work toward courteous environments. The numbering is cardinal, so do not let the order imply priority. It is better, as you read them, to jump around and loop back through them a few times at random so that the feeling of flow that they are intended to convey may come across. They do not constitute a handbook of "how-to's"; I would prefer that they be considered a set of attitudes.

1. The environment and its users interact in a set of physical parameters shared in common: e.g. touch-and-movement or e.g. a change of acoustical properties which allow the participant's own sounds to change as they are returned to him and change as he moves and listens differently. Parameters which do not have this shared intimacy should be influenced by those that do. Thus, in the environments to which we have become accustomed a shift of our visual attention does not change the light; our act of passive listening does not change the sound.

2. The control of each parameter must be looped back upon itself --- simply at least, but with more complex interconnections as the facilities for self-organizing control (1) are augmented. What happens is twofold: the environment acquires an exploratory behavior of its own and that behavior is related to the spread of what has happened and of what will happen. You will not understand if you think of time passage as a thin, straight line. The grammar of purposive behavior is not punctuated by the clock but is expressed in rates of change and rates of rates of change. (2,5).

3. Each loop behavior should possess a small amount of random variation. The time-grain or rhythm of these changes should be slow in comparison to the responsiveness of that loop to changes imposed from the outside. Totally random behavior is as discourteous as fixity and is likely to produce anxiety. A certain amount of redundant pattern or melody is pleasing and the slow variations lead to the delight of noticing the differences from anticipated patterns. These are the differences that make a difference.

4. At a more advanced stage, consideration must be given to decision processes and to learning processes. Decision implies the ability to shift abruptly between learned modes of successful behavior. Learning implies that an organism has the ability to acquire (slowly perhaps, but not by punch-card program) new modes or patterns that are successful in newer contexts. Let's keep it simple for now. Literature is available (14,7) on the how-to's of appropriate instrumentation for the modelling of these skills.

288

5. <u>N.B.</u> MEASUREMENTS ARE <u>NOT</u> TO BE MADE UPON THE OCCU-
PANTS OF AN ENVIRONMENT. THE ONLY MEASUREMENTS ALLOWED ARE
THE SELF-REFERENT ONES OF CONTROL SYSTEMS UPON THE PARAMETERS
THEY CONTROL. AWARENESS OF OCCUPANT PARTICIPATION IS BY WAY
OF THE CHANGES HE IMPOSES UPON ENVIRONMENTAL LOOP BEHAVIORS
BECAUSE OF HIS INTERVENTION IN THEM.

BIG BROTHER IS <u>NOT</u> WATCHING YOU.

HE IS ENJOYING <u>HIMSELF</u>.

SO JOIN THE FUN.

6. Beware at all times of limiting the degrees of freedom of any part of a living environment. Choose with care
but with courage. Leave every parameter as free and self-
organizing as possible. In the long run it really is not
a question of how much looseness and control flexibility you
can afford for the project. The playfulness of an environment allows its organizing game to evolve into what did not
exist for it before. Remember, we are increasing adaptability! (11)

ANTITHESIS

If you have been quietly nodding "of course" to the
above, let me introduce some recently published rebuttal.
Constantinos A. Doxiadis, in an article entitled: "Ekistics,
the Science of Human Settlements". (<u>Science</u>, Vol 170 No. 3956)
discusses optimization and bids to lay waste a couple of
myths. In challenging the "myth of the static plan" he
says in part: "We need a room with constant dimensions, a
home that gives us a feeling of permanency, a street and a
square which do not change and which are esthetically satisfying. Such considerations lead to the question, to what
extent can our environment be a constant one? The answer
is that, if there is a unit of optimum size such as a room,
a home, a community (up to the one of 1-kilometer radius),
this can and should be constant. In this way we can face a
world of changing dynamic cities by building them with constant physical units within which we can create quality ---
units meant for a certain purpose and containing a certain
desirable mixture of residences, cultural facilities, industry, and commerce." If you read on, it becomes only more
obvious that Mr. Doxiadis is searching for some kind of

immutable truth. I think that Paolo Soleri is doing roughly
the same in his quest of "Arcology", and the proposed urban
spaces only horrify me with their total lack of playfulness.
I do not wish to be the victim of such plans.

COUNTERPOINT

A courteous environment will enrich its occupants.
Wealth is synonymous with access to the activities and
processes that matter to you. Playfulness provides access to
what was previously noise, redundancy, or garbage. Your
wealth is enhanced by the opportunity to have it matter to
someone else: your children, your friends, or your extended
community. The "loop" way of thinking provides for recycling
your own energy. It is an "ecological think".

So far I have been talking about the kinds of disem-
bodied relationships that seem acceptable when one is discussing control and process and the properties of change. I do
not intend to be any more specific about them because this
publication is neither a catalogue nor a handbook nor a
technical dictionary. If I can inspire you to dig further
into the bibliography or to contact authors directly, I have
been successful. I do want to mention briefly the kinds of
materials that some of us suspect will be necessary for the
realization of truly responsive, unprogrammed, playful
environments.

I will not talk about form. That, my friend, will
come about when you, the materials, the control and energy
sources get together. I predict that in form and behavior
the resulting structures will be described sometimes as
artificial organisms --- and I daydream that architecture
may eventuate as a technological branch of zoology!

THE STUFF OF IT

What kinds of materials and energetics will you find
in a responsive, playful environment? Look around you. (17)
Consider the correlations you may observe between the structure of mechanisms and the adaptability of their behaviors.
Most of the man-made machines we see rely for their entele-
chies upon hard, rigid materials and the moving parts are
guided or confined either by smooth, sliding surfaces or
by rotation about shafts. The energy equations one could
write for them are expressible in terms of vector forces,

lengths, velocities, and other such linearly related variables with time as an explicit parameter. These are the mathematical manipulations with which we are taught to grapple in school and they are easy to transfer from one mechanism to the next both as descriptors and as design tools. We are in fact taught early that the opportunity to reduce mechanism to symbolic formulae is highly beneficial to the process of decision and design --- and the habit thenceforth programs us to seek solutions by way of such mechanisms. We somehow never quite shake free of them until biological systems hit us in the face with their fascinating modes of coping in their environments.

When we start trying to imitate biology (and move into the area now called Bionics), we find it strangely difficult so long as we attempt the imitation with rigid materials: so long, for example, as we describe a man's movements as if he were merely an animated skeleton. A breakthrough into the realm of soft materials, with thermodynamic energy relationships, suddenly puts you into a position to fulfill the desired biological paradigm within the frame of ordinary, non-living materials.

To date a few of us have been working and playing with thin plastic films and foams, and with compressed air and other expandables. I am not talking simply about inflatable blisters nor even double-wall structures which generally have been patterned after their post-and-beam counterparts. Artificial organisms as living environments may be made highly permeable to their surroundings while also being courteous to their occupants. Self-organizing controllers can maintain (for example) average light levels or favorable brightness differences in the context of the weather, time of day, and the difference between your mood and that mood which was anticipated. The radiation or absorption of heat in direct exchange with the surrounding can be made relevant to your activities and to the thermodynamic conditions available. The acoustic properties of the inner spaces can be caused to enhance the privacy of a tête-à-tête or the mutual involvement in a larger gathering. Walls that move to the touch --- relevant to the function of support or moving back in retreat --- that change in color and form: streamlining

themselves to the wind or shrinking down when unoccupied,
are all possible within the state-of-the-art technology.

No architect's prior commitments to a fixed design
could possibly serve so many functions so well. Let us
accede to the admission that any immutable structure is going
to deprive its occupants of that which should by now be their
birthright: the active use of a responsive environment as an
artist with his brush so as to convey an affirmative message
of their own participation.

I will grant that much use will still be made of
rigid members as surfaces and exoskeletons, as articulated
components, or as lively contrast in an otherwise plastic
system. Let us bear in mind, however, that each use of
these materials serves in some way to delimit a priori
the richness of response which that part of the system
could enjoy. Rigidity as protection against wind and weather
simply isn't necessary when a structure has the capability
of reshaping itself actively and in a manner relevant to
the maintenance of its inner integrity and intention. The
wolf that came huffing and puffing to the first two little
pigs would have had a discouraging time of it had the straw
or sticks been resilient plastic with self-organizing control
systems in command. They would then laugh at the third
little pig and wonder why he would want a house that was the
same in summer as in winter?

A PARTING WORD OF
WARNING

I come from an academic background which relies
heavily for its communications about its reality upon
drawings, schematics, models, verbal description, computer
graphics, and other ersatz displays. When I began grappling
with the complexities and responsiveness of "soft control
material" (as we call it) I found that metaphors and simulation games just don't work any more. You have to play with
reality itself: life size.

Do you realize how much the means of expression and
design that are available to you serve to preprogram what
you create with them? Get away from your drafting tables and
T-squares; throw out all that flat, stiff cardboard and balsa;
destroy the squareness and flatness of the spaces you live

and work in; and generally unprogram yourself from the habits
of thought seemingly demanded by steel and concrete and glass.
Get yourself some plastic film and an old vacuum cleaner you
can run backwards, find a heat-sealer or some tape or adhesive
and go at it. There really is no other way to metabolize
for yourself the properties and topologies of this new world
of responsive, self-organizing, evolutionary artificial
organisms. Get out there and DO IT.

Who was that wolf anyway? The bill collector? landlord? hunger? a general neurosis? Just how strong were the
myths of our childhood and how much effort is entailed in
throwing them off? Those who cringe in stone boxes may not
yet be aware that they are already dead.

BIBLIOGRAPHY

1. Barron, R.L. "Self-Organizing and Learning Control Systems" In: H.L. Oestreicher and D.R. Moore (eds.), Cybernetic Problems in Bionics New York, Gordon and Breach, 1968

2. Brodey, W.M. "The Clock Manifesto", Ann. N.Y. Acad. Sci. Vol. 138, 1967

3. ----------- "Soft Architecture: The Design of Intelligent Environments" Landscape Vol. 7, No. 1 (Autumn 1967)

4. ----------- "If You Can't Support the Revolution, Let the Revolution Support You" Innovation Issue No. 15, October 1970

5. ----------- "Information Exchange in the Time Domain" In: Gray, Duhl, and Rizzo (eds.) General Systems Theory and Psychiatry Boston, Little Brown, 1969

6. ----------- and Lindgren, N. "Human Enhancement: Beyond the Machine Age" (two articles) IEEE Spectrum September 1967 and February 1968

7. Gamba, A. "Optimum Performance of Learning Machines" Proc. IRE, Vol. 49, No. 1, January 1961

8. Gibson, J.J. The Senses Considered as Perceptual Systems Boston, Houghton Mifflin, 1966

9. Hermann, H. and Kotelly, J.C. "An Approach to Formal Psychiatry" Perspec. in Biol. and Med. Vol. 10, No. 2, (Winter 1967)

10. Johnson, A.R. "Organization, Perception, and Control in Living Systems" Indus. Mngmt. Rev. (MIT) Vol. 10, No.2, (Winter 1969)

11. ----------- "Performance as Controller of Performance" presented at 3rd Ann. Symp. Am. Soc. for Cybernetics, October 1969 (in press ?)

12. ----------- "Information Tools that Decision Makers Can Really Talk With" (Innovation #10) or in Decision Making in a Changing World, Princeton: Auerbach, 1971

13. ----------- "Dialogue and the Exploration of Context: Properties of an Adequate Interface", presented at 4th Ann. Symp. Am. Soc. for Cybernetics, October 1970 (in press ?)

14. Kilmer, W.L., McCulloch, W.S., and Blum, J. "Some Mechanisms for e Theory of the Reticular Formation" In: Mesarovic (ed.) Systems Theory and Biology New York: Springer Verlag, 1968

15. Pask, G. "Comments on Men, Machines, and Communication Between Them" Vision -- 67 ICCAS New York Univ. October 1967

16. Storm, H.O. "Eolithism and Design" Colorado Quarterly Vol. 1, No. 3 (Winter 1953)

17. WHOLE EARTH CATALOGUE Published semi-annually by the Portola Institute, Menlo Park, California

18. Wiener, N. The Human Use of Human Beings: Cybernetics and Society Boston, Houghton Mifflin, 1950

Edward Allen

Prospects for a Magic Building Machine

Construction of a stabilized earth house, 1963.

In 1963 I built a small, rather sculptural house of stabilized earth. Following a method I had seen depicted in a book on Cameroun building techniques, I built mud domes and vaults much as a potter builds a large vessel from a coil of clay. The process was a discouraging one in that it required an inordinate amount of labor, and the finished construction cracked profusely as it dried. But several aspects of the process were intriguing: I had been able to span horizontal space, using a wet material, without the use of any sort of temporary support. I had been able to work for days at a stretch without using a level, plumb line, or measuring device, because the flowing, one-piece construction required no attention whatsoever to exact measurements and close tolerances. I had discovered, in fact, a simple building system which could realize any floor plan, no matter how irregular, because it required neither rigid materials nor formwork. Furthermore, the basic construction operations I had carried out so painfully by hand -- mixing, transporting, molding into place-- were operations which could be assumed naturally by machines: power mixers, pumps and extruders.

Simultaneously with my experimentation with mud, the Dow Chemical Company was developing its Spiral Generation technique, in which strips of polystyrene foam are assembled into large domes by a moving heat-bonding head at the end of a rotating boom. Both they and I thought at that time that it would be a fairly simple matter to program the length of the boom to produce more complex shapes. Later, in the late 1960's, the Araneida Corporation attempted actual development of such a magic building machine, using an ingenious wet-extrusion device to place epoxy foam in successive courses. For unpublicized reasons, they did not carry the work into full-scale testing, and the project is apparently dead.

In 1969 and 1970, I and several students, working under small research grants from the M.I.T. Urban Systems Laboratory and Hercules, Inc., resumed work where I had left off six years earlier. Mud had several inadequacies as a material, so we looked for something else. Foam plastics were promising in several respects, but from having visited some experimental sprayed-foam houses, I was convinced that the inherent problems of poor fire resistance and low stiffness would be difficult to overcome. We chose instead to work with lightweight concrete,

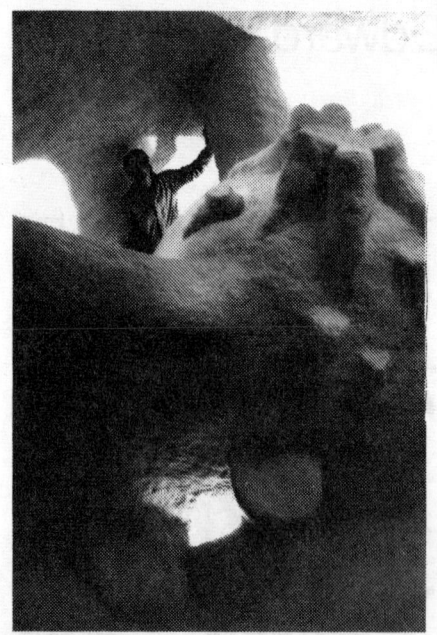

Interior of stabilized earth house.

Spiral generation of a polystyrene foam dome by the Dow Chemical Company.

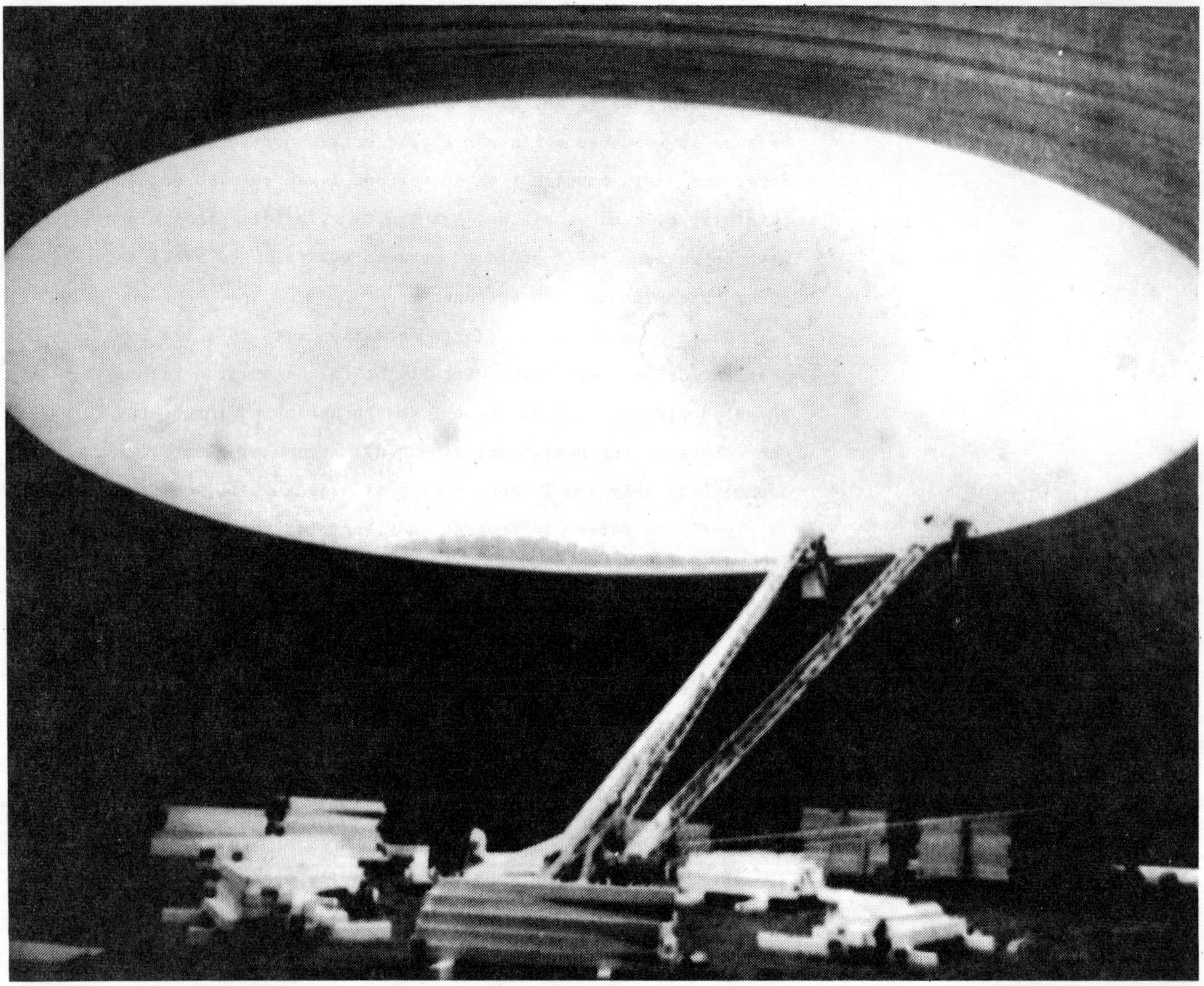

Extruding a course of concrete
wall in the laboratory.

The Araneida Corp. system:
a structure of filled epoxy
foam is generated from liquid
components by a moving arm.

WALL FORMATION OF A RECTANGULAR BUILDING

One version of an on-site
extrusion device.

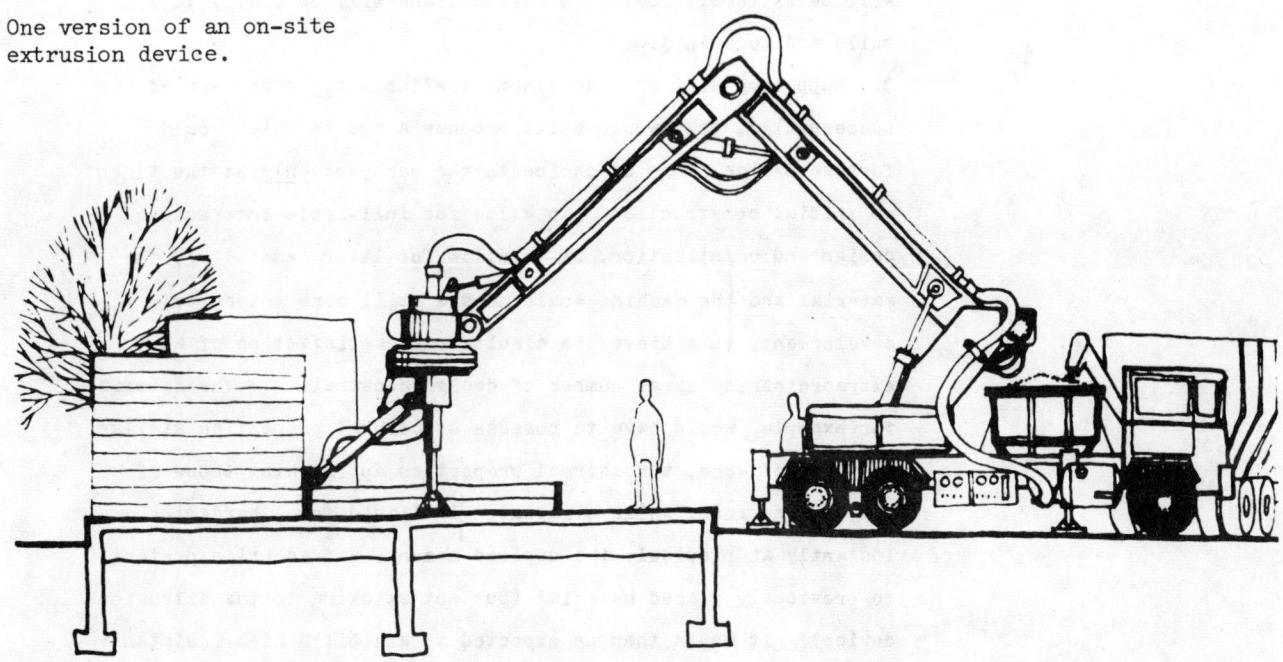

and we carried out a crash program to develop a concrete which
would simultaneously support and insulate. Using a Perlite
aggregate and air entrainment, we approached that ideal. In
May of 1970, using this material, we carried out a number of
successful pumping and extruding trials in the laboratory, and
developed our equipment to a stage where we felt we could make
our first attempts at actual construction, using a variable-
speed electric screw pump and a hand-guided extrusion device
which we had evolved.

Our on-site attempts to extrude concrete into a building
shell were failures. We were unable to control the consistency
of our rather complicated concrete mixture as closely as we had
in the laboratory. More development work was required, but our
research funding had expired, and the project became dormant.

In the course of our work, we had arrived at solutions to
several perplexing problems: Our device could form square corners.
We discovered a plausible method for building automatically any
desired floor plan, regardless of the complexity of its intersections.
We evolved a simple procedure for producing window and door
openings. But despite the progress we made, three basic difficulties remain:

1. The design and development of a suitable basic material for
use with an extrusion building device will require long and
expensive effort.

2. Any automated construction machine which we could project
will be extremely costly to develop, and will be costly to
build and to maintain.

3. Supposing that such development efforts <u>could</u> be carried out
successfully, they would still produce a system which could
furnish a wide range of choice to the occupant <u>only</u> at the time
of initial construction. To allow for full-scale interactive
design and construction, or to allow for later remodeling, the
material and the machine would need a still more intensive
development, to achieve the simultaneous satisfaction of an
extraordinarily large number of design constraints. The material,
for example, would have to possess structural properties similar
to those of wood, and thermal properties approaching those of
polystyrene foam, while originating in liquid form, hardening
instantly at precisely the desired moment, and adhering perfectly
to previously-placed material (but not sticking to the extrusion
device). It would then be expected to exhibit a high resistance
to fire, water, water vapor, ultraviolet attack, and biological
decay. Furthermore, it would have to revert to liquid form on
cue, ready to start the cycle over again, and it would have to
do all this at a cost competitive with those of conventional

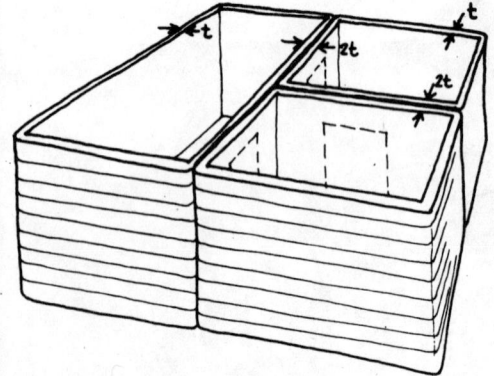

Extrusion patterns for producing
complex aggregations of rooms.

materials. A similarly discouraging list of requirements is associated with the design of the machine, and after all these problems were solved, one could then begin to deal with building code acceptance, labor union jurisdiction, and consumer attitudes.

It is tempting to note that we came very close to extruding a real house with a material and a machine which were only eight months in the making, and that Dow routinely builds satisfactory domes by a somewhat similar process. But we and Dow have done the easy parts, perhaps two percent of what must be done before a real house will be totally machine-built on its site.

Dow's domes are used as permanent formwork for a thin reinforced-concrete shell which becomes the actual load-bearing structure. Our lightweight concrete extruded shell was in itself the loadbearing structure, but required the application of surface coatings to achieve an attractive, waterproof enclosure. One can easily conceive of a third type of relationship between base material and coatings, in which the machine builds of a relatively tractable but structurally inadequate material such as plastic foam, and strong coatings are subsequently applied to both the interior and exterior faces to form a structural sandwich capable of carrying loads in bending. As a sequel to our work with extruded concrete, I decided in 1971 to explore this third type of relationship, to see where it might lead.

After much preliminary experimentation, two full-scale test structures, one indoors and one out, were built of hand-assembled slabs of polystyrene foam, and were then covered with expanded metal mesh and about one-eighth inch of latex-modified Portland cement plaster on both interior and exterior. Floors and roofs spanning up to twelve feet were made with a six-inch foam core. Walls were four inches thick. Subsequent load tests showed acceptable bending strength and deflection characteristics, and both visual quality and tactile "feel" were good, but several serious problems arose:

1. Flat surfaces of ordinary low-density foam are incapable of carrying the loads of workmen, tools, materials, and wind which are normally encountered before the strength-giving coatings can be applied and cured.

2. Over a period of months, the foam core of any horizontal spanning element creeps appreciably in shear, allowing the element to deflect progressively and permanently.

3. The thin concrete coatings which are sufficient for structural purposes do little to insulate the vulnerable foam core from fire. At least triple the thickness of coating would be required for adequate fire protection.

Finishing and making rigid a polystyrene foam structure by means of a thin coating of structural portland cement plaster.

A prototype plastic foam block, with glass fiber reinforcing for the faces and a metal shear channel across one edge.

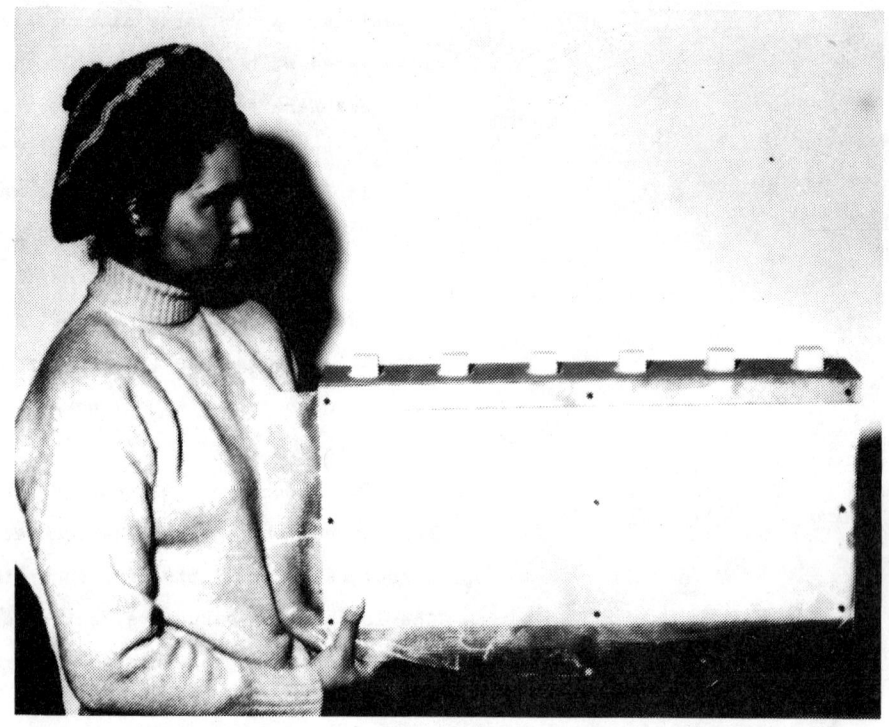

The first and second problems could be met effectively by
the insertion of metal shear reinforcing, such as sheet-metal
channels, through the foam, but any hypothetical machine which
might place the foam would have to be made several times more
complicated in order to be able to place ~~the~~ _{this} metallic reinforcing
as well. The third problem might be solved ~~more easily~~ by the
application of an intumescent paint.

In recent months, I have been investigating an offshoot of
this coated-foam idea which involves the factory production of
tongue-and-groove blocks and planks of plastic foam for later
manual assembly by the homeowner. These elements would come with
integral metal shear reinforcing, and with reinforcing meshes
for the coatings already in place. A standard wall block,
4" x 12" x 24", together with its reinforcing, would weigh only
about four pounds, and could be put in place even by a child,
who is undoubtedly the most magical of all building machines.
A twelve-foot roof or floor plank one foot wide could weigh
less than twenty pounds. In theory, the homeowner could order a
few cartons of blocks and planks, then use them to design his
house full-scale on the site, ripping down and rebuilding as he
desires, changing openings, ceiling heights, and wall locations
at will. The blocks are designed so that interlocking right-
angle junctions may be made at any four-inch interval without
cutting. Construction is so rapid and effortless that a person
working alone could build the first version of his bedroom
in less than half an hour, then spend the rest of the day experimenting with variations and changes.

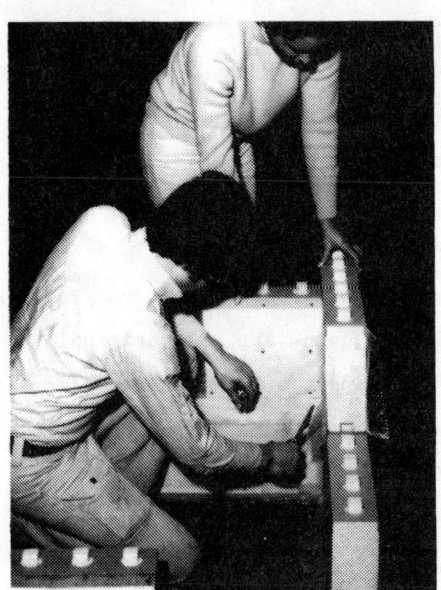

Building with foam blocks.

When a satisfactory arrangement was achieved the owner would open a can of premixed coating, pour some into a roller pan, and roll a continuous coating onto all exposed surfaces, inside and out. The next day he would apply a second coat. When the coatings were cured, he would insert windows, doors, and surface-applied wiring, and move in.

The attraction of this vision is that it combines design and construction into a single, undifferentiated process which requires no machinery, and little physical strength, skill, or capacity for abstract visualization on the part of the owner-builder. It takes the most instinctive of all building processes, that of piling up material to make walls, and puts it at the disposal of anyone who cares to work with it.

In reality, parts of this vision are working, and other parts are not. Our prototype blocks are delightfully easy to assemble and dismantle. But adequate temporary strength against wind and construction loads has not been achieved. The crudely-formulated roll-on coatings which we have concocted are applied only with the expenditure of a good deal of labor per square foot, but they hide the block joints and reinforcing joints well, and produce automatically a uniformly-textured, pleasing surface. Structural tests have given mixed results, depending on the coating formulation used. The problem of fire resistance remains. In my opinion, several years of adequately-funded developmental work could produce a simple, marketable system of lightweight blocks and planks, but the work will not be easy. Where the magic building machine will be almost impossible to develop because of its complexity, the block will be difficult to develop because ultimately it must be so simple.

I expect that a magic building machine (or at least a sleight-of-hand one) will be produced before this century is out. I doubt that I will produce it, not just because I doubt my ability to produce it, but because I no longer think I will be interested in the venture. A machine equal to the task will be so big, so complicated, and so expensive that it may well be more of a barrier than a boon to the would-be designer-builder.

On the other hand, although I foresee a considerable near-future expansion of conventional self-help housebuilding, a majority of American families will still be without ready means of assuming control of their own immediate environments, unless a middle technology can be found which neither exacts the sweat, blood, and gold of conventional construction methods, nor boxes in the human spirit, as do the products of the cyclical outbreaks of Detroit mentality which plague the housing field.

Concerning Responsive Architecture

Nicholas Negroponte

There are really three kinds of responsiveness. There's a responsive <u>design</u> technology that people are talking about--participation, advocacy planning. There is a responsive <u>building</u> technology. And the third is responsive architecture itself. I think the three are reasonably different, and that quite often they are confused.

There is the interesting thing about middlemen. Yesterday someone was appalled to see Wolf Hilbertz using the same original starting point, a return to nature, to support a very technocratic view, where he a few days earlier had used the same images to point to the opposite extreme. What I think they have in common is that both are interested in getting rid of middlemen. Architecture is <u>plagued</u> with middlemen. The architect is traditionally the middleman between somebody and his need for design. The builder is the middleman between that and the building itself. In fact, what advocacy planning is all about is getting rid of the first middleman, and responsive building technology is about getting rid of the second. What I would like to talk to you about today is neither; it is the third, responsive architecture itself, the removal of all middlemen.

It's interesting that this kind of picture [of a Greek island village] appears in almost all documents that try to talk about responsive building technology. It doesn't matter if it's an Aegean island or an African village. It's the referral to an indigenous architecture. I happen to live here three months a year. There's something very interesting about this place. This little village of six hundred residents, only twenty of whom are people like me who have come from the outside--of the 580 permanent residents, most of them hate it. They really dislike living here. This is very important, and most people don't talk about it. Most of these residents would give an arm and a leg to get out, for other reasons. They would like to live in Boston's Prudential Center skyscraper, at least that's what they think. On the other hand, we want to live in <u>their</u> village. The meanings that we ascribe to this village are very, very superficial, compared to the ones of people who actually live there. It's just too easy to try to emulate the evolution that has taken place here. This town evolved, in a very true definition of the word "evolution." Materials were very limited, and they had to paint things white primarily

---

Nicholas Negroponte is an Associate Professor of Architecture at M.I.T., and is the father of The Architecture Machine.

for thermal control. Now, even on this remote island, people have access to a very broad range of materials and lots of different colors of paint. By law, now, if I want to paint my window sill, I have three colors to choose from: a brown, a dark blue, and a dark red, all special paints made by a special manufacturer. My wife didn't know this, and painted one yellow, and the police came. So all of a sudden, to preserve an indigenous architecture which has evolved over a long period has become a matter of legal control.

I want to tell you about something which happened last fall at a conference in England on design participation. I found it a very interesting conference. There were two groups represented: One group was the design methodologists, who said, "We need more information, we want to know what people want. We want the psychologists, sociologists, and anthropologists to tell us more about what people want. We want people to fill out more questionnaires. We want to know _more_, so we can design better buildings."

The other side were the "Advocacy Planner" types, saying, "No, it's really a matter of political mobility. We're going to get people to be _heard_. We're going to help them to affect the design of their environments."

I felt that the architects holding both those views were being as paternalistic as ever. I think that most efforts in participatory design and most efforts in responsive design and responsive building technology are paternalistic.

When I graduated from architecture school in 1965, I sincerely felt that I knew better how other people ought to live. In design classes we talked about life styles, and about educating people, and it was _so_ paternalistic.

My topic is computers, really, and computers are terribly paternalistic too. Our efforts in computer-aided design were, in fact, taking this paternalism and amplifying it, trying to get a computer system which would not only amplify your fantastic design abilities, but would sit there and ring bells and blow whistles to tell you when you had contradicted yourself, and make in a sense a slicker paternalism than existed before. After a while we realized that wasn't what we wanted, that we wanted the machine to participate much more overtly in the design process, and so its role had to be intelligent. But this could become in some sense a surrogate architect, which is equally paternalistic, and in some sense you'd be no better off than you were before.

The idea I really want to talk about is having a physical environment which has knowledge about _you_. I'm talking about a responsive building that knows _me_ . . . . If you walk into our machine room down the hall, you'll notice that the door is all jimmy-rigged. It's a simple effort to make a door that will recognize people. The major intent is--Gordon Pask gave us the name--it's a _you_-sensor. This is where I lose a lot of my colleagues, simply because a lot of them feel that you don't need a you-sensor, that an environment can be responsive without knowing _you_. I don't think that is true. I think a you-sensor is very important.

One of the examples I use frequently is what happens when I go home at night and ask my wife to "get the whatchamacallit and put it you-know-where." She knows _exactly_ what I'm talking about, and she'll _get_ the correct whatchamacallit, and she'll _put_ it you-know-where. If I were to make the same request to you, you wouldn't be able to handle it. She has, first of all, a terribly good predictive model of me. She also can handle this model in context. The whatchamacallit may be in fact an umbrella, and it's raining, and she knows it's raining. The whatchamacallit may be shoes that have come back from the repair place, and she knows that we're going out that night, and that those are the shoes I normally wear when I go out. It can be lots of different things at lots of different times. I think that's crucial.

The other question I bring up in connection with responsive or intelligent environments is to ask people to think of examples where the physical environment _does_ respond to you . . . The one example that interests me is elevators. There are some elevator computing systems recently installed in New York which are absolutely fantastic, that are self-organizing systems that cater to the pattern of the building . . . Aside from these systems, in the day-to-day environment, I know of no other examples of intelligent environments.

I want to show you where I live in Boston. I want to talk about the idea of what is, in this case, a very passive memory. Each of these things, I don't care if it's the candlestick here, or the paintings in the background, were objects that I went through some process to buy. This thing I remember buying at an auction, and writing a rubber check because I overbid. There"s a whole story behind each one of these that you don't know that I know, and hence, they have that much extra meaning to me, besides being nice objects. They're a form of memory. But it's a very passive

memory. This fireplace has nothing to do with me, so in some sense
it doesn't mean too much more to me than it would to you, but
obviously somebody, sometime, has a great interest in it. The tray
came from a great-great grandfather, and it has a little more meaning than the fireplace, but perhaps not as much as these other
things I've participated in much more overtly, in the acquisition
and so forth. I'd like to call all this passive memory. Active
memory, unfortunately, I can't give you any examples of, except the
boy standing in the background there, who has a very active memory . . .

A physical environment that knew me could be more active.
Unfortunately, this is where I get into trouble, because I cannot
think of a very convincing architectural response that an environment could make, given the fact that it knew me as well as my wife
knows me. If I had an environment which could understand all my
intentionalities as well as my wife can, what would it do? Unfortunately, examples such as floors that can tell how many people
are walking on them, and doors that recognize people, usually
end up driving second-rate light shows, or doing very banal things
in directing the physical environment, that are so unconvincing
that a lot of people have good reason to be suspicious of the idea
of responsive environments. I'm afraid I can't give you any better
examples. I sometimes use Sean Wellesley-Miller's design of a . . .
sculpture exhibit which counts the number of people that go into it
and come out of it, and inflates or deflates additional sections
of the building, depending on how many people are in the exhibit.
This is one of the few examples of a genuine architectural response--
the more people that enter the damn thing, the bigger it gets.
There are operational kinds of responses that are more convincing,
but they haven't much to do with architecture. Informational types
of responses, for example: I hate reading newspapers and looking
at news on the television, but I would love to have some sort of
device which knew me well enough to synopsize the news each night,
and tells me if there happens to be something interesting on television today or tomorrow, without having to read TV Guide. There
are lots of little applications of a surrogate "me" that I think of
in the context of synopsizing and presenting information, but again,
this is not an architectural response . . .

Conclusion

Richard Bender, an architect who teaches at the University of California in Berkeley, has done some rather significant real-world work toward involving inhabitants in the design and construction of their dwellings, as evidenced by a lecture he had given in Cambridge some weeks earlier. As last speaker at The Shirt-Sleeve Session, he confronted a straggling band of word-weary, image-inundated participants, and he did a very selfless thing. Rather than talk about his own work, he focussed our attention on what we had and hadn't discussed previously:

"It's interesting that during these three days we all agreed that housing is a process, not a product. It's not a commodity, but an activity; a verb. We're all good guys; we're on the right side.

"On the other hand, I was really sad, because if we're the good guys, we're not doing anything. In many of the presentations we've heard how many people really need to be connected into some system of making a dwelling . . . but our real involvements are at a really primitive scale. Most of us were talking about making some house for ourselves with a little gimmick in it.

"If we really want to be effective, how do we reach out beyond this conference? . . . how can all our energies add up to more? We've talked about a lot of good things, but they're not going to be useful until they get hooked into a larger system. What are some institutions we might build to help people to create and change their own environments?"

In response to his questioning, a large number of suggestions were offered. Some of the more coherent ones:

We need something which is to the house-maker what the library is to the scholar--perhaps an expanded, service-oriented lumberyard, furnishing not only materials of construction, but also instruction sheets for various construction operations, design and planning instructions and assistance, and help in making materials takeoffs.

Mass-media efforts like the "Stud House" cartoon should be nationally syndicated in popular newspapers.

More shelter-focussed publications of the <u>Whole Earth Catalog</u> type should be made available; perhaps even a monthly magazine specifically aimed at people who work on their own dwellings, with how-to advice on design, contracting, legal and financial problems, and construction operations.

Power groups (such as the Office of Economic Opportunity) should be identified within the federal bureaucracy, and a major effort should be mounted to get such groups working on occupant-designed and occupant-built housing.

As these and other points were being debated, sherry corks were popping at a buffet to one side of the room. Bits of the discussion were taken up by small groups nearer the buffet, and over a period of a few minutes, the arguments and agreements were dispersed among twos and threes of sherry-sipping participants. Without any conscious, formal word of benediction, The Shirt-Sleeve Session on Responsive Housebuilding Technologies had come to an end-- or had we subconsciously <u>avoided</u> calling an end? By letter, telephone, and published manifesto, the discussions still go on, and in various parts of the country, the experiments continue. Maybe it's not over yet.